Studies in Computational Intelligence 1119

Series Editor

Janusz Kacprzyk, *Polish Academy of Sciences, Warsaw, Poland*

The series "Studies in Computational Intelligence" (SCI) publishes new developments and advances in the various areas of computational intelligence—quickly and with a high quality. The intent is to cover the theory, applications, and design methods of computational intelligence, as embedded in the fields of engineering, computer science, physics and life sciences, as well as the methodologies behind them. The series contains monographs, lecture notes and edited volumes in computational intelligence spanning the areas of neural networks, connectionist systems, genetic algorithms, evolutionary computation, artificial intelligence, cellular automata, self-organizing systems, soft computing, fuzzy systems, and hybrid intelligent systems. Of particular value to both the contributors and the readership are the short publication timeframe and the world-wide distribution, which enable both wide and rapid dissemination of research output.

Indexed by SCOPUS, DBLP, WTI Frankfurt eG, zbMATH, SCImago.

All books published in the series are submitted for consideration in Web of Science.

Jonathan Garibaldi · Christian Wagner ·
Thomas Bäck · Hak-Keung Lam · Marie Cottrell ·
Kurosh Madani · Kevin Warwick
Editors

Computational Intelligence

Springer

Editors
Jonathan Garibaldi
Jubilee Campus
University of Nottingham
Nottingham, UK

Thomas Bäck
Advanced Computer Science
Leiden University
Leiden, The Netherlands

Marie Cottrell
Paris 1 Panthéon-Sorbonne
SAMM University
Paris, France

Kevin Warwick
Vice Chancellors Office
Coventry University
Coventry, UK

Christian Wagner
Computer Science
University of Nottingham
Nottingham, UK

Hak-Keung Lam
King's College London
Strand, UK

Kurosh Madani
University of Paris-EST Créteil (UPEC)
Créteil, France

ISSN 1860-949X ISSN 1860-9503 (electronic)
Studies in Computational Intelligence
ISBN 978-3-031-46220-7 ISBN 978-3-031-46221-4 (eBook)
https://doi.org/10.1007/978-3-031-46221-4

This Springer imprint is published by the registered company Springer Nature Switzerland AG
The registered company address is: Gewerbestrasse 11, 6330 Cham, Switzerland

Paper in this product is recyclable.

Preface

The present book includes extended and revised versions of a set of selected papers from the International Joint Conference on Computational Intelligence-IJCCI 2020 and 2021 that were exceptionally held as online events due to COVID-19.

IJCCI 2020 received 65 paper submissions from 29 countries, of which 6% were included in this book. IJCCI 2021 received 67 paper submissions from 27 countries, of which 10% were included in this book.

The papers were selected by the event chairs, and their selection is based on a number of criteria that include the classifications and comments provided by the program committee members, the session chairs' assessment and also the program chairs' global view of all papers included in the technical program. The authors of selected papers were then invited to submit a revised and extended version of their papers having at least 30% innovative material.

The purpose of IJCCI is to bring together researchers, engineers and practitioners on the areas of fuzzy computation, evolutionary computation and neural computation. IJCCI is composed of three co-located conferences, each specialized in at least one of the aforementioned main knowledge areas.

The papers selected to be included in this book contribute to the understanding of relevant trends of current research on computational intelligence, including: Applications: Image Processing and Artificial Vision, Pattern Recognition, Decision Making, Industrial and Real-World Applications, Financial Applications, Neural Prostheses and Medical Applications, Neural-Based Data Mining and Complex Information Process, Deep Learning, Convolutional Neural Networks, Complex Artificial Neural Network-Based Systems and Dynamics, Learning Paradigms and Algorithms, Industrial, Financial, Medical, and Other Applications Using Fuzzy Methods, Fuzzy Logic in Robotics, Fuzzy Decision Analysis, Multi-Criteria Decision Making and Decision Support, Fuzzy Control and Evolutionary Search and Meta-Heuristics.

We would like to thank all the authors for their contributions and also to the reviewers who have helped ensuring the quality of this publication.

October 2021

Jonathan Garibaldi
Christian Wagner
Thomas Bäck
Hak-Keung Lam
Marie Cottrell
Kurosh Madani
Kevin Warwick

Organization

Conference Chair

Served in 2020

Kurosh Madani — University of Paris-EST Créteil (UPEC), France

Served in 2021

Juan Julian Merelo — University of Granada NIF Q1818002F, Spain

Program Co-chairs

ECTA

Served in 2020
Juan Julian Merelo — University of Granada NIF Q1818002F, Spain

Served in 2021
Thomas Bäck — Leiden University, Netherlands

FCTA

Served in 2020 and 2021
Jonathan Garibaldi — University of Nottingham, UK
Christian Wagner — University of Nottingham, UK

NCTA

Served in 2020
Thomas Bäck — Leiden University, Netherlands

Served in 2021

H. K. Lam	King's College London, UK
Marie Cottrell	Université Paris1, France

ECTA Program Committee

Served in 2020

Anca Andreica	Babes-Bolyai University, Romania
Ben Paechter	Edinburgh Napier University, UK
Chih-Chin Lai	National University Kaohsiung, Taiwan, Republic of China
Christine Zarges	Aberystwyth University, UK
Clara Pizzuti	National Research Council of Italy-CNR, Italy
Frédéric Saubion	University of Angers, France
Gary Parker	Connecticut College, USA
Giovanni Iacca	University of Trento, Italy
Hisao Ishibuchi	Southern University of Science and Technology, China
Julian Miller	University of York, UK
Khairul Kasmiran	Universiti Putra Malaysia, Malaysia
Krzysztof Trojanowski	Uniwersytet Kardynała Stefana Wyszyńskiego, Poland
Lutz Hamel	University of Rhode Island, USA
Marcus Randall	Bond University, Australia
Mario Giacobini	University of Torino, Italy
Matthieu Basseur	University of Angers, France
Miqing Li	University of Birmingham, UK
Mohamed Arezki Mellal	M'Hamed Bougara University, Algeria
Mohammed Salem	University of Mustapha Stambouli Mascara, Algeria
Pablo Mesejo Santiago	Universidad de Granada, Spain
Paolo Sibani	University of Southern Denmark, Denmark
Ricardo Landa	CINVESTAV-IPN, Mexico
Richard Allmendinger	University of Manchester, UK
Riyaz Sikora	University Texas Arlington, USA
Rui Mendes	University of Minho, Portugal
Sabri Arik	Istanbul University-Cerrahpasa, Turkey
Shengxiang Yang	De Montfort University, UK
Tan Tse Guan	Universiti Malaysia Kelantan, Malaysia
Tatiana Tambouratzis	University of Piraeus, Greece

Viviana Mariani	Pontifical Catholic University of Parana, Brazil
Wenjian Luo	Harbin Institute of Technology, Shenzhen, China
William Buckley	California Evolution Institute, USA

Served in 2021

Amelia Zafra	University Córdoba, Spain
Andy Connor	Auckland University of Technology, New Zealand
Carmelo Militello	National Research Council (CNR), Italy
Dariusz Jakóbczak	Technical University of Koszalin, Poland
David Greiner	Universidad de Las Palmas de Gran Canaria, Spain
Diarmuid O'Donoghue	Maynooth University, Ireland
Donato D'Ambrosio	University of Calabria, Italy
Enrique Carmona	Universidad Nacional de Educación a Distancia, Spain
Francisco Gallego-Durán	University of Alicante, Spain
Gianluigi Folino	ICAR-CNR, Italy
Heiko Hamann	University of Lübeck, Germany
Jerzy Balicki	Warsaw University of Technology, Poland
Luca Manzoni	University of Trieste, Italy
Marco Tomassini	University of Lausanne, Switzerland
Mario Garza-Fabre	Cinvestav, Mexico
Miguel Melgarejo	Universidad Distrital Francisco José de Caldas, Colombia
Philip Mckinley	Michigan State University, USA
Robiah Ahmad	University of Technology Malaysia, Malaysia
Shigeru Obayashi	Tohoku University, Japan
Tao Gong	Education Testing Service, USA
Vincent Cicirello	Stockton University, USA
Wei Fang	Jiangnan University, China
Xinchao Zhao	Beijing University of Posts and Telecommunications, China

Served in 2020 and 2021

Adrian Bekasiewicz	Gdansk University Technology, Poland
Adrien Goeffon	University of Angers, France
Ahmed Kheiri	Lancaster University, UK
Alexander Brownlee	University of Stirling, UK
Andres Faina	IT University of Copenhagen, Denmark
Andrzej Skowron	Institute of Mathematics UW, Poland

Clara Pizzuti	National Research Council (CNR), Italy
Conor Ryan	University of Limerick, Ireland
Dalila B. M. M. Fontes	Faculdade de Economia and LIAAD-INESC TEC, Universidade do Porto, Portugal
Daniel Porumbel	Conservatoire National des Arts et Métiers, Paris (CNAM), France
David A. Pelta	University of Granada, Spain
Dominik Sobania	Johannes Gutenberg University Mainz, Germany
Eduardo Rodriguez-Tello	Cinvestav, Mexico
Fabio Caraffini	Swansea University, UK
Francesco Fontanella	Università di Cassino e del Lazio Meridionale, Italy
Francisco Chicano	University of Málaga, Spain
Gareth Howells	University of Kent, UK
Hui Wang	Nanchang Institute of Technology, China
Iwona Karcz-Duleba	Wroclaw University of Science and Technology, Poland
J. Manuel Colmenar	Universidad Rey Juan Carlos, Spain
Janos Botzheim	Eötvös Loránd University Faculty of Informatics, Hungary
Jean-Marc Montanier	Faurecia, France
José Fonseca	UNINOVA, Portugal
José Ribeiro	Instituto Politécnico de Leiria, Portugal
José Santos	University of A Coruña, Spain
Juan Luis Jimenez Laredo	University of Le Havre, France
Luis Nunes	Instituto Universitáario de Lisboa (ISCTE-IUL) and Instituto de Telecomunicações (IT), Portugal
Marc Ebner	Ernst-Moritz-Arndt-Universität Greifswald, Germany
Moshe Sipper	Ben-Gurion University, Israel
Mustafa Misir	Istinye University, Turkey
Nasimul Noman	University of Newcastle, Australia
Nicolas Jozefowiez	University of Lorraine, France
Nuno Leite	Instituto Superior de Engenharia de Lisboa, Portugal
Oussama Hamid	University of Nottingham, UK
Paola Festa	University of Napoli, Italy
Paola Pellegrini	French Institute of Science and Technology for Transport, France
Paolo Cazzaniga	University of Bergamo, Italy
Pauline Haddow	The Norwegian University of Science and Technology, Norway

ECTA Additional Reviewers

Served in 2020

FCTA Program Committee

Served in 2020

Giovanna Castellano	University of Bari, Italy
József Dombi	University of Szeged, Institute of Informatics, Hungary
Javier Montero	Complutense University of Madrid, Spain
John Macintyre	University of Sunderland, UK
José Luis Verdegay	University of Granada, Spain
Khairul Kasmiran	Universiti Putra Malaysia, Malaysia
Leonilde Varela	University of Minho, School of Engineering, Portugal
Li-Pei Wong	Universiti Sains Malaysia, Malaysia
Lucia Vacariu	Technical University of Cluj Napoca, Romania
Luis Martinez Lopez	University of Jaén, Spain
Michael Vrahatis	University of Patras, Greece
Mitsuharu Matsumoto	The University of Electro-Communications, Japan
Olympia Roeva	Institute of Biophysics and Biomedical Engineering, Bulgarian Academy of Sciences, Bulgaria
Parag Pendharkar	Pennsylvania State University, USA
Pawel Myszkowski	Wroclaw University of Technology, Poland
Philippe Thomas	Université de Lorraine, France
Rahul Caprihan	Dayalbagh Educational Institute, India
Robert Schaefer	AGH University of Science and Technology, Poland
Schütze Oliver	CINVESTAV-IPN, Mexico
Tatiana Tambouratzis	University of Piraeus, Greece
Thomas Baeck	Leiden University, Netherlands
Thomas Hanne	University of Applied Arts and Sciences Northwestern Switzerland, Switzerland
Vesa Niskanen	Univ. of Helsinki/VM University, Finland
Vilém Novák	University of Ostrava, Czech Republic
Vladik Kreinovich	University of Texas at El Paso, USA
Wladyslaw Homenda	Warsaw University of Technology, Poland

Served in 2021

Alper Basturk	Erciyes University, Turkey
Carlos Travieso-González	Universidad de Las Palmas de Gran Canaria, Spain
Chih-Cheng Hung	Kennesaw State University, USA
Frank Klawonn	Ostfalia University of Applied Sciences, Germany
László Kóczy	Budapest University of Technology and Economics, Hungary

Martina Dankova	University of Ostrava, Czech Republic
Penousal Machado	University of Coimbra, Portugal
Salvatore Vitabile	University of Palermo, Italy
Slawomir Zadrozny	Polish Academy of Sciences, Poland

Served in 2020 and 2021

Ahmed Bufardi	Independent Researcher, Switzerland
Alexander Hošovský	Technical University of Kosice, Slovak Republic
Chilukuri Mohan	Syracuse University, USA
Christel Kemke	University of Manitoba, Canada
Christopher Hinde	Loughborough University, UK
Chung-Hsing Yeh	Monash University, Australia
Cleber Zanchettin	Federal University of Pernambuco, Brazil
Colm Riordan	National University of Ireland, Galway, Ireland
Corrado Mencar	University of Bari, Italy
Daniel Sánchez	University of Granada, Spain
Dat Tran	University of Canberra, Australia
Edwin Lughofer	Johannes Kepler University, Austria
Etienne Kerre	Ghent University, Belgium
Fernando Bobillo	University of Zaragoza, Spain
France Cheong	RMIT University, Australia
Francisco Lupianez	Univ. Complutense de Madrid, Spain
Hao Ying	Wayne State University, USA
Hazlina Hamdan	Universiti Putra Malaysia, Malaysia
Ivo Bukovsky	Czech Technical University in Prague, Czech Republic
Jesús Alcaláa-Fdez	University of Granada, Spain
Joaquim Reis	ISCTE, Portugal
José Molina	Universidad Carlos III de Madrid, Spain
Jose de Jesus Rubio	Instituto Politecnico Nacional, Mexico
Katsuhiro Honda	Osaka Metropolitan University, Japan
Miguel Sanz-Bobi	Comillas Pontifical University, Spain
Mu-Chun Su	National Central University, Taiwan, Republic of China
Oussama Hamid	University of Nottingham, UK
Pablo Carmona	University of Extremadura, Spain
Patrick Siarry	University Paris 12 (LiSSi), France
Pavel Krömer	VSB Ostrava, Czech Republic
Pedro Coelho	State University of Rio de Janeiro, Brazil
Radu-Emil Precup	Politehnica University of Timisoara, Romania
Rainer Heinrich Palm	Örebro University, Sweden

Roseli Romero	University of São Paulo, Brazil
Stefka Fidanova	Bulgarian Academy of Sciences, Bulgaria
Wei-Chiang Hong	Asia Eastern University of Science and Technology, Taiwan, Republic of China
Yoshikazu Fukuyama	Meiji University, Japan

NCTA Program Committee

Served in 2020

Arfan Ghani	Coventry University, UK
Barry Nichols	Middlesex University, UK
Benoit Frenay	University of Namur, Belgium
Davide Bacciu	University of Pisa, Italy
Elena Marchiori	Radboud University, Netherlands
Erzsébet Merényi	Rice University, USA
Eyad Elyan	Robert Gordon University, UK
Jan Faigl	Czech Technical University in Prague, Czech Republic
Jan Mares	University of Chemistry and Technology, Czech Republic
Juan Julian Merelo	University of Granada NIF Q1818002F, Spain
Leslie Smith	University of Stirling, UK
Lin-Ching Chang	The Catholic University of America, USA
Maojiao Ye	Nanjing University of Science and Technology, China
Micheal Spratling	King's College London, UK
Miltos Alamaniotis	University of Texas at San Antonio, USA
Neel Mani	Amity University, India
Patricio Orio	Universidad de Valparaíso, Chile
Sreela Sasi	Gannon University, USA
Stefan Glüge	ZHAW School of Life Sciences and Facility Management, Switzerland
Stephane Pinel	Mailchimp, USA
Tatiana Tambouratzis	University of Piraeus, Greece
Vladik Kreinovich	University of Texas at El Paso, USA
William Buckley	California Evolution Institute, USA

Served in 2021

Alessio Martino	Department of Business and Management, LUISS University, Italy
Daniel Vasata	Faculty of Information Technology, Czech Technical University in Prague, Czech Republic
Fabio Scotti	Universita degli Studi di Milano, Italy
Gary Parker	Connecticut College, USA
Hongjian Liu	Anhui Polytechnic University, China
Junjie Fu	Southeast University, China
Mohamed Fakhr	Arab Academy for Science and Technology & Maritime Transport, Egypt
Oksana Pomorova	University of Lodz, Poland
Sivaramakrishnan Rajaraman	National Library of Medicine, USA
Yoshihiko Horio	Tohoku University, Japan

Served in 2020 and 2021

Abbas Fotouhi	Cranfield University, UK
Alfredo Vellido	Universitat Politècnica de Catalunya, Spain
Andrzej Skowron	Institute of Mathematics UW, Poland
Antonello Rizzi	Università di Roma “La Sapienza”, Italy
Artur Ferreira	ISEL-Instituto Superior de Engenharia de Lisboa and IT-Instituto de Telecomunicações, Portugal
Barry Bentley	Cardiff Metropolitan University, UK
Friedhelm Schwenker	University of Ulm, Germany
Gang Li	Deakin University, Australia
George Rudolph	Utah Valley University, USA
Gilles Bernard	PARIS 8 University, France
Jean-Jacques Mariage	Laboratoire d’Informatique Avancée de Saint-Denis and Université Paris 8, France
Jose de Jesus Rubio	Instituto Politecnico Nacional, Mexico
Khairul Kasmiran	Universiti Putra Malaysia, Malaysia
Kostantinos Demertzis	Democritus University of Thrace, Greece
Mark Oxley	Air Force Institute of Technology, USA
Monica Bianchini	University of Siena, Italy
Nicoletta Nicolaou	University of Nicosia Medical School, Cyprus
Norikazu Takahashi	Okayama University, Japan
Oussama Hamid	University of Nottingham, UK
Petr Hajek	Faculty of Economics and Administration, University of Pardubice, Czech Republic

Philipp Hoevel	University College Cork, Ireland
Philippe Thomas	Université de Lorraine, France
Sarangapani Jagannathan	Missouri University of Science and Technology, USA
Vincenzo Piuri	Università degli Studi di Milano, Italy

Invited Speakers

2020

Erdal Kayacan	Aarhus University, Denmark
Sanaz Mostaghim	Otto-von-Guericke-Universität Magdeburg, Germany
Kalyanmoy Deb	Michigan State University, USA
M. Verleysen	Machine Learning Group, Université Catholique de Louvain, Belgium

2021

Susana M. Vieira	University of Lisbon, Portugal
Joseph Rynkiewicz	Université de Paris 1 Panthéon-Sorbonne, France
Carlos C. Coello	CINVESTAV-IPN, Mexico
Barbara Hammer	Bielefeld University, Germany

Contents

Evolutionary Optimization of Roles for Access Control in Enterprise Resource Planning Systems

Simon Anderer[1(✉)], Bernd Scheuermann[1], and Sanaz Mostaghim[2]

[1] Hochschule Karlsruhe, Karlsruhe, Germany
{Simon.Anderer,Bernd.Scheuermann}@h-ka.de
[2] Otto-von-Guericke Universität Magdeburg, Magdeburg, Germany
Sanaz.Mostaghim@ovgu.de

Abstract. Role Based Access Control is a common concept in cyber security and authorization management of today's business landscapes. The corresponding (NP-complete) optimization problem, which consists in finding an optimum set of roles and their assignment to users, is known as the Role Mining Problem (RMP). Since real-world use cases often involve large-scale problems with a high number of users and permissions, it is desirable to find fast solution strategies for the RMP. This paper introduces to the addRole-EA, an evolutionary algorithm for the RMP. The addRole-EA is enhanced by advanced mutation methods and an in-depth performance analysis is carried out.

Keywords: Role Mining · Evolutionary Algorithm · Access Control

1 Introduction

As companies and organizations continue to fall victim to cyber crime, data privacy and security are more important than ever to protect their information systems. At this, it is not only attacks from the outside, but also erroneous and fraudulent behavior of employees, which can result in substantial damage [27]. It is therefore important to ensure, that only selected users can access sensitive data. For this purpose, companies and organizations implement access control systems. Due to the huge size of many companies and the large number of potential permissions to be issued, it is no longer reasonable to assign permissions directly to users, as this would quickly lead to very complex and unmanageable access control schemes. To address this issue, *Role Based Access Control (RBAC)* [22], which comprises the aggregation of permissions into roles and the subsequent assignment of roles to users, is one of the most common means to implement access control, resulting in reduced complexity and thus increased comprehensibility and manageability.

Conventionally, the definition of the roles, which is called *role engineering*, is carried out in a top-down manner. For this purpose, the entire company structure must be analyzed, which requires extensive use of human labor, especially in huge companies with a large number of users. For this reason, a bottom-up approach, where roles

J. Garibaldi et al. (Eds.): IJCCI 2021, SCI 1119, pp. 1–23, 2023.
https://doi.org/10.1007/978-3-031-46221-4_1

are derived more or less automatically from the user-permission assignment, which is easily available [16] in most enterprises, is increasingly being used for role engineering in recent times. The corresponding optimization problem is called the *Role Mining Problem (RMP)* and was shown to be NP-complete [25].

This paper provides an introduction to the addRole-EA [1], with a particular focus on examination and analysis of its mutation methods. For these, new variants are developed and evaluated to finally derive an advanced version of the addRole-EA.

The remainder of this paper is organized as follows: Sect. 2 introduces to the Role Mining Problem, gives a formal definition and presents some of its variants. Section 3 discusses the current state of the art considering algorithms and solution strategies for the RMP. Section 4 gives an overview of the addRole-EA in its original version and evaluates its different mutation methods. In Sect. 5, new variants of the mutation methods are presented and evaluated, while Sect. 6 aggregates the insights gained in the previous sections to derive and evaluate an advanced version of the addRole-EA. Finally, Sect. 7 concludes the paper and presents paths for future research.

2 Problem Description

In this section, the Role Mining Problem and some of its variants are presented. A first definition of the RMP was given in [25] as minimum biclique cover problem. In contrast to that, this paper introduces the RMP as binary matrix decomposition problem (cf. [1]), using the following definitions:

- $U = \{u_1, u_2, ..., u_m\}$ a set of $m = |U|$ users
- $P = \{p_1, p_2, ..., p_n\}$ a set of $n = |P|$ permissions
- $R = \{r_1, r_2, ..., r_k\}$ a set of $k = |R|$ roles
- $UPA \in \{0,1\}^{m \times n}$ the user-permission assignment matrix, where $UPA_{ij} = 1$ implies, that permission p_j is assigned to user u_i
- $UA \in \{0,1\}^{m \times k}$ the user-role assignment matrix, where $UA_{ij} = 1$ implies, that role r_j is assigned to user u_i
- $PA \in \{0,1\}^{k \times n}$ the role-permission assignment matrix, where $PA_{ij} = 1$ implies, that role r_i contains permission p_j.

Based on these definitions, the *Basic Role Mining Problem* can defined. Given a set of users U, a set of permissions P and a user-permission assignment UPA, find a minimal set of Roles R, a corresponding user-role assignment UA and a role-permission assignment PA, such that each user has exactly the set of permissions granted by the UPA matrix:

$$\textbf{\textit{Basic RMP}} = \begin{Bmatrix} \min & |R| \\ \text{s.t.,} & \|UPA - UA \otimes PA\|_1 = 0. \end{Bmatrix} \tag{1}$$

where $\|.\|_1$ denotes the L_1-norm for matrices and $\otimes$ the Boolean Matrix Multiplication:

$$(UA \otimes PA)_{ij} = \bigvee_{l=1}^{k} (UA_{il} \wedge PA_{lj}). \tag{2}$$

Figure 1 shows an example of the schematic representation of the UPA, UA and PA matrix, as used throughout the remainder of the paper to illustrate the developed methods and results. The matrices in the figure are based on 3 users, 6 permissions and 4 roles. Black cells indicate 1's, white cells represent 0's.

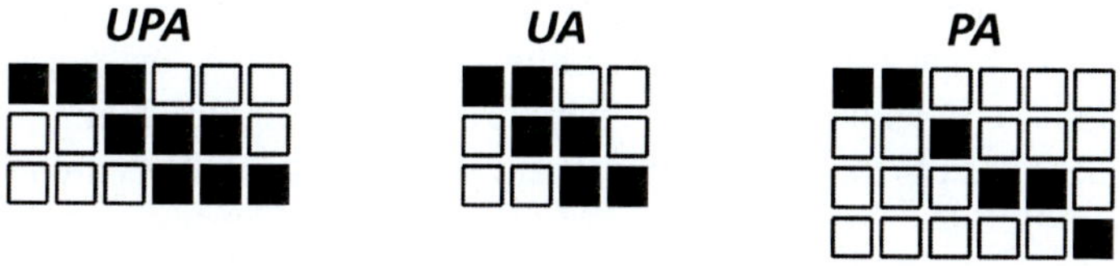

Fig. 1. Schematic representation of the UPA, UA and PA matrix.

A tuple $\pi := \langle R, UA, PA \rangle$ can be considered one possible solution to the underlying Role Mining Problem. If the constraint in (1) is satisfied, the corresponding solution is denoted *0-consistent*. Further variants of the RMP can be obtained relaxing the 0-consistency constraint, tolerating deviations between the user-permission assignment $UA \otimes PA$ and the original UPA matrix. The number of deviations which is given by $\|UPA - UA \otimes PA\|_1$ can then be considered as further optimization criteria in role miming problems. Based on that, a selection of common variants of the RMP is described in Table 1.

Table 1. Different variants of the Role Mining Problem based on [21].

RMP-Variant	Objective	Constraint
Basic RMP	$\lvert R\rvert \rightarrow min$	$\|UPA - UA \otimes PA\|_1 = 0$
Edge RMP	$\|UA\|_1 + \|PA\|_1 \rightarrow min$	$\|UPA - UA \otimes PA\|_1 = 0$
Min. Noise RMP	$\|UPA - UA \otimes PA\|_1 \rightarrow min$	$\lvert R\rvert = k$ constant
δ-approx. RMP	$\lvert R\rvert \rightarrow min$	$\|UPA - UA \otimes PA\|_1 \leq \delta$

Additionally, more specific variants of the RMP result from the consideration of business-driven aspects of role mining e.g. administrative costs [7], the temporal Role Mining Problem [5] or the meaningfulness of roles [17,28]. A survey on the different variants of the RMP and role mining in general can be found in [16].

3 Related Work

The Role Mining Problem and its different variants are well-studied problems, such that many solution techniques have been developed in the last years.

An obvious approach in this context is permission grouping. At this, a set of candidate roles is obtained by grouping permissions to roles, usually based on the intersection of permission sets of different users. These candidate roles are then ranked based on different prioritization functions and assigned to users. One of the first role mining tools is *ORCA* [23]. Further examples of role mining methods using permission grouping are *Simple Role Mining Algorithm* [6], *CompleteMiner* and *FastMiner* [26], *Pair Count* [18] and *Constrained Role Miner* [13].

Another approach consists in mapping the RMP to other, well-known problems in data mining, like binary integer programming problems [14], the Set Cover Problem [12], the Minimum Biclique Cover Problem [10], the Minimum Tiling problem [25] or the Discrete Basis Problem [15]. At this, mainly greedy algorithms are used as solution strategy.

Further approaches are based on graph optimization e.g. *GO* [29] or formal concept analysis e.g. *HierarchicalMiner* [17].

Since it was shown to be NP-complete, evolutionary algorithms are a straightforward solution approach for the RMP. In [24], the structural change in organizations is examined and different methods are proposed to respond to changing business environments. These provide a good basis of evolutionary operators for evolutionary algorithms in the RBAC context. In [8,9,19,20] evolutionary algorithms are applied as solution strategy for the Role Mining Problem. One main drawback of the proposed algorithms, however, is the design of its mutation and crossover methods, which result in violating the 0-consistency constraint, such that they are not applicable for the Basic Role Mining Problem. In addition, the proposed evolutionary algorithms are evaluated on randomly generated data or use case data, which is not publicly accessible. Thus, performance comparison is hampered.

A detailed description of many of the mentioned techniques for the RMP is provided by [16].

4 The AddRole-EA

In the course of this section, the addRole-EA [1] is described and evaluation results are presented. It is characterized by a new method of addition and consequential deletion of roles to UA and PA, such that the 0-consistency constraint holds at all times. Special focus is given to its mutation methods, as these will be further analyzed and used as basis for the development of new mutation variants in Sect. 5.

4.1 Presentation of the AddRole-EA

At first, a pre-processing procedure is described, which reduces the dimension of the initial UPA matrix without loss of information. Thereafter, the different methods and components of the addRole-EA are presented. The following listing provides a top-level description of the addRole-EA:

```
Input:  Users U, Permissions P, User-Permission Assignment UPA
Output: User-Role Assignment UA, Role-Permission Assignment PA

Begin addRole-EA
   Pre-Processing of initial UPA();
```

```
    Initialization(Population);
    Evaluation(Population);

    Repeat (Until stopping condition met)
        select(p1,p2)
        Crossover(p1,p2,c1,c2);
        Mutation(c1);
        Mutation(c2)
        Evaluation of Fitness(c1,c2);
        Replacement(Population,c1,c2);
    End Repeat.

    Post-Processing of resulting UA and PA.
End addRole-EA.
```

Pre-Processing. In many cases, the dimension of the initial user-permission assignment UPA can be reduced by the deletion of redundant data. For this purpose, the following methods, which are based on [12], are applied:

(PP1): Deletion of Empty Rows and Columns.
Deletion of permissions that are assigned to no user and deletion of users that have an empty permission set. This equals the deletion of columns and rows of the UPA matrix that contain zero-elements only.
(PP2): Aggregation of Permissions.
Aggregation of permissions that are assigned to exactly the same users. This corresponds to fusing all identical columns to a single column.
(PP3): Aggregation of Users 1.
Aggregation of all users that have exactly the same permission sets. This corresponds to fusing all identical rows to a single row.
(PP4): Aggregation of Users 2.
Deletion of users whose permission set is equal to the union of other users' permission sets.

Chromosome Encoding and Initialization. Initially a seed individual is generated, where, $UA = UPA$ and $PA = I_n$ (I_n denotes the $n \times n$ identity-matrix). For this seed individual, the 0-consistency constraint is fulfilled, see Fig. 2 (left). Since the UPA, UA and PA matrices are populated very sparsely in practice, it is reasonable to use a sparse representation, instead of the classical representation as binary matrices. This results in huge savings of memory space, since binary sparse matrices only store the position of the one-elements in each row, while zero-elements, which constitute the majority of the matrices' elements, are omitted. The sparse encoding of the seed individual is shown in Fig. 2 (right).

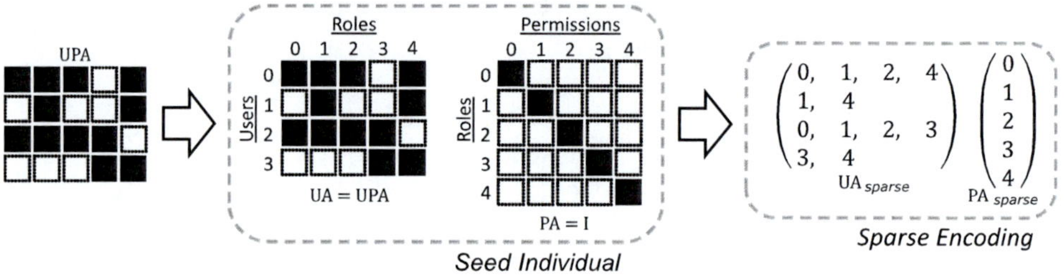

Fig. 2. Representation of the seed individual in the addRole-EA.

Subsequently, the initial population is generated. Each individual in this initial population is created by applying a random sequence of mutation operators to the seed individual. The different mutation operators are described in detail in Sect. 4.1

Evaluation of Fitness. In this paper, the Basic Role Mining Problem is considered (see Sect. 2). Therefore, the number of roles $|R|$ of an individual constitutes the fitness value.

Crossover. The crossover method of the addRole-EA consists in reciprocal exchange of roles between individuals. At this, two parent individuals are selected. One child individual is then obtained by the selection of a subset of all roles of the first parent and the addition of those roles to the second parent using the addRole-method (for the second child individual vice versa). An overview of the operation principle of the crossover methods is given in the following listing:

```
Input:  Individual p1, p2
Output: Individual c1, c2

Begin Crossover
   Choose role-selection method (Si) from (S1-3);
   Initialize c1 (copy of p1);
   Initialize c2 (copy of p2);
   Select set of roles R1 from p1 by (Si);
   Select set of roles R2 from p2 by (Si);
   addRole(c1,R2);
   addRole(c2,R1)
End Crossover.
```

The following selection methods determine how roles are selected for exchange:

(S1): Random-Role Selection.
To select the roles, which are to be exchanged, (S1) selects a subset of each parent's role set randomly based on uniform distribution.

(S2): User-Role Set Selection 1.
Instead of choosing random roles, (S2) selects one user randomly. The roles for exchange are then determined by the roles assigned to this user in the respective parent individuals.

(S3): User-Role Set Selection 2.
This method operates similar to (S2), except that the user, whose roles are to be exchanged, is not selected randomly. Instead the user is chosen, which has the highest difference in the number of assigned roles in the two parent individuals.

Mutation. Mutation of the addRole-EA consists in choosing a role-creation method, creating a new role and adding it to the individual, see the following listing:

```
Input:  Individual c
Output: mutated Individual c

Begin Mutation
   Choose role-creation method (Mi) from (M1-5) randomly;
   Apply (Mi) to individual c to create r_new;
   addRole(c, {r_new});
End Mutation.
```

Hence, the mutation methods differ in the methods used to create a new role (one of (M1)-(M5)). At this, only roles are created, that can be assigned to at least one user. An example of the different methods to create a new role r_{new} is provided in Fig. 3.

(M1): Intersection of Permission Sets.
A role can be considered good if it can be assigned to many users. Therefore, to create r_{new}, this method intersects the permission sets of a random number of randomly chosen users.

(M2): Permission Set setminus Union of Permissions of Roles.
To gather permissions of a user, which are not covered by good roles yet, (M2) randomly selects a user and some of the user's current roles. The new role r_{new} is then obtained from all permissions of the selected user, which are not contained in the union of the permissions of the selected roles.

(M3): Splitting of Roles.
Reducing the number of permissions contained in a role, increases the probability that this role can be assigned to more users. Furthermore, it can be desirable for roles to not share common permissions. Therefore, (M3) selects two random roles and then obtains r_{new} from all permissions of the first role, which are not contained in the second role.

(M4): Permission Set of a User.
It is possible that there are users having a unique permission set with little or no intersections with other users' permission sets. In this case, it can be reasonable to create a role containing all of the user's permissions. Thus, (M4) simply copies all permissions of a random user into r_{new}.

(M5): Merging of Roles.
Creating one role from two roles may directly decrease the total number of roles. For this purpose, a user and two of the user's current roles are selected randomly. The new Role r_{new} is then obtained as the union of the permissions of the selected roles.

The AddRole-Method. The addRole-method, which is the principle method of the algorithm, consists mainly of two steps. In step 1, a given role r_{new} is added to the PA matrix as new row and assigned to all users, whose permission sets contain the permissions of r_{new}, by appending a new column to the UA matrix. In step 2, all roles that became obsolete by adding the given role are removed from the UA and PA matrices. For this purpose, all roles are identified, whose permission sets have non-empty intersection with the permissions set of r_{new}. For each of these roles, it is analyzed, whether it is possible to delete the role without causing deviations due to the previous addition of r_{new}. If that is the case, the considered role can be deleted. This ensures, that the

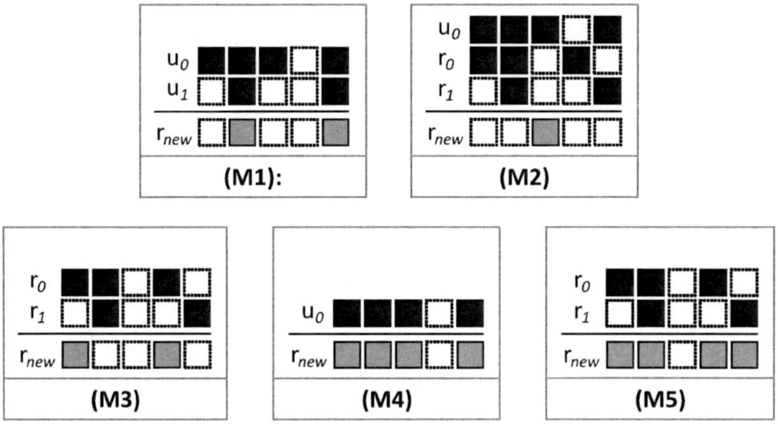

Fig. 3. Example of role-creation for the different mutation methods.

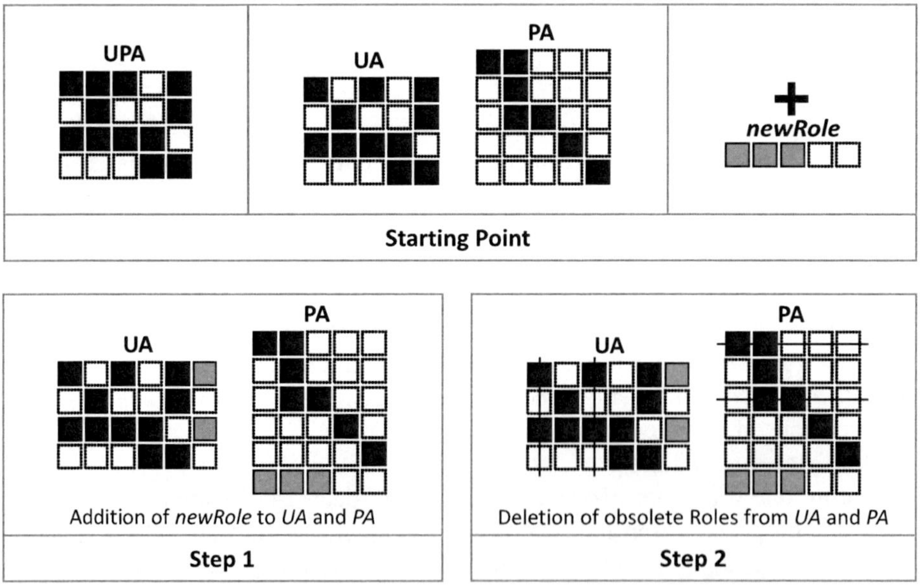

Fig. 4. The addRole-method (example).

0-consistency constraint holds at all times. An example of the operating principle of the addRole-method can be found in Fig. 4.

Replacement. The addRole-EA is a steady-state evolutionary algorithm. For replacement an elitist selection scheme is applied. At this the elitism rate determines the number of individuals that survive based on the quality of their fitness function. Subsequently, the individuals needed to fill the population are selected randomly from the remaining individuals.

Post-Processing. After optimization, the obtained role concepts must be extended to match the original problem specifications before pre-processing. For this purpose, the aggregation of users and permissions, carried out by (PP1-4), has to be undone in a post-processing step.

4.2 Evaluation

This section starts with introducing and analyzing the benchmark instances used for evaluation purposes. Then, the performance of the addRole-EA is examined and compared to the results obtained by other role mining algorithms. In addition, as a basis for the methods and results presented in the further sections of this paper, the mutation methods of the addRole-EA are analyzed in more detail.

Benchmark Data. Until today, the instances of the *HPLabs* benchmark [10] are commonly used to investigate and compare methods and algorithms for the Role Mining Problem. However, it has been shown that applying the presented pre-processing steps (see Sect. 4.1), which remove obvious redundancies from the data, significantly reduces the size of its instances [1]. Table 2 shows the comparison between the original sizes of the instances and their sizes after executing these four pre-processing steps. At this, the role lower bound is derived from [10].

Table 2. Pre-processing *HPLabs* benchmark data.

Instance	Original Size		Size after (PP1-4)		Role Lower	
	Users	Prms.	Users	Prms.	Bound	Δ_{HPLabs}
America large	3,485	10,127	430	1,354	390	40
America small	3,477	1,587	225	349	172	53
APJ	2,044	1,164	475	578	453	22
EMEA	35	3,046	34	263	34	0
Healthcare	46	46	16	19	14	2
Domino	79	231	20	38	20	0
Firewall 1	365	709	71	86	64	7
Firewall 2	325	590	10	11	10	0

There are two trivial solutions for the RMP [2], one of which is obtained from $UA = I_m$ and $PA = UPA$ and contains the same number of roles and users $|R| = m$. The other trivial solution is given by $UA = UPA$ and $PA = I_n$ and contains as many roles as permissions $|R| = n$. It is noticeable, that the number of users after pre-processing is always smaller than the number of permissions after pre-processing for all instances of the *HPLabs* benchmark. Thus, the low values for Δ_{HPLabs}, which is defined as the difference between the number of users after pre-processing and the role lower bound, show that the instances of the *HPLabs* benchmark leave little room for proper role optimization. In three cases (*EMEA*, *Domino* and *Firewall 2*), the optimal solution is even already attained by applying solely the pre-processing procedure.

To address this issue, new benchmark instances, which are more robust against the four steps of the pre-processing procedure, were created and published as *RMPlib* [2]. In its current state, *RMPlib* comprises three benchmarks:

- The *PLAIN_x*-benchmark: a collection of synthetically generated instances obtained from Boolean matrix multiplication of randomly created of UA and PA matrices.
- The *COMP_x*-benchmark: a collection of synthetically generated instances reflecting the component structure typically found in user-permission assignments obtained from SAP's enterprise systems.
- The *RW_x*-benchmark: a collection of instances derived from real-world use case data.

Again, the values for Δ_{RMPlib}, which is defined as difference between the minimum number of roles of the two trivial solutions and a role bound, was examined. Since no lower bound on the number of roles is known for the instances of *RMPlib*, the role bound applied for the definition of Δ_{RMPlib} represents the number of roles, which was used for the creation of the benchmark instance and thus serves as natural upper bound on the minimum number of roles. It is shown, that the values for Δ_{RMPlib} are comparatively high and range from 16 to 600 [2]. This implies that the instances of the *PLAIN_x*-benchmark are well suited for the evaluation of role mining algorithms. An overview of all instances of the *RMPlib* benchmarks and corresponding key figures can be found in the documentation of *RMPlib* [2].

Evaluation. To evaluate the effectiveness of the addRole-EA, it was repeated 20 times on each instance of the *HPLabs* benchmark as well as the *PLAIN_small* instances of *RMPlib* with different random seeds. The values of the different parameters of the addRole-EA applied in these evaluation tests were adopted from [1] and are summarized in Table 3.

Table 3. Parameters used for evaluation tests adopted from [1].

Population Size	20
Mutation Rate	1.0
Crossover Rate	0.1
Elitism Rate	0.7
Max. Number of Iterations	100,000
Max. Number of Iterations without Improvement	10,000

Table 4 shows an overview of the evaluation results of the addRole-EA in comparison with the best results of other known algorithms for role-mining based on the instances of the *HPLabs* benchmark.

Table 4. Evaluation of addRole-EA based on instances of *HPLabs* benchmark.

	America large	America small	APJ	EMEA	Healthcare	Domino	Firewall 1	Firewall 2
others$_{best}$	400	193	453	34	14	20	65	10
addRole$_{best}$	400	184	453	34	14	20	64	10
addRole$_{avg.}$	401.85	187.15	453.1	34	14	20	64.95	10
addRole$_{SD}$	0.93	1.71	0.30	0.00	0.00	0.00	0.22	0.00
$Dev_{avg.}$	+0.46%	−3.03%	0.02%	0.00%	0,00%	0.00%	−1.54%	0.00%

The values others$_{best}$, addRole$_{best}$ and addRole$_{avg.}$ represent the number of roles, which is the objective function of the Basic Role Mining Problem (see Sect. 2). For each benchmark instance, addRole$_{best}$ denotes the overall best fitness value obtained by one of the individuals, addRole$_{avg.}$ represents the average fitness value of the best individual after convergence over the 20 evaluation test runs, while addRole$_{SD}$ denotes the corresponding standard deviation. Furthermore, the deviation $Dev_{avg.}$ between the average results obtained from the application of the addRole-EA and the best solution found by the other role-mining algorithms (others$_{best}$) is shown in each case. The values of the best solutions found so far were adopted from [16]. It can be noted, that for each *HPLabs* benchmark instance, there exists at least one run of the addRole-EA, where the best number of roles achieved was smaller or equal the best number of roles achieved by any other algorithm. In two of the instances (*America small* and *Firewall 1*) the best solution could even be improved. Furthermore, the attained average results range from −3.03% to +0.46% compared to the best results found in literature. This emphasizes the high performance of the addRole-EA considering the *HPLabs* benchmark.

Also on the *PLAIN_x*-instances of *RMPlib*, the addRole-EA provides good results (see Table 5). Since no comparable results are available in current literature at this point, the results obtained are compared with the given role bound of each instance. For better overview, the results for the *PLAIN_small* instances are shown as an example. Results on the other instances of the *RMPlib* benchmarks and further analyses can be found in the documentation of *RMPlib* [2]. Again, the best solution of the addRole-EA addRole$_{best}$, the average solution addRole$_{avg.}$ the standard deviation addRole$_{SD}$ and the deviation $Dev_{avg.}$ between the average results obtained from the application of the addRole-EA and the given role bounds are shown in each case.

Table 5. Evaluation of the addRole-EA based on instances of *RMPlib.*

	PLAIN_small_01	PLAIN_small_02	PLAIN_small_03	PLAIN_small_04	PLAIN_small_05	PLAIN_small_06	PLAIN_small_07	PLAIN_small_08
Role Bound	25	25	25	25	50	50	30	50
addRole$_{best}$	24	27	27	28	49	50	33	50
addRole$_{avg.}$	24.65	30.05	29.8	32.8	49.8	50.25	39.2	52.5
addRole$_{SD}$	0.48	1.83	1.54	2.84	1.03	0.43	3.71	1.63
$Dev_{avg.}$	−1.40%	20.20%	19.20%	31.20%	−0.40%	0.50%	30.67%	5.00%

Evaluation of Mutation Methods. In [1], it has already been investigated how often the application of a certain mutation method leads to a local improvement, in order to show the general functionality of the mutation methods of the addRole-EA. At this, local improvement means, that the number of roles of an individual after mutation is lower than the number of roles of the corresponding individual before mutation. Here, in contrast, the relation of the mutation methods and global improvements is examined. For this purpose, the cases in which the number of roles of an individual obtained from the application of one of the mutation methods falls below the number of roles of the currently global best solution is counted. The percentage distribution of global improvements among the different mutation methods of the addRole-EA is shown in Fig. 5. All results presented in this section are based on the evaluation of the same test runs as in Sect. 4.2.

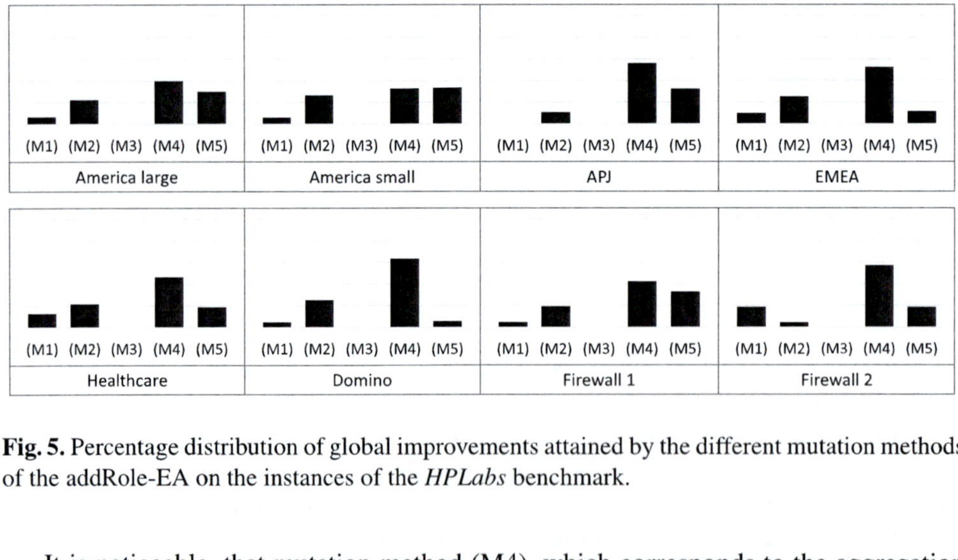

Fig. 5. Percentage distribution of global improvements attained by the different mutation methods of the addRole-EA on the instances of the *HPLabs* benchmark.

It is noticeable, that mutation method (M4), which corresponds to the aggregation of all permissions of one user in one role, leads to global improvements in an above-average number of cases. This corresponds to the observation of Sect. 4.2, that the number of users after pre-processing and the optimal solution of the benchmark instances are very close for the instances of the *HPLabs* benchmark. In addition, it can be seen, that mutation method (M3), which reduces the permission set of a given role by the permissions of another role, leads to rare global improvements.

The instances of the *RMPlib* benchmark, however, reveal a different pattern (see Fig. 6). Here, mutation methods (M1), which creates a new role from the intersection of permission sets of different users, (M2), which derives a role from all permissions of one user setminus the union of the permissions of some of the user's roles and especially (M5), which corresponds to the merging of two roles, perform particularly strong, while (M4) and again (M3) perform below average. This can be explained by the fact, that considering the instances of the *RMPlib* benchmark, unlike the instances of the *HPLabs* benchmark, most users have more than only one role. Hence, these results suggest, to focus on the further development and investigation of (M1), (M2) and (M5). In addition, the question arises as to whether omitting (M3) or (M4) might improve, respectively fasten the optimization process as it hardly contributes to the global improvement of the solutions of the addRole-EA on the *RMPlib* benchmark instances.

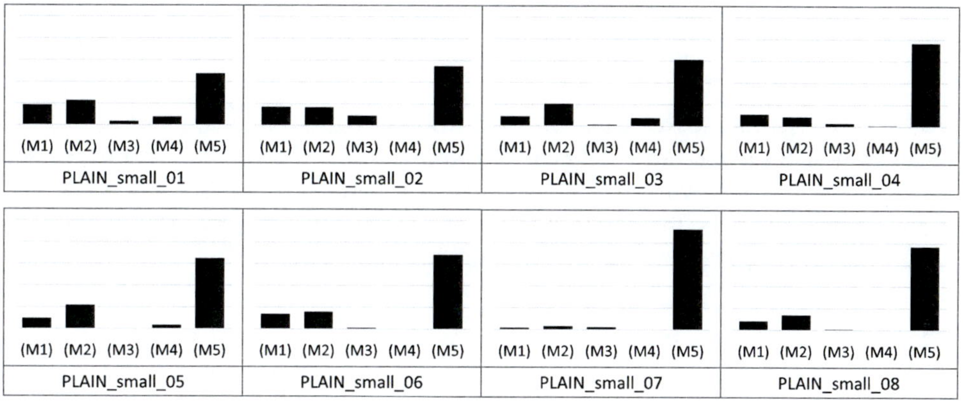

Fig. 6. Percentage distribution of global improvements attained by the different mutation methods of the addRole-EA on instances of the *RMPlib* benchmark.

5 New Mutation Methods for the AddRole-EA

Building on the results of Sect. 4.2, new variants of the mutation methods of the addRole-EA are presented and evaluated in this section. Since (M4) does not allow the development of further variants and (M3) is a candidate for being omitted, only variants of (M1), (M2) and (M5) are investigated in more detail.

5.1 (M1): Intersection of Permission Sets

One way to create new roles is to intersect the permission sets of different users. The users, whose permission sets are intersected for this purpose, are chosen randomly in the current version of the addRole-EA. However, this repeatedly leads to the creation of roles containing no permission due to the fact that the permission sets of randomly chosen users can be very heterogeneous and thus have empty intersections. In contrast, it can be assumed, that similar users have a large intersection considering their permission sets and are thus assigned similar roles. Hence, the user selection of the three new variants of (M1), which are introduced in the following, is based on the users' reciprocal similarity. Subsequently, these variants are compared to the current random selection (M1). In order to calculate the similarity of two users, the Jaccard-coefficient used:

$$J(u_i, u_j) = \frac{P(u_i) \cap P(u_j)}{P(u_i) \cup P(u_j)}, \tag{3}$$

where $P(u_i)$ is the set of permissions assigned to user u_i. Based on this, the new variants of (M1) can now be described. In each of them, a random number $k_{int.}$ is drawn to determine the number of users, whose permission sets are to be intersected. Subsequently, the first user u_{init} is chosen randomly, while the other $(k_{int.} - 1)$ users are selected based on the similarity to the first user depending on the chosen variant.

(M1v1): Deterministic Selection of Most Similar Users. It is straightforward from the definition of the Jaccard-coefficient that similar users share many common permissions. It is therefore very likely, that these also have similar roles. To obtain these roles, in this variant of (M1), the permission set of u_{init} is intersected with the permission sets of the $(k_{int.} - 1)$ users, which have highest similarity to u_{init}.

(M1v2): User Selection Based on Roulette Wheel Selection. The deterministic selection of the most similar users is prone to result in the repeated selection of the same users and thus in the repeated creation of the same roles. Moreover, this approach hampers the creation of roles shared by users who have rather low similarity, which can nevertheless be important in the optimization of the role concept. It can therefore be reasonable to include a random component in the selection process. This approach is based on the well-known *Roulette Wheel Selection (RWS)* [11], here applied to the remaining set of users $U_r := U \setminus \{u_{init}\}$ to select $(k_{int.} - 1)$ users. For this purpose, in a first step, so-called *slotsizes* are calculated for each of the remaining users:

$$slotsize(u_i) := \frac{J(u_i, u_{init})}{\sum_{u \in U_r} J(u, u_{init})}. \tag{4}$$

In a second step, the corresponding distribution function is calculated:

$$F(u_i) := \sum_{j \leq i} slotsize(u_j). \tag{5}$$

In a third step, the remaining users can be selected. To select a user for intersection, a random number $r_{RWS} \in [0, 1)$ is drawn. At this, user u_i, is selected if the following criteria is fulfilled:

$$F(u_{i-1}) \leq r_{RWS} < F(u_i). \tag{6}$$

To conclude the selection process, this is repeated $(k_{int.} - 1)$ times to select all remaining users.

(M1v3): User Selection Based on Stochastic Universal Sampling. Another possibility to add a random component to the selection process is *Stochastic Universal Sampling (SUS)* [3,4]. In order to select the users for intersection, the same slotsizes and distribution function is used as for *RWS* (see Eqs. (4) and (5)). *SUS* draws a random number $r_{SUS} \in \left[0, (k_{int.} - 1)^{-1}\right)$. At this, user u_i is selected if the following condition is fulfilled:

$$F(u_{i-1}) \leq r_{SUS} + \frac{s}{k_{int.} - 1} < F(u_i), \quad s \in \{0, 1, ..., (k_{int.} - 2)\}. \tag{7}$$

This selection condition provides users with lower similarity values a fair chance to be selected, while *RWS* may repeatedly prefer users with high similarity values compared to u_{init}.

Evaluation. To evaluate the different variants of (M1), the addRole-EA was tested on three different benchmark instances of *RMPlib* (*PLAIN_small_02*, *PLAIN_small_05* and *PLAIN_medium_01*). Due to their structure, the instances of the *HPLabs* benchmark were not considered as suitable benchmark instances for this test setup. In order to focus on the performance of the different variants of (M1), all mutation and crossover methods were deactivated except for the considered variant of (M1). These tests were performed 20 times for each of the three variants as well as the basic version of (M1). The values of all other parameters were adopted from [1]. Figure 7 shows the course of the average number of roles from all 20 test runs against iterations.

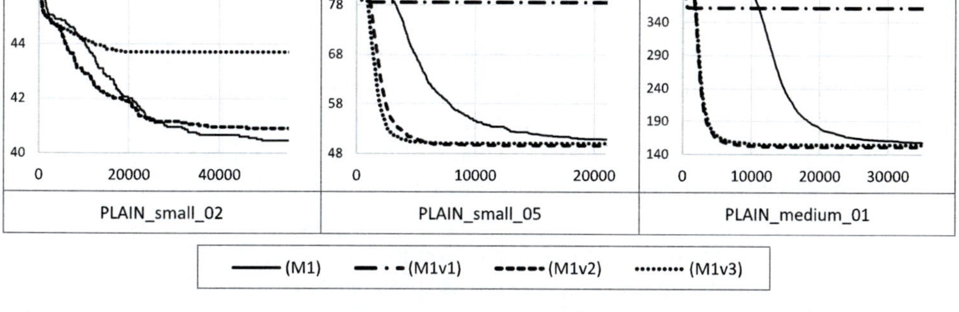

Fig. 7. Comparison of the different variants of (M1).

Figure 7 first indicates, that, (M1v1) performs quite poorly, on all three benchmark instances, since hardly any roles can be reduced. This is probably due to the fact that the same users are selected repeatedly by deterministic selection and therefore only a limited number of roles is created. (M1v2) and (M1v3) perform about equally well on *PLAIN_small_05* and *PLAIN_medium_01* and clearly better than the standard selection (M1). For *PLAIN_small_02*, however, the situation is different: At the beginning of the optimization process, (M1v2) still reduces the number of rolls the fastest, but is eventually surpassed by the standard mutation method (M1). Moreover, (M1v3) performs significantly worse in this case compared to the other two benchmark instances.

5.2 (M2): Permission Set Setminus Union of Permissions of Roles

Also for mutation method (M2) several variants were created and evaluated. In contrast to the standard selection of (M2), the roles, that constitute the basis for the union of permissions, which is taken from a users permission set to create a new role, are selected based on their size or popularity.

(M2v1): Role Selection Based on Role Size. The goal of mutation method (M2) is to find permissions of a user that are not covered by already existing roles. For this purpose, a random number k_{union} of roles of one user is selected. Subsequently, all permissions contained in one of the selected roles are removed from the user's permission set to create the new role. In (M2v1), the k_{union} currently largest roles, in terms of the number of contained permissions, of a user are selected.

(M2v2): Role Selection Based on Popularity. In this variant of (M2), it is not a role's size, but the popularity of a role, which is used as selection criterion for the k_{union} roles. The popularity of a role $pop(r_j)$ in this sense, refers to the number of users who are assigned a given role r_j.

$$pop(r_j) : R \mapsto \mathbb{N}, \quad pop(r_j) := \sum_{i=1}^{m} UA_{ij}. \tag{8}$$

Hence, to create the new role, rather those permissions are used, that are not part of a role assigned to many users, which can thus be considered a potentially good role regarding the optimization process.

Evaluation. To evaluate the different variants of (M2), the addRole-EA was tested on the three different benchmark instances, with (M2) respectively one of its variants activated as only mutation method and crossover switched off. Again, each test setup was performed 20 times. Figure 8 shows the course of the average number of roles from all 20 test runs against iterations. It can bee seen, that both new variants perform significantly worse than the standard operator (M2) on all three benchmark instances. This leads to the conclusion that prioritized selection may hamper the optimization process compared to the pure random selection in (M2). Nevertheless, it can be observed that (M2v2) performs above average, in some first few iterations, which shows the general strength of the popularity approach. After some iterations, however, this variant is overtaken by the standard version of (M2), which is probably due to the fact, that, at this point, the role concepts proposed by the addRole-EA change less, which leads (M2v2) to repeatedly create the same roles.

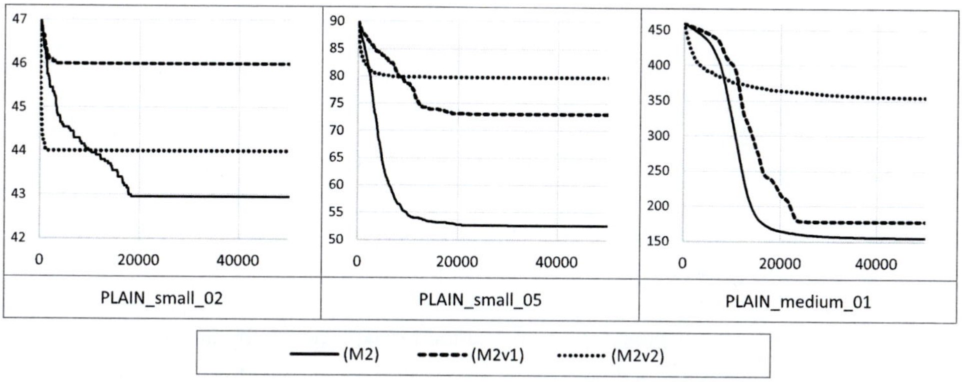

Fig. 8. Comparison of the different variants of (M2).

5.3 (M3): Splitting of Roles

In Sect. 4, it was shown that mutation method (M3) hardly contributes to the role mining process of the addRole-EA. Therefore, instead of studying possible variants of (M3), the effects of omitting (M3) on the performance of the optimization process are investigated at this point.

Evaluation. To investigate the performance of the addRole-EA with and without (M3), the standard version of the algorithm, as well as a variant in which (M3) is deactivated, was tested 20 times on the three benchmark instances.

Figure 9 shows, that the addRole-EA with (M3) deactivated leads to good results more rapidly. As, in addition, omitting (M3) provides similar or even better results on each of the three benchmark instances, it might be worthwhile to consider excluding (M3) completely from the mutation methods of the addRole-EA. However, if the obtained role concepts are to consist mainly of roles that have no intersections, this method may be a tool to achieve this objective.

5.4 (M4): Permission Set of a User

Similar to (M3), the analysis part of the previous section revealed that (M4) hardly contributes to the reduction of roles on the instances of *RMPlib*. Therefore, it is reasonable to examine the effect of omitting (M4) on the optimization process.

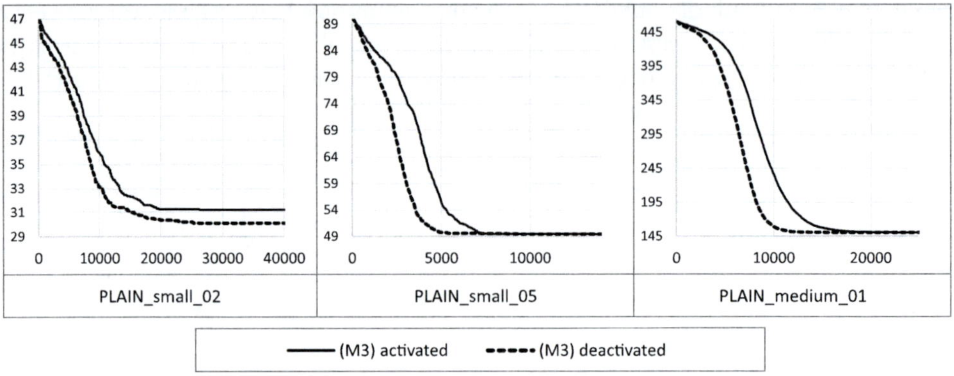

Fig. 9. Comparison of addRole-EA with (M3) activated/deactivated.

Evaluation. For this purpose, again, the standard version of the addRole-EA, as well as a variant in which (M4) is deactivated, was tested 20 times on each of the considered benchmark instances (Fig. 10).

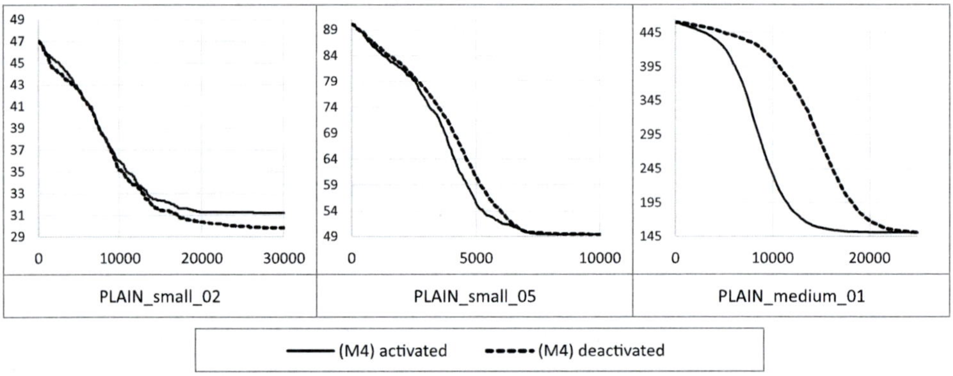

Fig. 10. Comparison of addRole-EA with (M4) activated/deactivated.

In contrast to (M3), where the deactivation of the mutation method accelerated the optimization process, omitting (M4), although having little contribution in terms of global improvements, resulted in slower convergence on two of the three benchmark instances considered (*PLAIN_small_05* and *PLAIN_medium_01*).

5.5 (M5): Merging of Roles

As shown in Sect. 4, (M5) is the mutation method that causes the most global improvements on the examined instances of the *RMPlib* benchmark. Also on the instances of the *HPLabs* benchmark, (M5) belongs to the rather successful methods considering global improvement. It may therefore be beneficial to investigate whether this method can be further improved. For this purpose, three different variants of (M5) are presented and evaluated subsequently.

(M5v1): Merging of Smallest Roles. It is clear that maintaining many small roles, in terms of the number of contained permissions, typically involves a large total number of roles. While large roles, covering multiple permissions of several users, rather contribute to a small total number or roles. Hence, this variant of (M5) merges the two smallest roles of a randomly selected user.

(M5v2): Merging of Roles with Largest Intersection. Whenever two roles are very similar, they may be assigned to the same users. If that is the case, also the union of the contained permissions can be assigned to those users. In order to reflect this, again, two roles of one randomly chosen user are selected and merged. At this, the first role is selected randomly, while the second role is chosen in such a way, that its permission set has the largest intersection with the permission set of the first role.

(M5v3): Merging of Roles with Highest Popularity. Another approach to select the two roles, which are to be merged, is to consider popularity as selection criterion. Again, a user and one of its roles are chosen randomly. As a second role, among all roles assigned to the selected user, the role having the highest joint popularity with the first role is selected. At this, the joint popularity of two roles r_j and r_l is defined (similar to Eq. 8) as:

$$pop(r_j, r_l) : (R \times R) \mapsto \mathbb{N}, \quad pop(r_j, r_l) := \sum_{i=1}^{m} UA_{ij} \cdot UA_{il}. \tag{9}$$

Evaluation. Evaluation of the different variants of (M5) was conducted on the three chosen benchmark instances of *RMPlib* with (M5) activated as only mutation and crossover method. Again, each test setup was performed 20 times (Fig. 11).

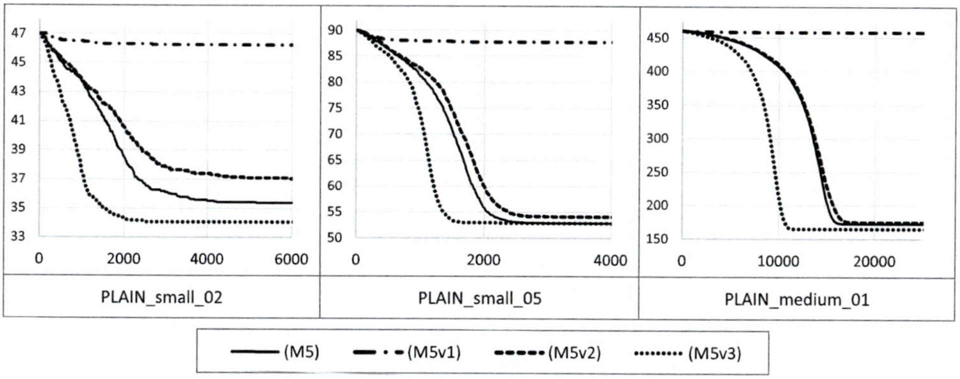

Fig. 11. Comparison of the different variants of (M5).

The evaluation of the different variants of (M5) leads to similar results on each of the three selected benchmark instances. It can be clearly shown, that variant (M5v1) hardly helps to reduce the total number of roles. Also (M5v2) performs worse than the pure random selection of the original version of mutation method (M5). Of all presented variants of (M5), variant (M5v3) clearly performs best, as it produces similar or even better results in significantly fewer iterations.

6 Evaluation

In order to provide a final evaluation based on findings in Sect. 4 and Sect. 5, the addRole-EA is executed using the best performing variants of the mutation operators. In concrete terms this means replacing (M1) by (M1v2) using Roulette Wheel Selection. (M2) is maintained in its original version, since none of the presented variants could outperform (M2). Mutation method (M3) is completely omitted in the course of the algorithm. (M4) is kept unchanged, as this method can play an important role on instances of different structure (e.g. *HPLabs* benchmark). Thus, it is continued to simply copy the permission set of a random user to create a new role. Considering (M5), it was shown, that, for merging two roles, it is worthwhile to analyze, which roles are assigned to the same users. Therefore, the original version of (M5) is replaced by (M5v3).

The performance of this advanced addRole-EA was tested on *PLAIN_small_02*, *PLAIN_small_05* and *PLAIN_medium_01* and compared to the performance of the original version of the addRole-EA. At this, each test setup was repeated 20 times. Table 6 gives an overview of the results.

Table 6. Performance comparison of advanced addRole-EA and original version.

	PLAIN_small_02		PLAIN_small_05		PLAIN_medium_01	
	original	advanced	original	advanced	original	advanced
Number of roles	30.05	29.30	49.80	49.80	150.40	151.50
Last improvement at iteration	12,605.20	9,920.65	4,132.05	3,414.75	12,504.80	5,922.85
Time (ms) per iteration	17.31	14.18	7.81	7.70	110.38	94.43

Figure 12 shows the development of the number of roles against iterations for the advanced and original version of the addRole-EA. It can be seen that the advanced version of the addRole-EA provides similar results in terms of the number of roles. On *PLAIN _small_02* the advanced version performs slightly better, while on *PLAIN_medium_01* the original version leads to a slightly better result considering the number of roles obtained. However, considering the number of iterations needed, the time per iteration and the course of the number of roles against iterations, Table 6 and Fig. 12 show, that the advanced version of the addRole-EA significantly accelerates the optimization process. Especially on the comparatively larger benchmark instance *PLAIN_medium_01*, the number of iterations needed by the advanced version is reduced to substantially less than 50% compared to the number of iterations needed by the original version.

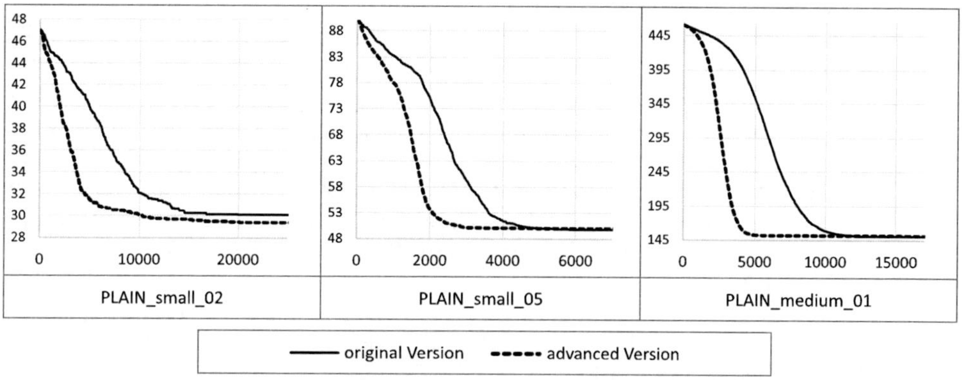

Fig. 12. Performance comparison of advanced addRole-EA and original version.

7 Conclusion and Future Works

In this paper, an advanced version of the addRole-EA for the Role-Mining Problem was presented. For this purpose, the contribution of the different mutation methods of the original version of the addRole-EA to the global improvement of the individuals was analyzed. Based on the results obtained, new variants of the existing mutation methods were proposed and evaluated. For each mutation method, the best performing variant was then integrated into an advanced version of the addRole-EA. Evaluation shows, that the advanced version attains similar results in terms of the number of roles in significantly fewer iterations.

Even if it could be shown, that the advanced version of the addRole-EA leads to good results much faster on synthetic benchmark instances, it remains an open question, whether the same is valid for real-world data. Such real world instances may have structures that could be fundamentally different from synthetic benchmark instances, for example considering problem size or in terms of the distribution of the user-permission assignment. Furthermore, running and managing an industrial enterprise system also involves dynamic changes in the user permission assignment caused by dynamically occurring events like new employees joining a company, employees leaving a company, position changes and permission requests, to which the addRole-EA must be adapted.

References

1. Anderer, S., Kreppein, D., Scheuermann, B., Mostaghim, S.: The addrole-EA: a new evolutionary algorithm for the role mining problem. In: Proceedings of the 12th International Joint Conference on Computational Intelligence, IJCCI 2020, Budapest, Hungary, November 2–4, 2020, pp. 155–166. SCITEPRESS (2020). https://doi.org/10.5220/0010025401550166

2. Anderer, S., Scheuermann, B., Mostaghim, S., Bauerle, P., Beil, M.: RMPlib: a library of benchmarks for the role mining problem. In: Proceedings of the 26th ACM Symposium on Access Control Models and Technologies. SACMAT 2021, New York, NY, USA, pp. 3–13. Association for Computing Machinery (2021). https://doi.org/10.1145/3450569.3463566
3. Baker, J.E., et al.: Reducing bias and inefficiency in the selection algorithm. In: Proceedings of the Second International Conference on Genetic Algorithms, vol. 206, pp. 14–21 (1987)
4. Baker, J.E.: Adaptive selection methods for genetic algorithms. In: Proceedings of an International Conference on Genetic Algorithms and Their Applications, vol. 1. Hillsdale, New Jersey (1985)
5. Bertino, E., Bonatti, P.A., Ferrari, E.: TRBAC: a temporal role-based access control model. In: Proceedings of the Fifth ACM Workshop on Role-Based Access Control. RBAC 2000, New York, NY, USA, pp. 21–30. Association for Computing Machinery (2000). https://doi.org/10.1145/344287.344298
6. Blundo, C., Cimato, S.: A simple role mining algorithm. In: Shin, S.Y., Ossowski, S., Schumacher, M., Palakal, M.J., Hung, C.C. (eds.) Proceedings of the 2010 ACM Symposium on Applied Computing - SAC 2010, New York, New York, USA, pp. 1958–1962. ACM Press (2010). https://doi.org/10.1145/1774088.1774503
7. Colantonio, A., Di Pietro, R., Ocello, A.: A cost-driven approach to role engineering. In: Wainwright, R.L., Haddad, H.M. (eds.) Proceedings of the 2008 ACM symposium on Applied computing - SAC 2008, New York, USA, pp. 2129–2136. ACM Press (2008). https://doi.org/10.1145/1363686.1364198
8. Dong, L.J., Wang, M.C., Kang, X.J.: Mining least privilege roles by genetic algorithm. Appl. Mech. Mater. **121–126**, 4508–4512 (2011). https://doi.org/10.4028/www.scientific.net/AMM.121-126.4508
9. Du, X., Chang, X.: Performance of AI algorithms for mining meaningful roles. In: 2014 IEEE Congress on Evolutionary Computation (CEC), pp. 2070–2076. IEEE (2014). https://doi.org/10.1109/CEC.2014.6900321
10. Ene, A., Horne, W., Milosavljevic, N., Rao, P., Schreiber, R., Tarjan, R.E.: Fast exact and heuristic methods for role minimization problems. In: Ray, I., Li, N. (eds.) Proceedings of the 13th ACM Symposium on Access Control Models and Technologies - SACMAT 2008, New York, New York, USA, pp. 1–10. ACM Press (2008). https://doi.org/10.1145/1377836.1377838
11. Holland, J.: Adaptation in Natural and Artificial Systems. University of Michigan Press, Ann Arbor', Cité page 100 (1975)
12. Huang, H., Shang, F., Liu, J., Du, H.: Handling least privilege problem and role mining in RBAC. J. Comb. Optim. **30**(1), 63–86 (2015). https://doi.org/10.1007/s10878-013-9633-9
13. Kumar, R., Sural, S., Gupta, A.: Mining RBAC roles under cardinality constraint. In: Jha, S., Mathuria, A. (eds.) ICISS 2010. LNCS, vol. 6503, pp. 171–185. Springer, Heidelberg (2010). https://doi.org/10.1007/978-3-642-17714-9_13
14. Lu, H., Vaidya, J., Atluri, V.: Optimal Boolean matrix decomposition: application to role engineering. In: 2008 IEEE 24th International Conference on Data Engineering, pp. 297–306. IEEE (2008). https://doi.org/10.1109/ICDE.2008.4497438
15. Lu, H., Vaidya, J., Atluri, V.: An optimization framework for role mining. J. Comput. Secur. **22**(1), 1–31 (2014). https://doi.org/10.3233/JCS-130484
16. Mitra, B., Sural, S., Vaidya, J., Atluri, V.: A survey of role mining. ACM Comput. Surv. **48**(4), 1–37 (2016). https://doi.org/10.1145/2871148
17. Molloy, I., et al.: Mining roles with semantic meanings. In: Ray, I., Li, N. (eds.) Proceedings of the 13th ACM Symposium on Access Control Models and Technologies - SACMAT 2008, pp. 21–30, New York, New York, USA. ACM Press (2008). https://doi.org/10.1145/1377836.1377840

18. Molloy, I., Li, N., Li, T., Mao, Z., Wang, Q., Lobo, J.: Evaluating role mining algorithms. In: Carminati, B., Joshi, J. (eds.) Proceedings of the 14th ACM Symposium on Access Control Models and Technologies - SACMAT 2009, New York, New York, USA, pp. 95–104. ACM Press (2009). https://doi.org/10.1145/1542207.1542224
19. Saenko, I., Kotenko, I.: Genetic algorithms for role mining problem. In: 2011 19th International Euromicro Conference on Parallel, Distributed and Network-Based Processing, pp. 646–650. IEEE (2011). https://doi.org/10.1109/PDP.2011.63
20. Saenko, I., Kotenko, I.: Design and performance evaluation of improved genetic algorithm for role mining problem. In: 2012 20th Euromicro International Conference on Parallel, Distributed and Network-based Processing, pp. 269–274. IEEE (2012). https://doi.org/10.1109/PDP.2012.31
21. Saenko, I., Kotenko, I.: Using genetic algorithms for design and reconfiguration of RBAC schemes. In: Unknown (ed.) Proceedings of the 1st International Workshop on AI for Privacy and Security - PrAISe '16, New York, New York, USA, pp. 1–9. ACM Press (2016). https://doi.org/10.1145/2970030.2970033
22. Sandhu, R.S., Coyne, E.J., Feinstein, H.L., Youman, C.E.: Role-based access control models. Computer **29**(2), 38–47 (1996). https://doi.org/10.1109/2.485845
23. Schlegelmilch, J., Steffens, U.: Role mining with orca. In: Ferrari, E., Ahn, G.J. (eds.) Proceedings of the tenth ACM Symposium on Access Control models and Technologies - SACMAT 2005, New York, New York, USA, pp. 168–176. ACM Press (2005). https://doi.org/10.1145/1063979.1064008
24. Suganthy, A., Chithralekha, T.: Role-evolution in role-based access control system. Int. J. Emerging Res. Manag. Technol. **6**(7), 223–227 (2018). https://doi.org/10.23956/ijermt.v6i7.215
25. Vaidya, J., Atluri, V., Guo, Q.: The role mining problem. In: Lotz, V., Thuraisingham, B. (eds.) Proceedings of the 12th ACM symposium on Access control models and technologies - SACMAT 2007, New York, New York, USA, pp. 175–184. ACM Press (2007). https://doi.org/10.1145/1266840.1266870
26. Vaidya, J., Atluri, V., Warner, J., Guo, Q.: Role engineering via prioritized subset enumeration. IEEE Trans. Dependable Secure Comput. **7**(3), 300–314 (2010). https://doi.org/10.1109/TDSC.2008.61
27. Verizon: Data breach investigations report 2019. Comput. Fraud Secur. **2019**(6), 4 (2019). https://doi.org/10.1016/S1361-3723(19)30060-0
28. Xu, Z., Stoller, S.D.: Algorithms for mining meaningful roles. In: Atluri, V., Vaidya, J., Kern, A., Kantarcioglu, M. (eds.) Proceedings of the 17th ACM symposium on Access Control Models and Technologies - SACMAT 2012, New York, New York, USA, pp. 57–66. ACM Press (2012). https://doi.org/10.1145/2295136.2295146
29. Zhang, D., Ramamohanarao, K., Ebringer, T.: Role engineering using graph optimisation. In: Lotz, V., Thuraisingham, B. (eds.) Proceedings of the 12th ACM Symposium on Access Control Models and Technologies - SACMAT 2007, New York, New York, USA, pp. 139–144. ACM Press (2007). https://doi.org/10.1145/1266840.1266862

Behavioural Modelling of Digital Circuits in SystemVerilog Using Grammatical Evolution

Michael Tetteh[1(✉)], Conor Ryan[1(✉)], and Douglas Mota Dias[1,2(✉)]

[1] Biocomputing and Developmental Systems Group, University of Limerick, Limerick, Ireland
{michael.tetteh,conor.ryan,douglas.motadias}@ul.ie
[2] UERJ, Rio de Janeiro State University, Rio de Janeiro, Brazil
douglas.dias@uerj.br

Abstract. Digital circuit design is a very complex and time consuming task which requires a great deal of skill. It has been greatly aided by the use of Hardware Description Languages (HDLs) and powerful logic simulators. HDLs permit circuit designers to design circuits at a very abstract level. These are then tested before being committed to hardware. We present Automatic Design of Digital Circuits (ADDC), a system that employs Grammatical Evolution, SystemVerilog (an HDL) and Icarus Verilog (simulator) to design conventional circuits. ADDC is easily configurable to use with different HDLs and logic simulators. ADDC designs circuits at a higher abstraction level compared with previous works due to the use of HDLs, which are very expressive. Constructs such as `if-else`, `always` procedural block, `generate for-loop` (or synthesizable `for-loop`) and `switch-case` are readily available aiding designers to behaviourally describe circuits. In addition, due to the expressiveness of HDLs, ADDC evolved solutions are quite human interpretable. ADDC is tested using three combinational and two sequential circuits. We show that ADDC is successful on all five benchmark problems. In addition, we show that the introduction of simple domain knowledge into grammars has a major impact on evolutionary performance. Furthermore, Probabilisitc Tree Creation 2 initialization routine performed better on most digital circuit benchmark problems but not all compared to Sensible Initialization.

Keywords: Grammatical evolution · Digital circuit design · Combinational circuit · Sequential circuit · Evolvable hardware · Hardware description languages · Lexicase selection · Sensible Initialization · Probabilistic Tree Creation 2 (PTC 2)

1 Introduction

Circuit design is a highly skilled and demanding job, a slow and expensive process [17], with minor design errors costing millions of dollars to fix. Circuit design has evolved

The authors are supported by Research Grant 16/IA/4605 from the Science Foundation Ireland and by Lero, the Irish Software Engineering Research Centre. The second author is partially financed by the Coordenação de Aperfeiçoamento de Pessoal de Nível Superior - Brasil (CAPES) - Finance Code 001, and FAPERJ.

J. Garibaldi et al. (Eds.): IJCCI 2021, SCI 1119, pp. 24–43, 2023.
https://doi.org/10.1007/978-3-031-46221-4_2

from the use of vacuum tubes to the use of integrated circuits. This advancement has made digital circuits ubiquitous in the modern world but, as their complexity increases, so too does the difficulty associated with their design. In order to make the design task tractable, Hardware Description Languages (HDLs) were designed. HDLs aid designers in coping with complex circuits at an abstract level. These are then heavily tested using logic simulators to ensure functional correctness. Functionally correct circuits are then advanced through logic synthesis (logic gates and gates connections implementation of circuit) before fabrication takes place on a chip.

We present Automatic Digital Circuit Design Tool (ADDC) that uses GE and a simulator (Icarus Verilog) to automatically evolve behavioural models of circuits in SystemVerilog for a number of classic benchmarks. However, ADDC is easily configurable to use other HDLs and simulators other than SystemVerilog and Icarus Verilog respectively. We constrain ADDC to functional simulations only in this work with the facility to accommodate other features such as timing analysis moving foward.

Though there has been some related work, described in Sect. 2, most researchers evolve circuits at a lower level of abstraction (gate level). With the use of HDLs, which is akin to using high-level programming language (such as `C++` in software design) instead of Assembly Language, we are capable of evolving circuits at the highest abstraction level—behavioural level.

Grammar design has also shown that the use of appropriate operators from the target language can dramatically reduce the evaluation time of objective functions, in this case circuit desgins.

This work is an extension of the paper published under the title: "*Behavioural Modelling of Digital Circuits in System Verilog Using Grammatical Evolution*", publisehed in the proceedings of ECTA 2020 [19]. The initial work explored the benefit of evolving **combinational** circuits at a higher level abtraction through the use of GE with SystemVerilog and also investigated the impact of introducing simple domain knowledge into grammars and their impact on the evolutionary performance. We extend this work in two ways: we include two **sequential** circuit benchmark problems (both evolved at behavioural level, the same as the combinational circuit benchmarks) and determine the best GE initialization routine for this set of circuit design benchmark problems.

The contributions of this work are summarised as follows:

- Investigate the applicability of GE to circuit design and how best to set it up;
- Behavioural modelling of combinational and sequential circuits;
- We investigate the effect of exploiting information known about a problem to obtain a grammar variant for each of combinational circuit benchmarks;
- Identification of the most suitable GE initialization scheme for digital circuit benchmarks.

2 Background

Evolvable Hardware (EHW) is a field that deals with the design of electronic digital circuits using evolutionary algorithms. EHW is comprised of two major application domains: adaptive hardware (also known as Darwin Machines [4]) and electronic circuit design. In adaptive hardware, the hardware is capable of autonomous adaptation/reconfiguration of its architecture in response to changing environment, change in specification, etc. On the other hand, with conventional electronic circuit design, circuit specifications are known upfront and designed accordingly without the capability of evolving itself to changing factors. At present, there are three approaches used for circuit evolution: extrinsic [5] (simulation of circuits using simulators), intrinsic [25] (simulation of circuits in a target hardware such as Field Programmable Gate Arrays (FPGAs)) and mixtrinsic evolution (population of candidate circuits is divided in two, with each evaluated using either a simulator or an FPGA for example but not both) [23].

Digital circuit design can be done at different abstraction levels, namely, gate level, Register Transfer Level (RTL) and Behavioural level. Gate level designs deal with the use of logic gates to design circuits, while RTL designs, sometimes referred to as Functional Level, operates at a slightly higher level in that, instructions used to realize circuit functionality can be synthesized to a gate level representation. Behavioural level, the highest level of circuit design, is done using Hardware Description Languages (HDLs) such as Verilog/SystemVerilog, Very High-Speed Integrated Circuit Hardware Description Language (VHDL), etc. Behavioural models of digital circuits consist of programming constructs such as `if-else`, `switch-case`, `for-loops`, `bitwise, equality, logical, arithmetic` operators, etc. that describe the desired hardware functionality. However, not all programming constructs used in behavioural circuit models are synthesizable, thus they cannot be realized in hardware.

Mostly, HDLs are used to design modern complex circuits, usually by adopting a mixed-style approach. That is, designing circuits using a combination of the different design styles mentioned earlier. Most HDLs support mixed style designs of digital circuits. Circuit designs constructed using HDLs are subjected to numerous simulation and robust testing steps/cycles to ensure circuits are functionally correct and synthesizable. Many powerful tools, such as Vivado (Xilinx), Quartus (Intel/Altera), etc., are used by hardware designers to perform several forms of testing on circuit designs before being committed to hardware. However, there are some very resonable open source alternatives such as Icarus Verilog [28] (for simulation) and YoSys (for synthesis) [29].

Computer Aided Design (CAD) for digital circuits make use of these tools in a iterative manner, as shown in Fig. 1. Hardware designers rely on this to repeatedly perform functional simulations to ensure circuits are functionally correct before being transferred to the synthesis phase. At the synthesis phase, the object is to ensure the circuit is synthesizable. However, there are a number of reasons why a circuit may not be synthesizable; these vary from timing issues (components are too far apart, thus introducing some unexpected delays) to fitting issues (creation of a circuit is not possible given the available real estate), amongst others.

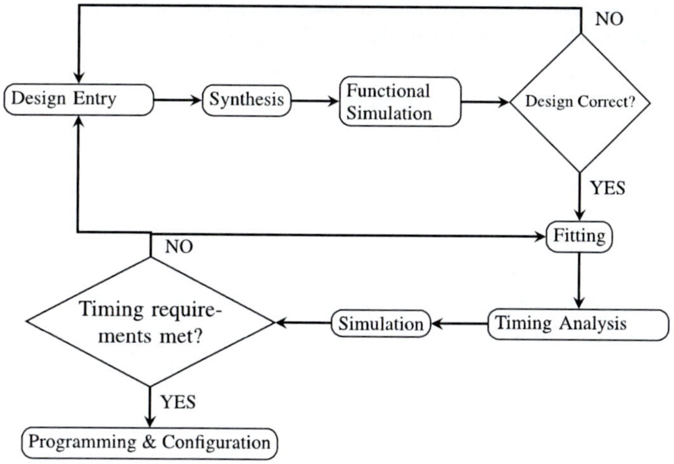

Fig. 1. The standard CAD flow for digital circuit design.

2.1 Related Work

Much work has been done in the application of EAs to both analog and digital circuit designs. However, we restrict the scope of this review to digital circuits, as this work falls under that category. Digital circuits fall under two main catergories: combinational and sequential circuits. Output(s) of a combinational circuit depends solely on its current input. Sequential circuits, in addition to combinational logic, use memory elements such as a flip flops and their output depends on both current and previous states of the circuits. The majority of EHW studies tackle the evolutionary design of digital circuits through the use of gate and functional evolution. Gate level evolution deals with the use of basic logic gates as the function set to design circuits, while functional level evolution in addition to logic gates use larger functional blocks such as multiplexers, adders and multipliers of varying bit width lengths, etc.

Cartesian Genetic Programming (CGP) is widely used in the evolutionary design of circuits. CGP is a GP variant which uses directed acyclic graphs to represent its programs instead of trees in conventional GP. CGP uses different types of genes in its genotype. A function gene encodes the address to a function in a lookup table a particular node must perform, connection genes encode addresses of the valid input sources for a node, usually in a feedforward manner [12], output genes holds addresses of nodes where program outputs are to be retrieved. CGP's representation makes it suitable for the low level evolution of circuits. Usually, the final output representation of CGP-evolved circuits are in the form of Boolean logic. CGP has been used to evolve combinational circuits such as adders and multipliers of small input sizes [20], seven segment display etc. Evolution of circuits such as the 1-bit adder, SR-latch and gated D-latch (sequential circuits) have been evolved at the gate level [2], although SR-latch results were not shown due to space limitations. The issue with gate-level evolution is its inability to scale to highly complex circuits from scratch [27]. Functional-level evolution was proposed by Murakawa et al. to increase scalability through the use of higher level functions such as multiplexers, small adders, subtractors and multipliers, etc. [13]. A 3-bit multiplier using binary multiplexers [27], 9- and 25-median approximate circuits [26] have all been evolved at the functional level.

In [7], CGP was used to evolve sequential circuits. CGP was extended to incorporate foward levels allowing feedback required for sequential circuits. A D Flip-flop was used in addition to logic gates for the evolution of sequential circuits. A custom-built translation system is then used to convert each evolved candidate sequential circuit into a VHDL program, which is then simulated using Xilinx software. All candidate circuits are subjected to timing analysis, an exercise which greatly increases the evaluation time. For example, it took approximately 2 days and 20 h to evolve a 4-bit up-counter [7]. Perhaps, a compromise, such as applying timing analysis to circuits only after they have either attained approximate functional correctness or fully functional correctness, may reduce experimental time.

2.2 Grammatical Evolution

GE evolves programs in arbitrary language specified in Backus-Naur Form (BNF). This feature makes GE attractive and it has been applied to many problems in different domains such as engineering, architecture, bio-informatics, software testing, explainable AI [1, 14, 18], etc. Despite its widespread adoption in many domains, digital circuit design hasn't received much attention from GE. However, the availability of good logic simulators makes it ideal to evolve digital circuits using GE with an HDL.

GE requires a genotype (a sequence of codons), a mapper and a valid subset of the target grammar designed for the problem. A codon is an integer equivalence of an 8-bit binary string decoded from a genome, though this can be extended if the need be. The defined grammar consists of rules which start with a non-terminal symbol with several productions which can expand into other non-terminals or terminals. The GE mapper uses a *modulo* rule to dictate which production belonging to a rule gets selected [16]. The process continues until one of the following happens: a valid phenotype is obtained (all non-terminals get expanded to terminals) or all codons in genome are exhausted leaving a number of non-terminals unexpanded (invalid phenotype). In case of the latter, wrapping counts are specified which determine the number of times the same genome can be reused to complete the mapping process. The syntactically correct program or program fragment in the target language is then evaluated and a fitness value assigned to the individual (Table 2).

3 Experimental Design

ADDC's circuit design flow is shown in Fig. 2. It captures only functional simulation, which is the current focus. The selected benchmark problems are briefly described in Sect. 3.1. For each benchmark problem we design a testbench which is used by the fitness evaluation function to determine the fitness score to assign to a candiate circuit. A testbench is analagous to regression testing in software programming. Essentially, it contains a set of test cases in addition to extra code to run the circuit under test on the testcases.

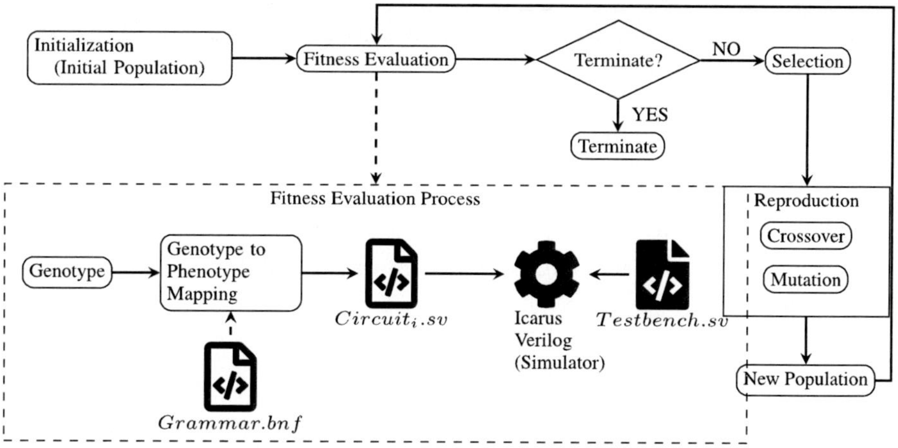

Fig. 2. GE Circuit Design Process.

Table 1. Experimental Run Parameters.

Parameter Type	Parameter Value	
	Set-up 1	Set-up 2
Initialization	Sensible Initialization	PTC2
Selection	Lexicase Selection	
No of generations	200	
Mutation rate	0.01	
Crossover rate	0.8	
Replacement rate	0.5	
No of runs	30	
Population size	2000	

Table 2. Number of Testcases Used for each problem. Sequential circuit benchmarks are marked with *.

Problem	Number of Testcases
11-Multiplexer	2,048
Seven Segment Display	16
Hamming Code (7,4) Decoder	128
*JK Flip-flop	10
*Mod-N Up-Down Counter	27

The parameters used for the experimental setup are shown in Table 1. Two different set-ups based on two initialization routines are used to run each experiments: Sensible Initialization and Probabilistic Tree Creation 2 (PTC 2). This is to enable as ascertain the most suitable GE initialization routine for digital circuit benchmark problems. The selection of PTC 2 initialization routine is based on evidence in [15], where PTC 2 adapted for use with GE consistently performed better compared to other initialization schemes on several symbolic regression and classification problems. Also, with or without domain knowledge, it offered the best solution. This particular PTC 2 implementation was based off the respective versions introduced in [6, 10]. It uses a pre-specified expansion budget to randomly select a non-terminal to expand, while favouring recursive productions with minimum expansion requirement less or equal to the remaining expansions. Additionally, the number of expansions required to fully expand outstanding non-terminals is budgeted for.

3.1 Benchmark Problems

We use three combinational and two sequential circuit benchmarks respectively. A variant grammar version for each combinational circuit is used to investigate the effect of the introduction of simple domain knowledge into grammar designs. The three combinational circuits benchmark considered are the 11-bit multiplexer, the Seven Segment Display and the Hamming Code (7,4) Decoder. We used the JK flip-flop and the Mod-N Up-Down counter for sequential circuits. Existing works mostly target the evolution up-counters [6,21,22]. In [7], 2-, 3- and 4- bit up-counters are evolved; 3-bit up-counters were also evolved in [21,22]. The selection of JK flip-flop as one of the sequential benchmark problems was partly motivated by its use as a memory element in the evolution of 3-bit up-counter in [22].

11-Bit Multiplexer. A multiplexer is a multi-input data device with a single output medium which selects a single input data via an address for transmission through the single output medium. The multiplexer problem is a classic GP benchmark problem. The 11-bit multiplexer has 8 data and 3 address/select input lines; it has been evolved though the use of both GP [9] and a classifier system [3]. We employ two grammars. Multiplexer Grammar A, in Listing 1, uses the following operators as its function set: bitwise-or, bitwise-and, ternary operator (`if-else`) and logical negation. Multiplexer Grammar B, on the other hand, uses the `ternary operator` as the only function ,as shown in Listing 2, which we speculate to be adequate to solve the 11-bit multiplexer problem. The data and address bits of the multiplexer are distinguished from each other using two separate rules, ⟨data-bit⟩ and ⟨address-bit⟩ respectively. These sort of constraints are easily imposable in GE, as it is simply a matter of grouping the two types of terminals under different rules. Achieving the same in standaard GP would be a more complex exercise as it requires the introduction of a new type that satisifes GP's closure principle. Both grammars use the `always block`, which is a proceedural construct which behaves in a similar manner like a while loop but executes statements within its body whenever the provided signals in the sentivity list change state.

Seven Segment Display. SSD is an electronic device consisting of seven segments or light emitting diodes (LEDs) for the display of digits (1 to 9) though there exist extensions that display letters from the English alphabet. The SSD specification considered in this work is limited to only digits. The device receives a 4-bit binary number (0000–1001) as input (digit to display) usually termed to as binary coded decimal

```
<design-module> ::= <begin-module><nextline>
                    <code-block><end-module>
                    <nextline>
   <code-block> ::= <always> <statement>
       <always> ::= "always@(*)" <nextline>
    <statement> ::= <output> "=" <expr>;
                    <nextline>
         <expr> ::= (<expr> & <expr>)
                  |(<expr> "|" <expr>)
                  |!(<expr>)|(<address-bit>
                  ? <expr> ":" <expr>)
                  |<data-bit>
       <output> ::= out
  <address-bit> ::= a0 | a1 | a2
     <data-bit> ::= d0 | d1 | d2 | d3 | d4
                  | d5 | d6 | d7
     <nextline> ::= "\n"
 <begin-module> ::= "module mux(output logic
                    out, input logic a0, a1,
                    a2, d0, d1, d2, d3, d4,
                    d5, d6, d7);"
   <end-module> ::= "endmodule"
```

Listing 1. 11-bit Multiplexer Grammar A [19].

```
<design-module> ::= <begin-module><nextline>
                    <code-block><end-module>
                    <nextline>
   <code-block> ::= <always> <statement>
       <always> ::= "always@(*)" <nextline>
    <statement> ::= <output> "=" <expr>;
                    <nextline>
         <expr> ::= (<address-bit> ? <expr>
                    ":" <expr>)
                  |<data-bit>
       <output> ::= out
  <address-bit> ::= a0 | a1 | a2
     <data-bit> ::= d0 | d1 | d2 | d3
                  | d4 | d5 | d6 | d7
     <nextline> ::= "\n"
 <begin-module> ::= "module mux(output logic
                    out, input logic a0, a1,
                    a2, d0, d1, d2, d3, d4,
                    d5, d6, d7);"
   <end-module> ::= "endmodule"
```

Listing 2. 11-bit Multiplexer Grammar B [19].

(BCD) which gets decoded into a 7-bit binary number. Each bit of the 7-bit binary number corresponds to a segment/LED of the device; a bit value of 1 and 0 indicates the corresponding segment's LED to be turned ON or OFF respectively, displaying the integer representation of the 4-bit binary number input. Both SSD Grammars in Listings 3 and 4 use `switch-case` construct as the only functional construct. Similar to the Muliplexer Grammars, the BCD and seven segment values are distinguished from each other using ⟨bcd-value⟩ and ⟨seven-segment⟩ rules respectively. To investigate the effect of domain knowledge/grammar design choices, SSD Grammar A in Listing 3 assumes no knowledge about the problem and, therefore, the valid bcd and seven-segment binary numbers are left to GE to deduce. This is a non-trivial task as not all the values are required. However, SSD Grammar B in Listing 4, has the bcd and seven-segment values explicitly provided for in the grammar, making it the least difficult and should record the best success rate.

Hamming Code (7,4) Decoder. Linear block codes are a category of codes capable of double error detection and single error correction. Thus, they have a minimum distance of three. Hamming Codes belong to this catergory of codes [11]. Hamming Code (7,4) Encoder encodes a 4-bit dataword by generating a three parity check bits. The 3 parity check bits and 4-bit dataword are merged into a 7-bit binary string termed as codeword either in a systematic or non-systematic encoding style before transmission. At the reception point, the Hamming Code (7,4) Decoder retrieves the transmitted dataword by generating a 3-bit binary number from the received codeword referred to as syndrome. Each bit serves as a parity check for the 4-bit dataword. Hamming Code (7,4) Decoder Grammars A and B are shown in Listings 7 and 9 respectively in the Appendix. Both grammars are designed to define a `function` that encapsulates instructions responsible for the generation of syndrome from the received codeword. For Hamming Code Decoders, if the syndrome is non-zero this means the codeword got corrupted during transmission. However, the integer equivalence of the non-zero syndrome is assummed to be the bit position in the codeword where the possible error occured. The bit in that position is flipped in an attempt to correct the codeword. The main difference between

```
<design-module> ::= <begin-module><nextline>
                    <code-block><next-line>
                    <end-module>
   <code-block> ::= <always><stmt>
       <always> ::= always@(bcd) <next-line>
         <stmt> ::= begin <nextline>
                    <switch-case><next-line>
                    end
  <switch-case> ::= case(bcd) <nextline>
                    <case-stmt> <next-line>
                    <default-case><nextline>
                    endcase
    <case-stmt> ::= <bcd-value> ":" <output>
                    "=" <seven-segment>;
                  | <bcd-value> ":" <output>
                    "=" <seven-segment>;
                    <nextline> <case-stmt>
 <default-case> ::= default ":" <output> "="
                    <seven-segment>;
    <bcd-value> ::= 4'b<bit><bit><bit><bit>
<seven-segment> ::= 7'b<bit><bit><bit><bit>
                    <bit><bit><bit>
     <nextline> ::= "\n"
       <output> ::= segment
          <bit> ::= 0 | 1
 <begin-module> ::= "module ssd(output logic
                    [6:0] segment, input
                    logic[3:0] bcd);"
   <end-module> ::= endmodule
```

Listing 3. Seven Segment Display Grammar A [19].

```
<design-module> ::= <begin-module><nextline>
                    <code-block><next-line>
                    <end-module>
   <code-block> ::= <always><stmt>
       <always> ::= always@(bcd) <next-line>
         <stmt> ::= begin <nextline>
                    <switch-case><next-line>
                    end
  <switch-case> ::= case(bcd) <nextline>
                    <case-stmt> <next-line>
                    <default-case><nextline>
                    endcase
    <case-stmt> ::= <bcd-value> ":" <output>
                    "=" <seven-segment>;
                  | <bcd-value> ":" <output>
                    "=" <seven-segment>;
                    <nextline> <case-stmt>
 <default-case> ::= default ":" <output> "="
                    <seven-segment>;
    <bcd-value> ::= 4'b0000 | 4'b0001
                  | 4'b0010 | 4'b0011
                  | 4'b0100 | 4'b0101
                  | 4'b0110 | 4'b0111
                  | 4'b1000 | 4'b1001
<seven-segment> ::= 7'b1111110 | 7'b0110000
                  | 7'b1101101 | 7'b1111001
                  | 7'b0110011 | 7'b1011011
                  | 7'b1011111 | 7'b1110000
                  | 7'b1111111 | 7'b1111011
                  | 7'b0000000
     <nextline> ::= "\n"
       <output> ::= segment
          <bit> ::= 0 | 1
 <begin-module> ::= "module ssd(output logic
                    [6:0] segment, input
                    logic[3:0] bcd);"
   <end-module> ::= endmodule
```

Listing 4. Seven Segment Display Grammar B [19].

Hamming Code (7,4) Decoder Grammar A and B is the ⟨expr⟩. Hamming Code (7,4) Decoder Grammar A contains no domain knowledge and is required to evolve expressions that generate the parity bits that form the syndrome. For that of Grammar B, we introduce some domain knowledge by restricting the generated expressions to use exactly four bits out of the 7 bit codeword which we know from the problem. As a result, we speculate Hamming Code (7,4) Decoder Grammar B to perform better than Grammar A. Both grammars use also use the `always` proceedural block.

JK Flip-flop. Flip-flops are bistable circuits used for the storage of information. A JK flip-flop has four possible states with no invalid states. No output change occurs when both inputs J and K have an input value of 0. However, when both J and K input values are 1, the current output value toggles. When $J = 0$, $K = 1$, the output is 0 and when $J = 1$, $K = 0$, the output is 1 at the clock transition. The JK flip-flop specification is captured in Table 3. Shown in Listing 5 is the JK flip-flop grammar designed and used for evolution. The grammar makes use of `always-block` and `switch-case` construct. The grammar uses the ⟨case⟩ rule to evolve the flip-flops input (J and K) and the ⟨statment⟩ rule to evolve the right output depending on the J and K input combinations.

Modulos-16 (Mod-16) Up-down Counter. A counter is a sequential logic device that increases or decreases its content by 1, or by some defined step value, when triggered by a clock signal. The mod-16 up-down counter is capable of counting up and down its content constrained to a range of 0 to 15. The specification for the mod-16 up-down counter considered in this paper is captured in the transition table in Table 4. The counter has an active low reset signal. The Mod-16 Grammar shown in Listing 6, uses the `always-block`, `switch-case` and two arithmetic operators ($-, +$) for the counting operations. The $\langle$case$\rangle$ rule is used to evolve to state of the counter and the $\langle$stmt$\rangle$ rule to evolve the appropriate output based on the state of the counter.

Table 3. JK Flip-flop Truth Table. ↑ represents the positive transition of the clock from 0 to 1; thus, the rising edge of the clock.

CLK	J	K	Q
↑	0	0	Q (no change)
↑	0	1	0
↑	1	0	1
↑	1	1	Q (toggle)

```
 <design-module> ::= <begin-module><new-line>
                     <code-block><new-line>
                     <end-module>
    <code-block> ::= <always-block><new-line>
                     <switch-case>
  <always-block> ::= "always @(posedge clk)"
   <switch-case> ::= "case ({j,k})" <new-line>
                     <case-statement><new-line>
                     endcase
<case-statement> ::= <case> ":" <statement>
                     <new-line><case-default>
                   | <case> ":" <statement>
                     <new-line><case-statement>
  <case-default> ::= default ":" <statement>
          <case> ::= "2'b"<bit><bit>
     <statement> ::= q "<=" <terminal>;
           <bit> ::= 0 | 1
        <lvalue> ::= q
      <terminal> ::= <bit> | <lvalue> | "~"<lvalue>
  <begin-module> ::= "module jk_ff(input j, input k,
                     input clk, output logic q);"
    <end-module> ::= endmodule
      <new-line> ::= "\n"
```

Listing 5. JK Flipflop Grammar.

Table 4. Mod-16 Up-Down Counter Transition Table. ↑ and ↓ represents *positive* and *negative* transitions of the clock respectively. The ? symbol represents *don't care* bits.

CLK	RESET	EN_LOAD	EN_UP	LOAD	UP_DOWN	Q (OUTPUT)
↑	0	?	?	?	?	0
↑	1	1	?	4-BIT NUMBER	?	LOAD
↑	1	0	1	?	1	Q + 1 (count up)
↑	1	0	1	?	0	Q − 1 (count down)
↑	1	0	0	?	?	Q (no change)
↓	?	?	?	?	?	?

```
<design-module> ::= <begin-module> <always> <endmodule>
       <always> ::= "always @(posedge clk)" <nextline> <switch-case>
         <expr> ::= <terminal> | <terminal> <op> <terminal> | <terminal> <op> <expr>
  <switch-case> ::= "casex  ({reset, en_load, enable_ud, up_ndown})" <nextline> <case-stmt>
                    endcase <nextline>
    <case-stmt> ::= <case> ":" <stmt> | <case> ":" <stmt> <case-stmt>
 <case-default> ::= default ":" "q<=" <expr>;
         <stmt> ::= "q<=" <expr>; <nextline>
         <case> ::= 4'b<bit><bit><bit><bit>
     <terminal> ::= 0 | 1 | q | load
          <bit> ::= 0 | 1 | ?
           <op> ::=  "-" | "+"
 <begin-module> ::= "module udcounter(input logic clk, up_ndown, en_load, enable_ud, reset,
                    input logic [3:0] load, output logic [3:0] q);" <nextline>
    <endmodule> ::= endmodule
     <nextline> ::= "\n"
```

Listing 6. Mod-16 Up-Down Counter.

4 Results and Discussions

In this section, we show results obtained from all the experiments conducted together with some discussions. We plot the average mean best per generation across the 30 independent runs conducted for each of the 5 selected benchmark problems with error bars. In each graph, we show results for the two experimental configurations: sensible initialisation and PTC2. These are shown in Figs. 3, 4, 5, 6, 7, 8, 9 and 10. In addition, we tabulate the success rate for each problem inclusive of the grammar variants where appropriate for each of the two experimental configurations as shown in Table 5.

Table 5. Success Rate.

Problem Grammar	Successful Runs using Sensible Initilization (out of 30)	Successful Runs using PTC2 (out of 30)
11–bit Multiplexer Grammar A	17	27
11–bit Multiplexer Grammar B	22	30
Seven Segment Display Grammar A	8	8
Seven Segment Display Grammar B	28	30
Hamming Code (7,4) Decoder Grammar A	22	26
Hamming Code (7,4) Decoder Grammar B	30	30
JK Flip-flop	30	30
Mod-16 Up-Down Counter	29	26

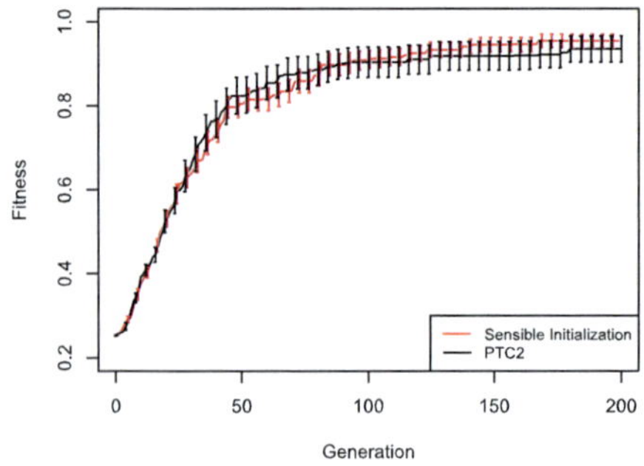

Fig. 3. Mean best across 30 independent runs using Hamming Code (7,4) Decoder Grammar A with error bars.

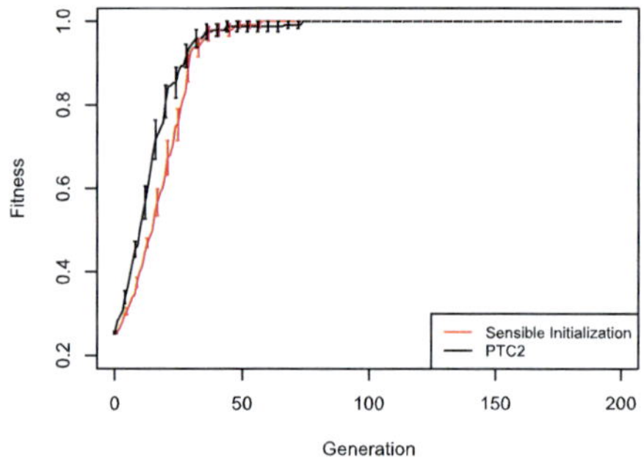

Fig. 4. Mean best across 30 independent runs using Hamming Code (7,4) Decoder Grammar B with error bars.

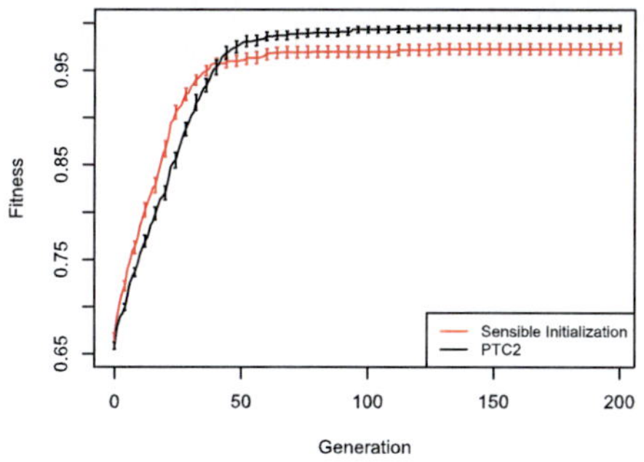

Fig. 5. Mean best across 30 independent runs using 11-bit Multiplexer Grammar A with error bars.

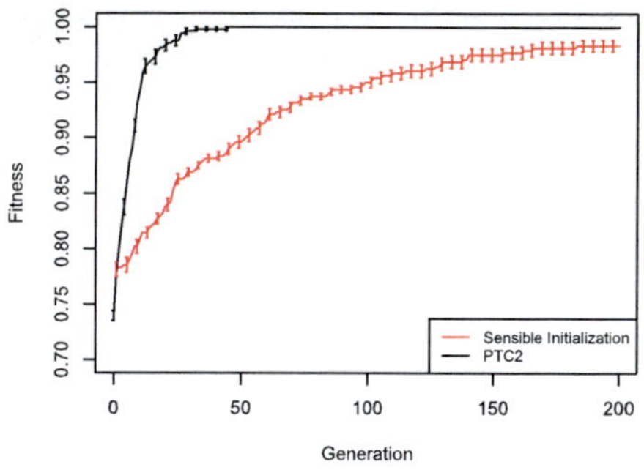

Fig. 6. Mean best across 30 independent runs using 11-bit Multiplexer Grammar B with error bars.

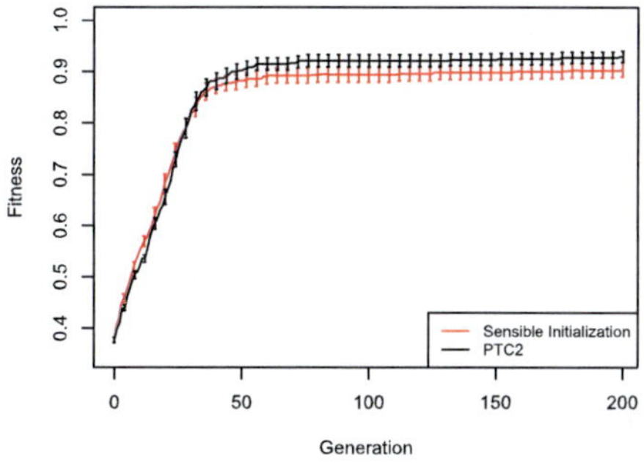

Fig. 7. Mean best across 30 independent runs using SSD Grammar A.

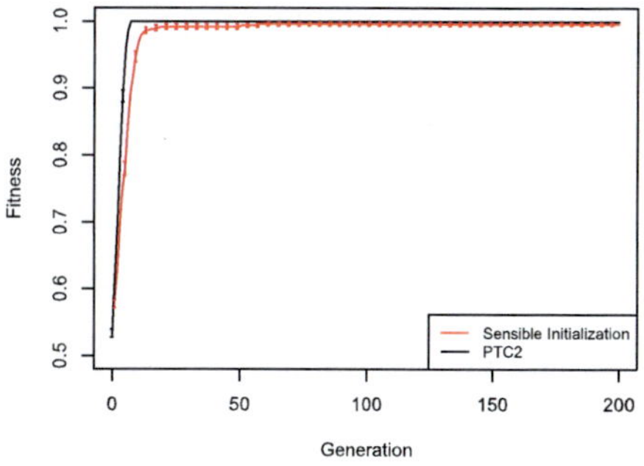

Fig. 8. Mean best across 30 independent runs using SSD Grammar B.

4.1 Success Rate

The success rate for all experiments are tabulated in Table 5 showing both the impact of grammar design and the effect of initialization routines on evolutionary performance. In [19], no optimal solutions were obtained using Grammar A variants (without domain knowledge) of 11-bit Multiplexer and SSD using sensible initialization. All these experiments were re-run following two minor modifications: refactoring of lexicase selection implementation to improve its runtime and assigning fitness score of zero to candidate circuits that take more than 10 s for simulation. These minor modifications greatly improved the performance of Sensible Initialization. We obtained 17 and 8 optimal solutions for 11-bit Multiplexer Grammar A and SSD GrammarA resptively.

Using PTC2, we obtained 27 and 8 optimal solutions using 11-bit Multiplexer Grammar A and SSD Grammar A respectively. Across all problems, PTC 2 obtained higher success rates except on the Seven Segment Grammar A, Hamming Code (7,4) Decoder Grammar B and JK Flip-flop where the tally was equal and on the Mod-16 up-down counter where Sensible initialization recorded three more successful runs. All Grammar B versions for all three combinational circuit benchmark problems performed better than their Grammar A counterparts because some simple domain knowledge was infused into these grammars.

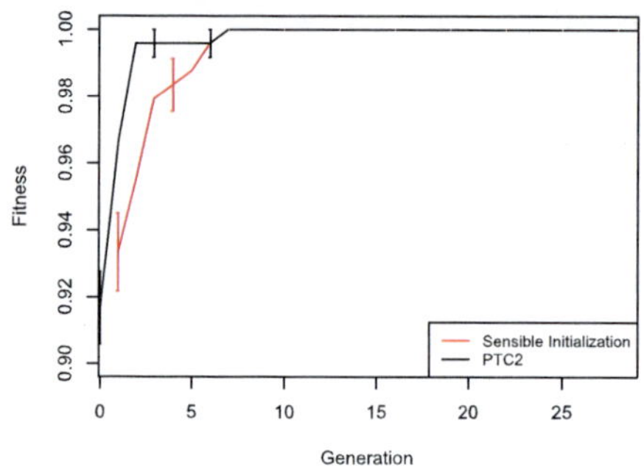

Fig. 9. Mean best across 30 independent runs for JK Flip-flop.

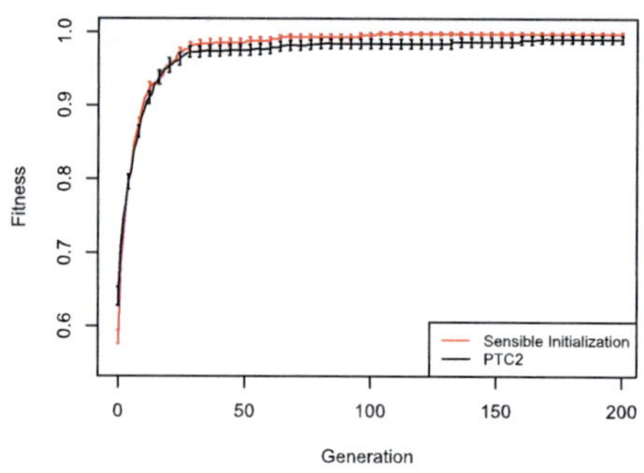

Fig. 10. Mean best across 30 independent runs for Mod-16 up-down counter.

4.2 Grammar Design

In order to assess the impact of grammar design choices and introduction of simple domain knowledge into grammars, two grammar variants are designed for each of the three combinational circuit benchmarks used: Hamming Code (7,4) Decoder, 11-bit Multiplexer and SSD. As speculated, the use of simple domain knowledge in some grammars had the intended influence.

The 11-bit Muliplexer Grammar A contained no domain knowledge and therefore contained more functions in its function set (ternary operator and bitwise operators). The 11-bit Multiplexer Grammar B used only a tenary operator, which based on the behavioural description of a multiplexer, we speculated to be sufficient for modelling the behaviour of a multiplexer. From Fig. 5 and Fig. 6, we observe that Multiplexer Grammar B outperformed Grammar A by a significant margin irrespective of the initialization routine used. The choice of highly expressive constructs/functions such as `if-else` at the behavioural level had the intended effect.

Recall that the SSD Grammar A version had to construct the BCD and segment-values from bits. The SSD Grammar B version had these values explicitly provided for in the grammar. As speculated, the Grammar B version performed better than Grammar A because Grammar A had the additional task of evolving the appropriate BCD and seven-segment values available in Grammar B.

The Hamming Code (7,4) Decoder Grammars differ only in the ⟨expr⟩ rule. The ⟨expr⟩ rule for Grammar A has recursive productions. As a result, evolution has to evolve arbitrary length expressions that generate the correct parity bit. In the case of Grammar B, we constrained the ⟨expr⟩ length of its final expression by limiting it to always select exactly 4 bits out of the 7-bit codeword, information we know from the Hamming Code (7,4) Decoder problem. This makes the Hamming Code (7,4) Decoder Grammar B more likely to solve the problem faster and with guarantee of no bloated expressions. From Fig. 3 and Fig. 4, we observe Hamming Code (7,4) Decoder Grammar B outperforms Grammar A even with the use of different initialization schemes. This shows despite the importance of good initialization schemes, they do benefit from concise grammars with simple domain knowledge infused into them. Furthermore, the use of grammars makes it possible to impose simple but generic constraints on the the search space though the introduction of simple domain knowledge into grammars.

Optimal solutions per grammar for all 5 benchmark problems are shown in the Appendix.

4.3 Higher Abstraction Levels

Modelling circuits in higher abstraction, such as behavioural modelling, is better suited for evolving complex circuits, but may not always guarantee optimized designs after synthesis. This is due to the fact that circuit opitmization is left to the mercy of synthesis tools, though modern synthesis tools are very good at optimization. A 1-bit full adder was evolved in RTL using GE with population size of 200 and a $100,000$ fitness evaluations budget [8]. Out of 35 independent experimental runs, only two optimal solutions were obtained. Using ADDC with the same grammar in [8] and population size but opting for 200 generations ($40,000$ maximum fitness evaluations), we obtained 8 out of 30 optimal solutions. These solutions were found on average after $4,600$ fitness evaluations. Detailed results of this experiment are not provided in this work as 1-bit full adder is trivial to evolve and Verilog/SystemVerilog's behavioural constructs were not exploited enough in the grammar design, greatly due to triviality of the chosen problem (1-bit adder). To investigate the latter claim, two grammar variants per combinational circuit design problem were used as described in Sect. 3.1, with focus on grammar design choices and the effect of introducing simple domain knowlegde about problems into grammar. Results as dicussed in Sect. 4.2 revealed that introducing simple information known about problems greatly improves the evolutionary search, thereby increasing the success rate. A recent study [24], exploited the parameterization construct in SystemVerilog to evolve very complex parameterized selective parity, adder, and muliplier combinational circuits using GE.

4.4 Impact of Initialization Schemes on Circuit Design Benchmark Problems

PTC2 offers the best performance on almost all experiments using grammars with and without domain knowledge introduced into them. However, experiments using sensible initialization obtained slightly better mean best fitnesses on the Hamming Code (7,4) Decoder grammars in Figs. 3 and 4 approximately after 80 and 65 generations. Sensible Initialization outperformed PTC2 marginally and quite distinctively on Hamming Code (7,4) Decoder Grammars A and B respectively. Similarly, sensible initialization slightly outperforms PTC 2 on mod-16 up-down counter. PTC2 obtained the worst starting mean best fitness scores in the initial generations for the 11-bit multiplexer (notably), SSD (slightly) and JK flip-flop problems but converged quicker to fitter and optimal solutions in general. This implies good fitness scores in intial generations does not necessarily guarantee a good overall performance which is consistent with findings in [15].

5 Conclusion and Future Work

In this work, ADDC was presented. ADDC designs digital circuit using GE, SystemVerilog and Icarus. ADDC is highly configurable; it can support other HDLs and simulators such as VHDL and ModelSim respectively. ADDC was tested on three combinational and two sequential benchmark problems. We designed two grammar variants per each combinational circuit – a version without domain knowledge and the other with simple domain knowledge infused into it. Results obtained revealed how grammars can be leveraged to dramatically reduce the search space such as separating out clearly distinct wires, for example input wires from output wires, and the inclusion of relevant functions only, such as in the case of the Multiplexer Grammar A.

Results obtained from the experiments reveal the impact of high-level constructs such as *Always* procedural block, `if-else`, `switch-case`, etc. on conventional circuit evolution. Due to the expressiveness of these programming constructs, they can replace many gates that would otherwise need to be evolved in gate-level evolution. This allows evolution to focus on evolving complex behaviours. All cases where we were able to test these features benefited from the introduction of simple domain knowledge.

Furthermore, experimental setups using PTC 2, overall had consistent results across all problems even on grammars without any domain knowledge infused into them. This makes PTC 2 more suitable for circuit design problems. The use of Lexicase selection made it easier to target poorly covered parts of the solution space. Results based on grammar design choices suggest Attribute Grammars may be more suited for more complex circuit benchmark problems.

Circuits evolved in this work were relatively complex and relevant in real-world applications. More complex circuits such as Hamming Code (127,120) or (63,57) Decoder will be looked at. These kind of problems may pose challenges but their solutions will greatly improve the robustness of ADDC. Such as the use of problem decomposition techniques to decompose circuits into evolveable blocks. Also, all circuits were evolved from scratch which is not usually the case in industry. Future works will target the use of IP blocks as modules to evolve bigger and more complex circuits. Adding these modules to our system is simply a matter of including them in the grammars. Finally, ADDC will require extension to support timing analysis.

Appendix

```
      <code-block> ::= <begin-module><declarations><func><new-line><always-block>
                       <new-line><end-module>
    <declarations> ::= <d-variable-type> <d-variable-name>; <new-line>
    <always-block> ::= "always @" (<input-codeword>) begin <new-line><output><new-line>
                       <syndrome> <new-line> <if-else> <new-line> end
         <if-else> ::= if(<condition>) "begin end" <new-line> "else begin" <new-line>
                       <output-codeword>[<d-variable-name>] "=" ~<input-codeword>
                       [<d-variable-name>]; end
       <condition> ::= <d-variable-name> "==" "3'b0"
          <output> ::= <output-codeword> "=" <input-codeword>;
        <syndrome> ::= <d-variable-name> "=" <func-call>
       <func-call> ::= <funct-name> (<input-codeword>);
            <func> ::= function <func-return-type><func-name> (<func-input>); <new-line>
                       begin <new-line><func-stmts> end <end-func>
    <func-stmts>   ::= syndrome[0] "=" <expr>; <new-line> syndrome[1]"="<expr>;<new-line>
                       syndrome[2] "=" <expr>; <new-line>
            <expr> ::= <sel-var> | <sel-var> <op> <expr>
         <sel-var> ::= <i-variable>[<index>]
<func-return-type> ::= "[2:0]"
       <func-name> ::= syndrome
      <func-input> ::= input <range> <i-variable>
           <range> ::= "[1:7]"
      <i-variable> ::= cw
           <index> ::= 1 | 2 | 3 | 4 | 5 | 6 | 7
              <op> ::= ^ | & | "|"
 <d-variable-type> ::= integer
 <d-variable-name> ::= syndrome_value
  <input-codeword> ::= i_codeword
 <output-codeword> ::= o_codeword
        <end-func> ::= endfunction
    <begin-module> ::= "module hc_decoder(input [1:7] i_codeword,
                       output logic [1:7] o_codeword);" <new-line>
    <end-module>   ::= endmodule
      <new-line>   ::= "\n"
```

Listing 7. Hamming Code (7,4) Decoder Grammar A [19].

```
module hc_decoder(input [1:7] i_codeword, output logic [1:7] o_codeword);
integer syndrome_value;
function [2:0] syndrome(input[1:7]cw);
begin
    syndrome[0]=cw[3]^cw[5]^cw[1]^cw[7];
    syndrome[1]=cw[3]&cw[3]^cw[7]^cw[2]^cw[6];
    syndrome[2]=cw[4]^cw[5]^cw[2]^cw[7]^cw[2]^cw[6];
end endfunction
always @(i_codeword)begin
    o_codeword=i_codeword;
    syndrome_value=syndrome(i_codeword);
    if(syndrome_value==3'b0) begin end
    else begin
        o_codeword[syndrome_value]=~i_codeword[syndrome_value]; end
    end
endmodule
```

Listing 8. Representative Optimal Solution obtained for Hamming Code (7,4) Grammar A.

```
        <code-block> ::= <begin-module><declarations><func><new-line><always-block><new-line>
                         <end-module>
      <declarations> ::= <d-variable-type> <d-variable-name>; <new-line>
      <always-block> ::= "always @" (<input-codeword>) begin <new-line><output><new-line>
                         <syndrome> <new-line> <if-else> <new-line> end
           <if-else> ::= if(<condition>) "begin end" <new-line> "else begin" <new-line>
                         <output-codeword>[<d-variable-name>] "=" ~<input-codeword>
                         [<d-variable-name>]; end
         <condition> ::= <d-variable-name> "==" "3'b0"
            <output> ::= <output-codeword> "=" <input-codeword>;
          <syndrome> ::= <d-variable-name> "=" <function-call>
         <func-call> ::= <funct-name> (<input-codeword>);
              <func> ::= function <func-return-type><func-name> (<func-input>); <new-line>
                         begin <new-line><func-stmts> end <end-func>
        <func-stmts> ::= syndrome[0] "=" <i-variable>[<index>] <op> <i-variable>[<index>]
                         <op><i-variable>[<index>] <op> <i-variable>[<index>]; <new-line>
                         syndrome[1] "=" <i-variable>[<index>]<op><i-variable>[<index>]<op>
                         <i-variable>[<index>] <op> <i-variable>[<index>]; <new-line>
                         syndrome[2] "=" <i-variable> [<index>] <op> <i-variable>[<index>]
                         <op> <i-variable>[<index>] <op> <i-variable>[<index>]; <new-line>
<func-return-type> ::= "[2:0]"
         <func-name> ::= syndrome
        <func-input> ::= input <range> <i-variable>
             <range> ::= "[1:7]"
        <i-variable> ::= cw
             <index> ::= 1 | 2 | 3 | 4 | 5 | 6 | 7
                <op> ::= ^ | & | "|"
   <d-variable-type> ::= integer
   <d-variable-name> ::= syndrome_value
    <input-codeword> ::= i_codeword
   <output-codeword> ::= o_codeword
      <end-function> ::= endfunction
      <begin-module> ::= "module hc_decoder(input [1:7] i_codeword,
                         output logic [1:7] o_codeword);" <new-line>
        <end-module> ::= endmodule
          <new-line> ::= "\n"
```

Listing 9. Hamming Code (7,4) Decoder Grammar B [19].

```
module hc_decoder(input [1:7] i_codeword, output logic [1:7] o_codeword);
integer syndrome_value;
function[2:0]syndrome(input[1:7]cw);
begin
    syndrome[0]=cw[1]^cw[7]^cw[3]^cw[5];
    syndrome[1]=cw[6]^cw[2]^cw[7]^cw[3];
    syndrome[2]=cw[6]^cw[7]^cw[5]^cw[4];
end endfunction
always @(i_codeword)begin
    o_codeword=i_codeword;
    syndrome_value=syndrome(i_codeword);
    if(syndrome_value==3'b0)begin end
    else begin
        o_codeword[syndrome_value]= ~i_codeword[syndrome_value]; end
end
endmodule
```

Listing 10. Representative Optimal Solution obtained for Hamming Code (7,4) Grammar B.

```
module mux(output logic out,
    input logic a0,a1, a2, d0,
    d1, d2, d3, d4, d5, d6, d7);

always@(*)
out=(a0?(a1?(a2?d0:(a2?d0:d1)):
    (a2?d2:d3)):(a0?(((a2?((d1&
    d6)|d2):(a2?((d1&d6)|d2):d3))
    |d6)&d6):(a1?(a0?d1:(a2?d4:
    d5)):(a0?!((a2?(d0&d0):
    d0)):(a2?d6:(a1?(a2?!(d6):
    d5):d7)))))));

endmodule
```

Listing 11. Representation Optimal Solution obtained for 11-bit Multiplexer Grammar A.

```
module mux(output logic out,
    input logic a0, a1, a2, d0,
    d1, d2, d3, d4, d5, d6, d7);

always@(*)
out=(a2?(a1?(a0?d0:(a2?d4:(a2?
    d4:d6))):(a0?d2:d6)):(a2?(a1?
    d2:(a1?(a1?d4:d7):d2)):(a1?
    (a2?(a0?d6:(a1?(a1?d7:d4):
    d7)):(a0?d1:(a2?d4:d5))):
    (a2?(a0?(a2?d3:d2):d6):
    (a1?(a0?d1:d5):(a1?d5:(a2?
    d7:(a2?(a0?d3:d7):(a0?
    d3:d7))))))))));

endmodule
```

Listing 12. Representation Optimal Solution obtained for 11-bit Multiplexer Grammar B.

```
module ssd(output logic[6:0]
segment, input logic[3:0] bcd);
always@(bcd)
begin
case(bcd)
    4'b0100:segment=7'b0110011;
    4'b0001:segment=7'b0110000;
    4'b1001:segment=7'b1111011;
    4'b0000:segment=7'b1111110;
    4'b1001:segment=7'b1011111;
    4'b0010:segment=7'b1101101;
    4'b0011:segment=7'b1111001;
    4'b0001:segment=7'b1111001;
    4'b0111:segment=7'b1110000;
    4'b0011:segment=7'b1011011;
    4'b0111:segment=7'b1111011;
    4'b0101:segment=7'b1011011;
    4'b1001:segment=7'b1111011;
    4'b0011:segment=7'b1011011;
    4'b0001:segment=7'b1011011;
    4'b1000:segment=7'b1111111;
    4'b0101:segment=7'b1111011;
    4'b1001:segment=7'b1111011;
    4'b0110:segment=7'b1011111;
    default:segment=7'b0000000;
endcase
end
endmodule
```

Listing 13. Representative Optimal Solution for SSD Problem.

```
module udcounter(input logic clk, up_ndown,
    en_load, enable_ud, reset, input
    logic [3:0] load, output logic [3:0] q);
always @(posedge clk)
casex({reset, en_load, enable_ud, up_ndown})
    4'b11??:q<=load;
    4'b0???:q<=0-load+load;
    4'b0???:q<=0-load+0+load-0;
    4'b??10:q<=1-1+q-1;
    4'b1?1?:q<=load+1+q-0-load;
endcase
endmodule
```

Listing 14. Representative Solution Obtained for Mod-16 Up-Down Counter Problem.

```
module jk_ff (input j, input k, input clk,
    output logic q);
always@(posedge clk)
case({j,k})
    2'b01:q<=0;
    2'b10:q<=1;
    2'b00:q<=q;
    2'b11:q<=~q;
    default:q<=q;
endcase
endmodule
```

Listing 15. Representation Solution Obtained for JK Flip-flop Problem.

References

1. Anjum, M.S., Ryan, C.: Seeding grammars in grammatical evolution to improve search based software testing. In: EuroGP (2020)
2. Cullen, J.: Evolving digital circuits in an industry standard hardware description language. In: Li, X., et al. (eds.) SEAL 2008. LNCS, vol. 5361, pp. 514–523. Springer, Heidelberg (2008). https://doi.org/10.1007/978-3-540-89694-4_52

3. Fredivianus, N., Prothmann, H., Schmeck, H.: XCS revisited: a novel discovery component for the eXtended classifier system. In: Deb, K., et al. (eds.) SEAL 2010. LNCS, vol. 6457, pp. 289–298. Springer, Heidelberg (2010). https://doi.org/10.1007/978-3-642-17298-4_30
4. de Garis, H.: Evolvable hardware genetic programming of a darwin machine. In: Albrecht, R.F., Reeves, C.R., Steele, N.C. (eds.) Artificial Neural Nets and Genetic Algorithms, pp. 441–449. Springer, Vienna (1993). https://doi.org/10.1007/978-3-7091-7533-0_64
5. de Garis, H.: An artificial brain ATR's CAM-Brain Project aims to build/evolve an artificial brain with a million neural net modules inside a trillion cell Cellular Automata Machine. New Gener. Comput. **12**(2), 215–221 (1994). https://doi.org/10.1007/BF03037343
6. Harper, R.: Ge, explosive grammars and the lasting legacy of bad initialisation. In: IEEE Congress on Evolutionary Computation, pp. 1–8 (2010)
7. Henson, B., Walker, J.A., Trefzer, M.A., Tyrrell, A.M.: Designing digital systems using cartesian genetic programming and VHDL. In: Stepney, S., Adamatzky, A. (eds.) Inspired by Nature. ECC, vol. 28, pp. 57–86. Springer, Cham (2018). https://doi.org/10.1007/978-3-319-67997-6_3
8. Karpuzcu, U.R.: Automatic verilog code generation through grammatical evolution. In: Proceedings of the 7th Annual Workshop on Genetic and Evolutionary Computation, GECCO 2005, pp. 394–397. ACM, New York (2005). https://doi.org/10.1145/1102256.1102346
9. Koza, J.R.: Genetic Programming: On the Programming of Computers by Means of Natural Selection. MIT Press, Cambridge (1992). https://mitpress.mit.edu/books/genetic-programming
10. Luke, S.: Two fast tree-creation algorithms for genetic programming. Trans. Evol. Comp **4**(3), 274–283 (2000). https://doi.org/10.1109/4235.873237
11. Miller, F.P., Vandome, A.F., McBrewster, J.: Hamming Code: Parity Bit, Two- out- of- Five Code, Hamming (7,4), Reed-Muller Code, Reed-Solomon Error Correction, Turbo Code, Low- Density Parity-Check Code, Telecommunication. Alpha Press, Linear Code (2009)
12. Miller, J.F., Thomson, P.: Cartesian genetic programming. In: Poli, R., Banzhaf, W., Langdon, W.B., Miller, J., Nordin, P., Fogarty, T.C. (eds.) Genetic Programming, pp. 121–132. Springer, Heidelberg (2000). https://doi.org/10.1007/978-3-642-17310-3_2
13. Murakawa, M., Yoshizawa, S., Kajitani, I., Furuya, T., Iwata, M., Higuchi, T.: Hardware evolution at function level. In: Voigt, H.-M., Ebeling, W., Rechenberg, I., Schwefel, H.-P. (eds.) PPSN 1996. LNCS, vol. 1141, pp. 62–71. Springer, Heidelberg (1996). https://doi.org/10.1007/3-540-61723-X_970
14. Murphy, A., Murphy, G., Amaral, J., MotaDias, D., Naredo, E., Ryan, C.: Towards incorporating human knowledge in fuzzy pattern tree evolution. In: Hu, T., Lourenço, N., Medvet, E. (eds.) EuroGP 2021. LNCS, vol. 12691, pp. 66–81. Springer, Cham (2021). https://doi.org/10.1007/978-3-030-72812-0_5
15. Nicolau, M.: Understanding grammatical evolution: initialisation. Genet. Program Evolvable Mach. **18**(4), 467–507 (2017). https://doi.org/10.1007/s10710-017-9309-9
16. O'Neill, M., Ryan, C.: Grammatical evolution. IEEE Trans. Evol. Comput. **5**(4), 349–358 (2001). https://doi.org/10.1109/4235.942529
17. Rabey, J.M., et al.: Digital Integrated Circuits, 2nd edn. Pearson, Boston (2003)
18. Ryan, C., O'Neill, M., Collins, J.J.: Handbook of Grammatical Evolution, 1st edn. Springer, Heidelberg (2018). https://doi.org/10.1007/978-3-319-78717-6
19. Ryan., C., Tetteh., M.K., Dias., D.M.: Behavioural modelling of digital circuits in system verilog using grammatical evolution. In: Proceedings of the 12th International Joint Conference on Computational Intelligence, vol. 1: ECTA, pp. 28–39. INSTICC, SciTePress (2020). https://doi.org/10.5220/0010066600280039
20. Sekanina, L., Walker, J.A., Kaufmann, P., Platzner, M.: Evolution of Electronic Circuits, pp. 125–179. Springer, Heidelberg (2011). https://doi.org/10.1007/978-3-642-17310-3_5

21. Soleimani, P., Sabbaghi-Nadooshan, R., Mirzakuchaki, S., Bagheri, M.: Using genetic algorithm in the evolutionary design of sequential logic circuits. CoRR abs/1110.1038 (2011). https://arxiv.org/abs/1110.1038
22. Soliman, A., Abbas, H.: Synchronous sequential circuits design using evolutionary algorithms. In: Canadian Conference on Electrical and Computer Engineering 2004 (IEEE Cat. No. 04CH37513), vol. 4, pp. 2013–2016 (2004). https://doi.org/10.1109/CCECE.2004.1347629
23. Stoica, A., Zebulum, R., Keymeulen, D.: Mixtrinsic evolution. In: Miller, J., Thompson, A., Thomson, P., Fogarty, T.C. (eds.) ICES 2000. LNCS, vol. 1801, pp. 208–217. Springer, Heidelberg (2000). https://doi.org/10.1007/3-540-46406-9_21
24. Tetteh, M.K., Mota Dias, D., Ryan, C.: Evolution of complex combinational logic circuits using grammatical evolution with SystemVerilog. In: Hu, T., Lourenço, N., Medvet, E. (eds.) EuroGP 2021. LNCS, vol. 12691, pp. 146–161. Springer, Cham (2021). https://doi.org/10.1007/978-3-030-72812-0_10
25. Thompson, A.: An evolved circuit, intrinsic in silicon, entwined with physics. In: Higuchi, T., Iwata, M., Liu, W. (eds.) ICES 1996. LNCS, vol. 1259, pp. 390–405. Springer, Heidelberg (1997). https://doi.org/10.1007/3-540-63173-9_61
26. Vasicek, Z., Sekanina, L.: Evolutionary approach to approximate digital circuits design. IEEE Trans. Evol. Comput. **19**(3), 432–444 (2015). https://doi.org/10.1109/TEVC.2014.2336175
27. Vassilev, V.K., Miller, J.F.: Embedding landscape neutrality to build a bridge from the conventional to a more efficient three-bit multiplier circuit. In: Proceedings of Genetic and Evolutionary Computation Conference. Morgan Kaufmann (2000)
28. Williams, S., Baxter, M.: Icarus verilog: open-source verilog more than a year later. Linux J. **2002**(99), 3 (2002). https://dl.acm.org/citation.cfm?id=513581.513584
29. Wolf, C.: Yosys open synthesis suite. https://www.clifford.at/yosys/

Crossover in Cartesian Genetic Programming: Evaluation of Two Phenotypic Methods

Roman Kalkreuth(✉)

Department of Computer Science, TU Dortmund University, Otto-Hahn-Straße 14, Dortmund, Germany
roman.kalkreuth@tu-dortmund.de
https://www.cs.tu-dortmund.de/

Abstract. The tree-based representation model of Genetic Programming is commonly used with subtree crossover as the predominant genetic operator. In contrast, Cartesian Genetic Programming (CGP) is widely used merely with mutation for genetic variation. Compared to comprehensive and advanced knowledge about crossover in the field of tree-based Genetic Programming, the state of knowledge in CGP appears to be generally weak. Even if CGP was officially introduced over twenty years ago, the role of recombination in CGP has been recently considered to be still an open question. Promising steps have been taken in the last years, but the outcomes show that more studies are needed to evaluate crossover operators in CGP on a wider set of benchmark problems. In this work, we compare algorithms that utilize two phenotypic crossover operators for CGP, called subgraph and block crossover, to the traditional mutation-only approach. The results of our experiments on well-known symbolic regression, Boolean function, and images operator design problems demonstrate that the use of crossover-based algorithms can outperform the mutation-only CGP approach on well-known benchmark problems.

Keywords: Cartesian genetic programming · Recombination · Subgraph crossover · Block crossover

1 Introduction

Genetic programming (GP) is a paradigm that enables the evolutionary inspired derivation of computer programs. First landmark work on GP has been done by Forsyth [4], Cramer [3] and Hicklin [5]. Later work by Koza [14–16] significantly popularized the field of GP. GP traditionally uses trees as program representation but is not limited to this representation model. Over two decades ago Miller, Thompson, Kalganova, and Fogarty proposed first work on Cartesian Genetic Programming (CGP) [7,23,24]. CGP is an encoding model based on a two-dimensional array of functional nodes. The standard CGP representation model offers an acyclic graph representation model which allows many graph-based applications such as electronic circuit design, image processing [29], and neural networks [33]. Compared to tree-based GP, CGP has pivotal advantages: 1) The maximum size of encoded solutions is fixed which saves CGP to some extent from *bloat*, a behavior that is characteristic to tree-based GP [17,18,21,31] 2) CGP encodes a directed acyclic graph (DAG) which allows the evolution of topologies [34].

J. Garibaldi et al. (Eds.): IJCCI 2021, SCI 1119, pp. 44–72, 2023.
https://doi.org/10.1007/978-3-031-46221-4_3

Standard integer-based CGP is predominantly used only with point mutation for genetic variation and with a $(1 + \lambda)$-strategy. Various standard genotypic crossover operators failed to improve the search performance of standard CGP in the past. Overall, the state of knowledge in CGP appears to be still poor when compared to the state of knowledge in other GP domains. The state of knowledge about crossover in CGP has been recently surveyed by Miller [27] and the role of crossover is still considered to be an open issue. Even if some progress has been made in recent years, comprehensive and advanced knowledge about crossover in CGP is still missing [27]. Steps forward on the crossover issue were made by the proposal of the subgraph crossover [11] and the block crossover [6] but both techniques have not been comprehensively evaluated and studied in the past. Therefore, this work follows up former work on the role of crossover operators in CGP by evaluating if the search performance of CGP algorithms that utilize crossover can be more effective than the commonly used $(1+\lambda)$-CGP on a set of well-known benchmark problems.

This work also surveys and analyzes relevant CGP literature which contributed to the knowledge about crossover in the past. Based on our experiments and their results, we formulate and analyze hypotheses that address important search performance dogmas in the field of CGP. The primary intention of this work is to shed more light on the role of the two phenotypic crossover operators in CGP. This works revises and extends the evaluation study of Kalkreuth [9] on this subject by including the block crossover technique in the experiments.

Section 2 of this work describes CGP as well as the subgraph and block crossover. Section 3 surveys previous work on crossover in CGP, outlines the need for a new evaluation and hypotheses to be examined are formulated. Section 4 is devoted to the experimental results and the description of our experiments. In Sect. 5 we discuss the results of our experiments and analyze the hypotheses that have been formulated in Subsect. 3.3. Finally, Sect. 6 gives a conclusion and outlines future work.

2 Preliminaries

2.1 Cartesian Genetic Programming

In contrast to tree-based GP, CGP represents a genetic program via genotype-phenotype mapping as an indexed, acyclic, and directed graph. In this way, CGP can be seen as an extension of the traditional tree-based GP approach. The CGP representation model is based on a rectangular grid or row of nodes. A general definition of a genetic program is given in Definition 1 and a specified definition of a cartesian genetic program (CP) is given in Definition 2. Each CP is encoded in the genotype of an individual and is decoded to its corresponding phenotype. Originally, the structure of the graph was represented by a rectangular grid of n_r rows and n_c columns, but later work focused on a representation with one row. The CGP decoding procedure processes groups of genes, and each group refers to a node of the graph, except the last one, which represents the outputs of the phenotype. Each node is represented by two types of genes that index the function number in the GP function set and the node inputs. These nodes are called *function nodes* and execute functions on the input values. The number of input genes depends on the maximum arity n_a of the function set.

Definition 1 (Genetic Program). *A genetic programm is an element of the Cartesian product* $\mathcal{T} \times \mathcal{F} \times \mathcal{E}$*:*

- $\mathcal{F}$ *is a finite non-empty set of functions*
- $\mathcal{T}$ *is a finite non-empty set of terminals*
- $\mathcal{E}$ *is a finite non-empty set of edges*

Definition 2 (Cartesian Genetic Program). *A cartesian genetic program is an element of the Cartesian product* $\mathcal{N}_\mathrm{i} \times \mathcal{N}_\mathrm{f} \times \mathcal{N}_\mathrm{o} \times \mathcal{F}$:

- $\mathcal{N}_\mathrm{i}$ *is a finite non-empty set of input nodes*
- $\mathcal{N}_\mathrm{f}$ *is a finite set of function nodes*
- $\mathcal{N}_\mathrm{o}$ *is a finite non-empty set of output nodes*
- $\mathcal{F}$ *is a finite non-empty set of functions*

A backward search is used to decode the corresponding phenotype. The decoding procedure is done for all output genes. An example of the backward search of the most popular one-row integer representation is illustrated in Fig. 1. The backward search starts from the program output and processes all nodes which are linked in the genotype. In this way, only active nodes are processed during evaluation. The genotype in Fig. 1 is grouped by its function nodes. The first (underlined) gene of each group refers to the function number in the corresponding function set in the figure.

The number of inputs n_i, outputs n_o, and the length of the genotype is fixed. Every candidate program is represented with $n_\mathrm{r} * n_\mathrm{c} * (n_\mathrm{a} + 1) + n_\mathrm{o}$ integers. Even when the length of the genotype is fixed for each candidate program, the length of the corresponding phenotype in CGP is variable, which can be considered as a advantage of the CGP representation.

CGP is traditionally used with a $(1 + \lambda)$ evolutionary algorithm (EA). The (1+λ)-EA is often used with a selection strategy called *neutrality*, which is based on the idea that genetic drift yields to diverse individuals having equal fitness. The genetic drift is implemented into the selection mechanism in a way that individuals that have the same fitness as the normally selected parent are determined, and one of these same-fitness individuals is returned uniformly at random.

A pseudo-code of the (1+λ)-EA is given in Algorithm 1. The new population in each generation consists of the best individual of the previous population and the λ created offspring. The breeding procedure is mostly done by a point mutation that swaps genes in the genotype of an individual in the valid range by chance. An example of a point mutation is given in Fig. 2. The figure shows the flip of the value of a connection gene, which rewires the corresponding function node. Another point mutation is the flip of the functional gene, which changes functional behavior of the corresponding function node.

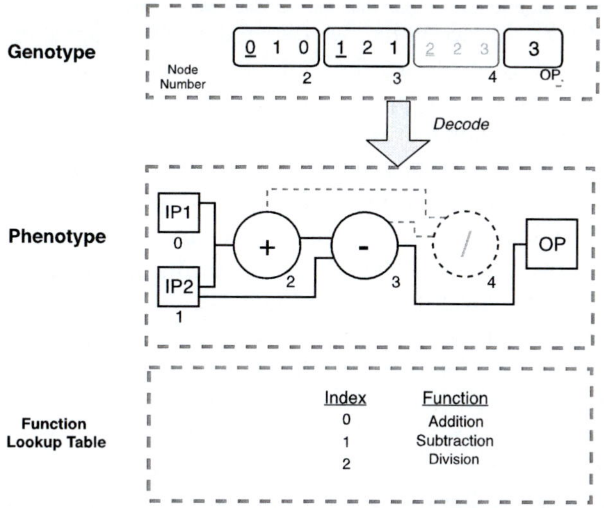

Fig. 1. Example of the decoding procedure of a CGP genotype to its corresponding phenotype. The nodes are represented by two types of numbers which index the number in the function lookup table (underlined) and the inputs (non-underlined) for the node. Inactive function nodes are shown in gray color. The identifiers IP1 and IP2 stand for the two input nodes with node index 0 and 1. The identifier OP stands for the output node of the graph. Figure has been taken from the work of Kalkreuth [9].

Algorithm 1. (1+λ)-EA.

1: initialize(P) ▷ *Initialize parent individual*
2: **repeat**
3: $Q \leftarrow$ breed(P) ▷ *Breed λ offspring by mutation*
4: Evaluate(Q) ▷ *Evaluate the fitness of the offspring*
5: **if** *at least one individual of Q has better fitness then P* **then**
6: $P \leftarrow$ best(Q) ▷ *Replace the parent by the best offspring*
7: **end if**
8: **until** *P meets termination criterion* ▷ *Until termination criteria not triggered*
9: **return** P

2.2 Advanced Crossover Operators for CGP

The two phenotypic crossover operators, which will be described in this section, differ from standard genotypic variation operators in a way that both operators directly operate on the phenotype of a CGP individual.

The Subgraph Crossover. The subgraph crossover technique for CGP was introduced by Kalkreuth et al. [11] and is inspired by Koza's two parent subtree crossover found in tree-based GP [15]. To recombine two directed acyclic graphs, the subgraph recombination is performed by respecting the CGP phenotype. Merely swapping parts of the genotype would be a disastrous approach according to the reportings by Clegg et al. [2]. The phenotype of each individual is represented by the active path of the graph and is determined through the evaluation process. Furthermore, the active path of a graph leads to the semantic value of a certain individual in CGP. As a consequence, the subgraph crossover only recombines the genetic material of the active paths. The basic idea of the subgraph crossover is that it should reduce the disruption which is caused by the genotypic one-point crossover in standard CGP.

For the description of the subgraph crossover procedure, let n_{i} be the predefined number of input nodes and let n_{f} be the predefined number of function nodes. In CGP, the input nodes are indexed from n_{i} to $n_{\mathrm{i}} - 1$ and the function nodes of each graph are indexed from 0 to $n_{\mathrm{i}} + n_{\mathrm{f}} - 1$. The nodes which lie between the input and output nodes are denoted as function nodes. The crossover is done with two parents which are denoted as P_1 and P_2. For the crossover procedure, the node numbers of the active function nodes are necessary. The node numbers of the active nodes of P_1 and P_2 are stored in two arrays M_1 and M_2. The active nodes are determined by the backward search in the evaluation procedure.

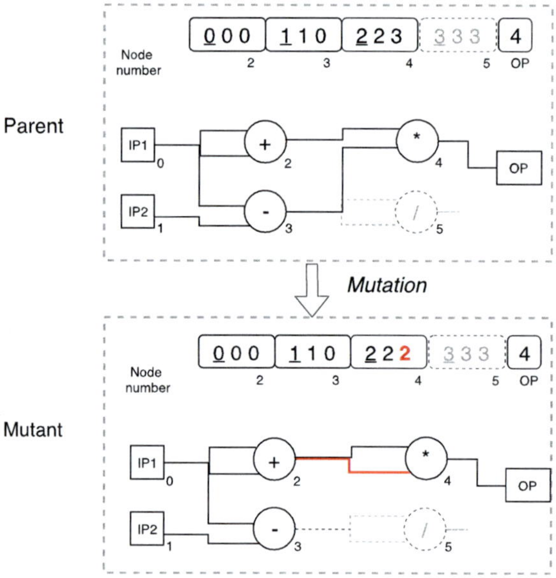

Fig. 2. Example of the standard point mutation operator in integer-encoded CGP. Genes of the genotype are selected by chance, and their values are randomly flipped within the legal range of possible values. The connection gene of node 4 is mutated from a value of 2 to a value of 3. This causes a rewiring of the second input of node 4 from the output of node 3 to the output node of node 2. Figure has been taken from the work of Kalkreuth [9].

To define one suitable crossover point, we define two possible crossover points C_{P1} and C_{P2} of the two parents. With information about the active nodes and the length of the path, we can choose two possible crossover points. The possible crossover points C_{P1} and C_{P2} are chosen by chance in the range of the active function nodes which are stored in M_1 and M_2. The possible crossover points may not be input or output nodes. A general crossover point C_P is defined by choosing the smaller crossover point from C_{P1} and C_{P2}. The reason for this is that the subgraphs of the parents, which will be placed in front of or behind the crossover point of the offspring's genome should be balanced. The representation of CGP allows active paths of an individual, which can start in the middle or back of the graph. The subgraph which will be placed in front of the crossover point has to start at more leading active nodes. If C_P is defined as the possible point C_{P1}, the subgraph of P_1 in front of C_P will be placed in front of C_P in the offspring genome. The subgraph behind C_P of P_2 will be placed behind C_P in the offspring genome.

Both sections are connected with a special step that we call *neighbourhood connect*. This step refers to the first active node of the section behind the crossover point which is connected to the last active node of the section in front of the crossover point. This is done by adjusting the connection gene of the first active node of the section behind the crossover point. To ensure that active nodes of the section behind the crossover point do not refer to inactive nodes of the section in front of the crossover point, we perform a step which we call *random active connect*. All connection genes of the active nodes of the section behind the crossover point are adjusted to the active nodes of the section in front of the crossover point, previous active nodes of S_2 or input nodes.

The crossover procedure produces a new genome that represents the offspring inheriting genetic parts of both parents. In the case that two offspring should be produced, the crossover procedure is performed twice with two different general crossover points. An example of the crossover procedure is illustrated in Fig. 3 and a definition is given in Definition 3.

Definition 3 (Subgraph Crossover). *The subgraph crossover is a genetic variation operator for CGP which exchanges and links subgraphs of function nodes between two selected parents, producing one or two offspring.*

The Block Crossover. The block crossover technique focuses on the one-dimensional representation of CGP where the number of rows is limited to one. The block crossover swaps the function genes between blocks of active function nodes which have been selected from each parent. Given the previously selected genotypes of two individuals serving as parents, the block crossover first determines a list of the active nodes by evaluating the genotype's active path. Afterward, the function nodes blocks are constructed concerning the predefined maximum block size B. The construction is done by randomly selecting a subset of active function nodes from each active function list. These random subsets of active function nodes serve as the blocks and are used for the recombination procedure. The procedure ensures that the blocks have the same size. The block crossover then iterates over the blocks and exchanges the function gene between all function nodes which are stored in the blocks. Since the blocks of the two selected

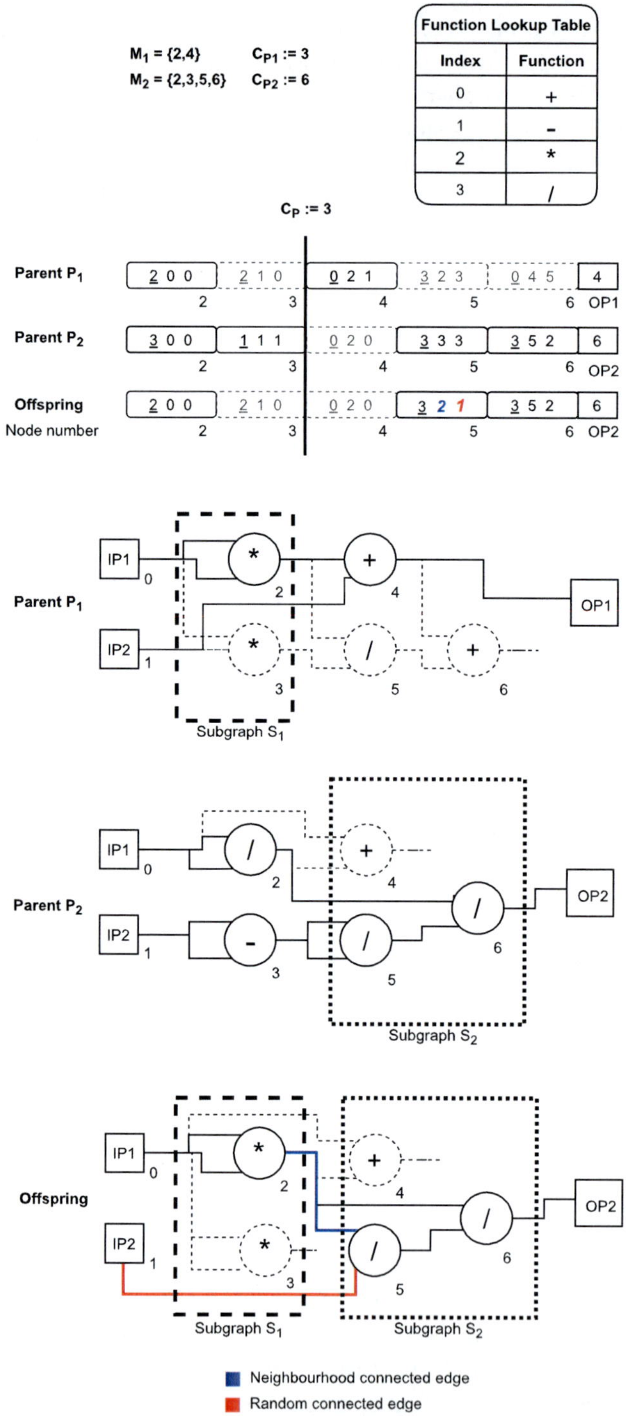

Fig. 3. The subgraph crossover procedure. Figure has been taken and revised from the work of Kalkreuth [9].

parents have the same dimension, the function genes of the nodes are swapped pairwise at each position.

First the active paths are determined and are stored in the lists M_1 and M_2. When at least one parent has no active function nodes, the crossover procedure is aborted. Otherwise, two blocks N_1 and N_2 containing randomly selected active function nodes from their respective lists are determined, This is done with respect to the maximum block size. If at least one parent has less active function nodes then the predefined maximum block size, the minimum size of active nodes is determined between the parents. Afterward, the minimum is taken as block size. To produce the first offspring O_1, the first parent P_1 is cloned, and the function genes of the nodes inside the block N_1, are replaced by the function genes of the nodes from block N_2. The second offspring O_2 is produced in the same way but instead of P_1 and N_2, the second parent $P2$ and the block N_1 are used. Figure 4 illustrates the crossover procedure. At the top of the figure, the active function nodes of the parents and the determined node block are listed. These listings are followed by the genotypes of the parents and the offspring. The function genes which have been swapped are highlighted in red and blue colors. Below the listing of the genotypes, the corresponding phenotypes are shown. The function nodes, which have been altered by the swapping process are also highlighted in red and blue colors. A definition of the block crossover is given in Definition 4.

Definition 4 (Block Crossover). *The block crossover is a genetic variation operator for CGP which exchanges blocks of function genes between two selected parents, producing two offspring.*

3 Review and Motivation

3.1 Previous Work on Crossover in CGP

According to Clegg et al. [2], the first attempts of recombination in standard CGP included the testing of different genotypic crossover techniques. For instance, the genetic material was recombined by swapping parts of the genotypes of the parent individuals or randomly exchanging selected nodes. Clegg et al. reported that all four techniques failed to improve the convergence of CGP and that merely swapping the integers (in whatever manner) in the CGP representation disrupts the search performance. Compared to running CGP with mutation only, the addition of these crossover techniques hindered the performance. The four methods were tested on the standard integer-based representation of CGP. In one of the first empirical studies in CGP, Miller [24] analyzed the computational efficiency on Boolean function problems. More precisely, Miller analyzed and studied the influence of population size on the efficiency of CGP. The key finding was that extremely low populations are most effective for the tested problems. The experiments of this study also revealed that the use of a genotypic crossover reduces the computational effort only marginally. This was the motivation for the introduction of a real-valued representation and new crossover technique for CGP by Clegg et al. The real-valued representation of CGP represents the directed graph as a fixed-length list of real-valued numbers in the interval [0, 1]. The genes are decoded to the integer-based

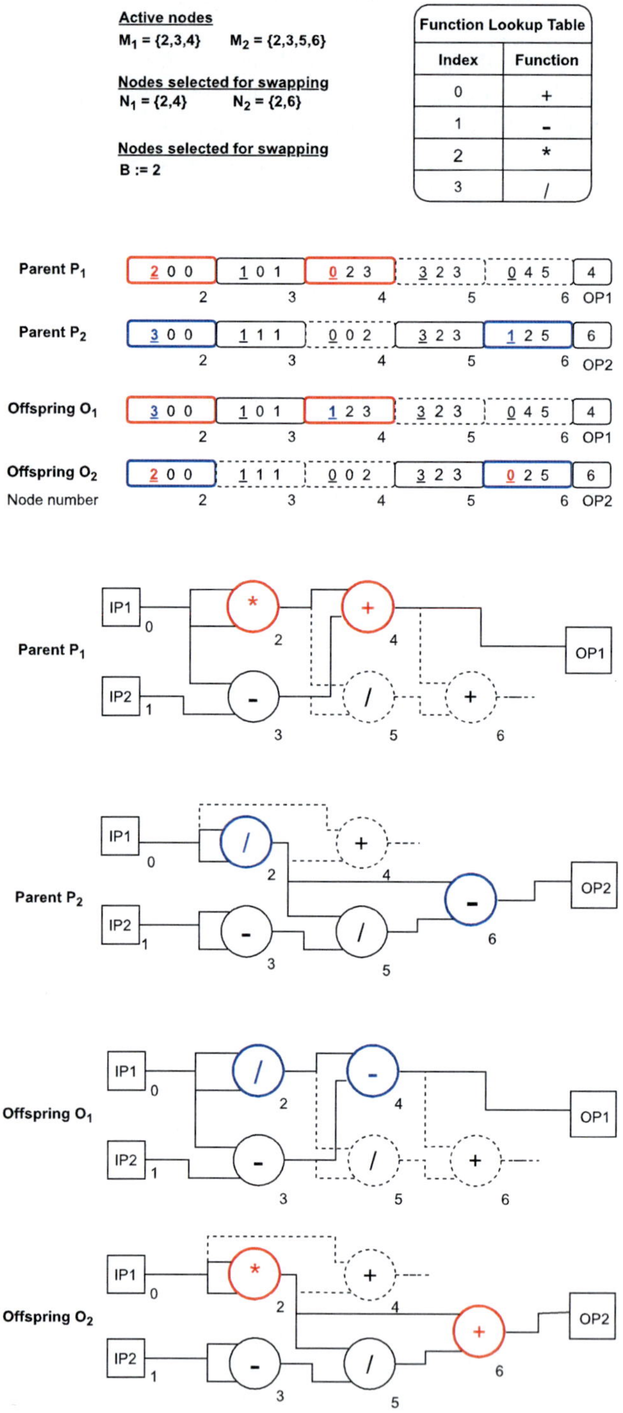

Fig. 4. The procedure of the proposed block crossover technique. Figure has been taken and revised from the work of Husa and Kalkreuth [6].

representation by their normalization values (number of functions or maximum input range). The recombination of two genotypes is performed by an arithmetic crossover with a random weighting factor, which can also be found in the field of real-valued Genetic Algorithms. Clegg et al. showed that the new representation in combination with crossover improves the convergence behavior of CGP. However, for the later generations, Clegg et al. found that the use of crossover in real-valued CGP disrupts the convergence on one of the two tested problems. Later work by Turner [32] presented results on three additional classes of computational problems, digital circuit synthesis, function optimization, and agent-based wall avoidance. On these problems, it was found that the real-valued representation together with the crossover operation performed worse than standard CGP.

Kalkreuth et al. [11] introduced and investigated a subgraph crossover in CGP. They evaluated the utility of the new crossover operator for a range of crossover rates on a suite of benchmark problems in circuit design, symbolic regression and image filter design. Often high crossover rates were beneficial and in all cases a nonzero crossover rate performed better than without crossover. However, they did not compare their results with the traditional $(1 + \lambda)$-CGP so it still remains unclear whether the subgraph crossover can significantly outperform mutational CGP.

Husa and Kalkreuth [6] proposed the block crossover and performed parameter sweeps with the $(1 + \lambda)$-strategy versus a genetic algorithm using block crossover on a suite of Boolean functions and symbolic regression problems. The results of the comparative study demonstrated that the traditional $(1+\lambda)$-CGP algorithm cannot be stated as the universally predominant algorithm for CGP. The outcome of the study gave significant evidence that the $(1 + \lambda)$-CGP cannot be considered as the most efficient CGP algorithm in the Boolean function domain, although it seems to be often a good choice. The experiments on the 1-resilient Boolean function demonstrated that the $(1 + \lambda)$-CGP may indeed be significantly inferior to the other CGP algorithms. In this way, the outcome of the study gave the first evidence, that it is possible for crossover operators to outperform the standard $(1 + \lambda)$ strategy. Sivla et al. [30] introduced a new form of crossover for CGP when a single genotype representation is used and the desired model has multiple outputs. The proposed method modifies the standard evolutionary strategy commonly adopted in CGP by combining the subgraphs of the best outputs of the parent and its offspring in order to generate a new fittest individual. Recently, Kalkreuth [8, 10] presented results on two phenotypic mutation techniques for CGP which are called *Insertion* and *Deletion*. The subgraph crossover was used within a $(\mu + \lambda)$ EA for a search performance comparison between CGP and a GP model called Evolving Graphs with Graph Programming (EGGP) [1], another *state-of-the-art* technique for evolving graphs. The search performance of the $(\mu + \lambda)$ EA with subgraph crossover and *Insertion* and *Deletion* mutation has been found very effective, when compared to the search performance of EGGP.

3.2 Motivation for a New Evaluation

Husa and Kalkreuth [6] pointed out that it is difficult to obtain well-performing parameter settings of crossover algorithms and complicates fair comparisons in CGP. The reason for that is that crossover-based CGP algorithms require the configuration of more parameters as the $(1+\lambda)$-CGP. Husa and Kalkreuth [6] also stated that merely relying on the results of the meta-evolution can lead to ineffective parameter settings which result in unfair comparisons. Since the meta-evolution is done by an evolutionary algorithm, it is possible that the meta-evolution evolves towards a local optimum, which can result in ineffective parameter settings. Therefore, solid and precise parameter tuning requires additional steps which will be described in the following section. Another reason for a new evaluation study is that comparisons on a bigger set of benchmark problems are needed to shed more light on the question of crossover in CGP. Husa and Kalkreuth [6] tested only 8 problems in two major GP benchmark fields. In a very recent survey about the status and the future of CGP, Miller [27] addressed open questions in CGP. Regarding the open question of crossover in CGP, Miller stated that crossover is still very underdeveloped in CGP and that the subgraph crossover needs to be comprehensively studied in the future.

3.3 Formulation of Hypotheses

With the evaluation study which will be presented in the experiments section, we want to shed more light on some important dogmas in the field of standard CGP, which we address with the formulation and analysis of the following hypotheses:

Hypothesis 1 (H1): *Crossover does not contribute to the search performance of integer-based standard CGP.*

Hypothesis 2 (H2): *Extremely large genotypes perform most effectively in CGP.*

Hypothesis 3 (H3): *Small populations perform most effectively in CGP.*

4 Evaluation

The primary purpose of this section is to present the result of an evaluation of the methods proposed in this work. To classify the significance of the presented methods and the respective algorithmic setup, the results are compared to the traditional $(1+4)$-CGP algorithm. Therefore, this evaluation study intends to investigate if the use of the presented crossover techniques can be considered as more effective than the use of the $(1+4)$-CGP algorithm. We will compare different evolutionary algorithms. Some of them are consequently equipped with the methods, which have been proposed in Sect. 2. A list of the algorithms which will be compared in this study is given in Table 1.

4.1 Experimental Setup

We performed experiments in the problem domains of symbolic regression, Boolean function learning, and image operator design. We also included more complex Boolean function problems. To evaluate the search performance of the tested algorithms, we measured the number of fitness evaluations until the CGP algorithm terminated successfully (*fitness-evaluations-to-success*) and the best fitness value, which was found after a predefined number of generations (*best-fitness-of-run*). In addition to the mean

Table 1. List of the CGP algorithms.

Algorithm	Description
$(1+4)$-CGP	Standard $(1+4)$-CGP algorithm
$(1+\lambda)$-CGP	Standard $(1+\lambda)$-CGP algorithm
$(\mu+\lambda)$-CGP (Subgraph)	$(\mu+\lambda)$-CGP algorithm with subgraph crossover
$(\mu+\lambda)$-CGP (Block)	$(\mu+\lambda)$-CGP algorithm with block crossover
Canonical-CGP (Subgraph)	Standard canonical EA with subgraph crossover
Canonical-CGP (Block)	Standard canonical EA with block crossover

values of the measurements, we calculated the standard deviation (SD) and the standard error of the mean (SEM). To classify the significance of our results, we used the Mann-Whitney-U-Test. The mean values are denoted $a^{\dagger}$ if the p-value is less than the significance level 0.05 and $a^{\ddagger}$ if the p-value is less than the significance level 0.01 compared to the $(1+4)$-CGP. Note that the mean values are **only** denoted with the significance level marker if the result of a certain algorithm is better than the result of the $(1+4)$-CGP. We performed 100 independent runs with different random seeds except the complex and computing-intensive Even-Parity 8 problem for which we performed only 30 runs. The elitism size was set to 1. The levels back parameter l was set to ∞. For the block crossover we used a block size of 2 function nodes.

4.2 Benchmarks

Symbolic Regression. The nine symbolic regression problems are selected from the work of McDermott et al. [20] and White et al. [36]. The functions of the problems are shown in Table 2. A training data set U[a, b, c] refers to c uniform random samples drawn from a to b inclusive and E[a, b, c] refers to a grid of points evenly spaced with an interval of c, from a to b inclusive. The respective function sets are shown in Table 3. The fitness of the individuals was represented by a cost function value. The cost function was defined by the sum of the absolute difference between the real function values and the values of an evaluated individual. We evaluated the simpler symbolic regression problems Koza 1, 2 & 3 with the *fitness-evaluation-to-termination* method. The other, more complex, symbolic regression problems were evaluated with the *best-fitness-of-run* method. We defined a maximum number of $8 \cdot 10^7$ fitness evaluations for these three experiments. The reason for our choice of these three problems is the fact that we can find an ideal solution more likely on average than the other more complex benchmark problems, which require a huge number of fitness evaluations to find an ideal solution. For the more complex problems, we measured the best fitness after a budget of 10^4 fitness evaluations.

Boolean Functions. In the Boolean domain, we chose six Parity-Even problems with $n = 3$ to 8 Boolean inputs. The goal was to find a program that produces the value of the

Table 2. List of symbolic regression benchmarks. Table has been taken from the work of Kalkreuth [9].

Problem	Objective Function	Vars	Training Set	Function Set
Koza-1	$x^4 + x^3 + x^2 + x$	1	U[−1, 1, 20]	Koza
Koza-2	$x^5 - 2x^3 + x$	1	U[−1, 1, 20]	Koza
Koza-3	$x^6 - 2x^4 + x^2$	1	U[−1, 1, 20]	Koza
Nguyen-4	$x^6 + x^5 + x^4 + x^3 + x^2 + x$	1	U[−1, 1, 20]	Koza
Nguyen-5	$\sin(x^2)\cos(x) - 1$	1	U[−1, 1, 20]	Koza
Nguyen-6	$\sin(x) + \sin(x + x^2)$	1	U[−1, 1, 20]	Koza
Nguyen-7	$\ln(x + 1) + \ln(x^2 + 1)$	1	U[0, 2, 20]	Koza
Keijzer-6	$\sum_i^x 1/i$	1	E[1, 50, 1]	Keijzer
Pagie-1	$1/(1 + x^{-4}) + 1/(1 + y^{-4})$	2	E[−5, 5, 0.4]	Koza

Table 3. Function sets for the symbolic regression benchmarks.

Function Set	Functions
Koza	$+, -, *, /, \sin, \cos, \ln(\lvert n \rvert), e^n$
Keijzer	$+, *, \frac{1}{n}, -n, \sqrt{n}$

Table 4. Function sets for the Boolean function problems.

Problems	Function Set
Parity-Even 3, 4, 5, 6, 7, 8	AND, OR, NAND, NOT, NOR
Adder-1, 2, 3, Subtractor-2, Multiplier-2, 3 Bit	AND, OR, NAND, NOR

Boolean even parity depending on the n independent inputs. The fitness was represented by the number of fitness cases for which the candidate solution failed to generate the correct value of the Even-Parity function.

We also investigated multiple output problems as the 1, 2 & 3,-Bit-Adder, 2-Bit-Subtractor, 2-Bit- and 3-Bit-Multiplier. To evaluate the fitness of the individuals on the multiple output problems, we defined the fitness value of an individual as the number of different bits to the corresponding truth table. When this number became zero, the algorithm successfully terminated. We evaluated the problems with the *fitness-evaluations-to-success* method. The function sets for the experiments with the Parity-Even problems and the multiple output problems are shown in Table 4.

Image Operator Design. We chose two noise reduction problems from the field of evolutionary image operator design [29]. For these problems, we wanted to minimize the difference between a filtered image I_f and a reference image I_r, which were based on an input image I_i. The images I_i and I_r were of size $N \times M$ pixels. We used the pixel values of a 3×3 kernel matrix as the inputs of the cartesian program. Figure 5 provides an example of an image operator evaluation procedure in CGP. For the filtered image we took a size of $(N-2) \times (M-2)$ pixels. The reason for this is the convolution problem on the edges of the input image which is often treated by ignoring the edges. Let

$$\text{MDPP} := \frac{\sum_{i=1}^{N-2} \sum_{j=1}^{M-2} |I_f(i,j) - I_r(i,j)|}{(N-2)(M-2)}$$

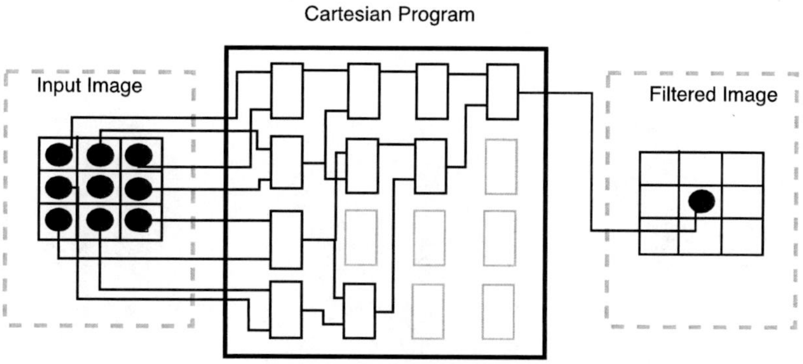

Fig. 5. Illustration of an image operator evaluation procedure in CGP.

Table 5. Function set for the image operator design problems.

Function	Description
OR	Bitwise OR
OR*	Bitwise OR with one inverted input
AND	Bitwise AND
NAND	Bitwise NAND
XOR	Bitwise XOR
ADD	Addition
ADD*	Addition with saturation
MAX	Maximum
MIN	Minimum
AVG	Average
ID	Indenity
INV	Inversion
SWAP	Swap upper and lower nibble
RIGHT SHIFT ONE	Right bit-shift by one
RIGHT SHIFT TWO	Right bit-shift by two
CONSTANT	Return of value 255

be the mean difference per pixel between the reference image I_r and the filtered image I_f. If MDPP is zero, the images I_f and I_r are identical, except the pixels on the edges. The image data consisted of the Lena test image set with an image size of 128×128 pixel, which includes the input images and reference image is shown in Fig. 6, 7 and 8. We evaluated all image operator design problems with the *fitness-evaluations-to-success* method. The function set for the image operator design problems is shown in Table 5. The goal of the resulting problem was to evolve a filter, which reduces the Salt & Pepper noise or Gaussian noise in reference to the Lena image without noise as shown in Fig. 8. We defined the termination criterion with a MDPP of less than or equal 3.

Fig. 6. Salt & Pepper noise.

Fig. 7. Gaussian noise.

Fig. 8. Original image.

4.3 Meta-optimization

We performed meta-optimization experiments intending to compare CGP algorithms in a fair way. Meta-optimization [22], can be described as a technique, which uses a certain optimization algorithm to tune the parameters of another optimization algorithm. Basically, meta-optimization is a simple way of finding performant parameters settings for a certain optimizing algorithm by using an overlaying optimizing algorithm which is called the meta-optimizer. Moreover, we tuned significant parameters for all used CGP algorithms on the set of benchmark problems. We used the meta-optimization extension package of the Java Evolutionary Computation Research System (ECJ) [19,28]. The parameters for the respective CGP algorithms are given in Table 6. The ranges for the number of nodes are oriented with the parameter settings found by Kaufmann and Kalkreuth [12,13]. Meta-evolution is very costly in terms of computational effort, which is necessary to find an optimal parameter setting. Furthermore, since GP benchmark problems can be very noisy in terms of finding the ideal solution, we used a common approach of fine-tuning. The meta-evolution is repeated several times, and the best settings are used to find the ideal algorithm setting. This is done afterward with manual fine-tuning of the respective parameters.

Tables 7, 8, 9, and 10 show the results of the meta-optimization experiments. The results in the Boolean domain reveal an effective parametrization pattern, characterized by the choice of very small population size and an extremely high number of function nodes. This finding also holds for the crossover-based algorithms. In the symbolic regression domain, the number of function nodes is, on average smaller compared to

Table 6. Parameter space explored by meta-optimization for the respective CGP algorithms. Table has been taken and revised from the work of Kalkreuth [9].

(1 + 4)-CGP		
Parameter	Description	Range
N	number of nodes	[10, 4000]
M	point mutation rate[%]	[1,20]
(1 + λ)-CGP		
λ	number of offspring	[1, 1000]
N	number of nodes	[10, 4000]
M	point mutation rate[%]	[1, 20]
Canonical-CGP		
N	number of nodes	[10, 4000]
M	point mutation rate[%]	[1, 20]
C	crossover rate[%]	[10, 100]
P	population size	[5, 1000]
T	tournament size	[2, 20]
(μ + λ)-CGP		
μ	number of parents	[2, 150]
λ	number of offspring	[1, 1000]
N	number of nodes	[10, 4000]
M	point mutation rate[%]	[1, 20]
C	crossover rate[%]	[10, 100]

the Boolean function domain. The results indicate that bigger population sizes seem to work more effectively in this problem domain. It can also be observed that the crossover rate for the symbolic regression problems is higher in average when compared to the rates determined in the Boolean domain. For the image operator design problems, very small population sizes and an extremely high number of function nodes seem to be a good choice which is similar to the findings in the Boolean domain.

4.4 Experiments

For the algorithm comparison, which we performed on the basis of the results of our meta-optimization experiments. We evaluated all problems with the respective parameter setting, which had been determined in the meta-optimization part of our study. For our comparison, we chose the traditional (1 + 4)-CGP as the baseline algorithm.

The results of our experiments in the symbolic regression domain are shown in Table 11 and Table 12. It is visible that the Canonical-CGP algorithm with subgraph crossover performs better than the (1 + 4)-CGP on all tested problems. Moreover, on

Table 7. Results of the meta-optimization for the symbolic regression problems. Table has been taken and revised from the work of Kalkreuth [9].

Problem	Algorithm	N	M	C	μ	λ	P	T
Koza-1	(1 + 4)-CGP	10	20	–	–	–	–	–
	$(1 + \lambda)$-CGP	10	20	–	–	8	–	–
	Canonical-CGP (Subgraph)	10	20	70	–	–	50	4
	Canonical-CGP (Block)	10	10	70	–	–	50	4
	$(\mu + \lambda)$-CGP (Subgraph)	10	20	70	4	16	–	–
	$(\mu + \lambda)$-CGP (Block)	10	20	70	4	16	–	–
Koza-2	(1 + 4)-CGP	10	20	–	–	–	–	–
	$(1 + \lambda)$-CGP	10	20	–	–	8	–	–
	Canonical-CGP (Subgraph)	10	20	70	–	–	50	4
	Canonical-CGP (Block)	10	10	70	–	–	50	4
	$(\mu + \lambda)$-CGP (Subgraph)	10	20	90	4	16	–	–
	$(\mu + \lambda)$-CGP (Block)	10	20	70	4	16	–	–
Koza-3	(1 + 4)-CGP	10	20	–	–	–	–	–
	$(1 + \lambda)$-CGP	10	20	–	–	8	–	–
	Canonical-CGP (Subgraph)	10	20	70			50	4
	Canonical-CGP (Block)	10	10	70	–	–	50	4
	$(\mu + \lambda)$-CGP (Subgraph)	10	20	70	1	8	–	–
	$(\mu + \lambda)$-CGP (Block)	10	20	70	4	16	–	–
Nguyen-4	(1 + 4)-CGP	120	10	–	–	–	–	–
	$(1 + \lambda)$-CGP	100	10	–	–	16	–	–
	Canonical-CGP (Subgraph)	220	9	90	–	–	50	5
	Canonical-CGP (Block)	100	5	90	–	–	50	7
	$(\mu + \lambda)$-CGP (Subgraph)	200	1	70	10	200	–	–
	$(\mu + \lambda)$-CGP (Block)	100	10	90	10	20	–	–
Nguyen-5	(1 + 4)-CGP	60	7	–	–	–	–	–
	$(1 + \lambda)$-CGP	60	7	–	–	16	–	–
	Canonical-CGP (Subgraph)	100	6	25	–	–	10	2
	Canonical-CGP (Block)	60	7	90	–	–	50	7
	$(\mu + \lambda)$-CGP (Subgraph)	300	5	70	10	250	–	–
	$(\mu + \lambda)$-CGP (Block)	60	5	75	10	20	–	–
Nguyen-6	(1 + 4)-CGP	100	10	–	–	–	–	–
	$(1 + \lambda)$-CGP	100	10	–	–	16	–	–
	Canonical-CGP (Subgraph)	20	20	90	–	–	50	7
	Canonical-CGP (Block)	60	7	90	–	–	50	7
	$(\mu + \lambda)$-CGP (Subgraph)	300	1	70	10	250	–	–
	$(\mu + \lambda)$-CGP (Block)	60	7	90	4	16	–	–
Nguyen-7	(1 + 4)-CGP	200	2	–	–	–	–	–
	$(1 + \lambda)$-CGP	1000	2	–	–	16	–	–
	Canonical-CGP (Subgraph)	500	3	70	–	–	250	7
	Canonical-CGP (Block)	200	5	90	–	–	50	7
	$(\mu + \lambda)$-CGP (Subgraph)	200	10	90	10	200	–	–
	$(\mu + \lambda)$-CGP (Block)	200	5	75	4	16	–	–
Keijzer-6	(1 + 4)-CGP	2000	3	–	–	–	–	–
	$(1 + \lambda)$-CGP	2000	5	–	–	16	–	–
	Canonical-CGP (Subgraph)	100	5	70	–	–	250	10
	Canonical-CGP (Block)	200	5	70	–	–	50	7
	$(\mu + \lambda)$-CGP (Subgraph)	700	5	70	10	250	–	–
	$(\mu + \lambda)$-CGP (Block)	200	5	75	4	16	–	–
Pagie-1	(1 + 4)-CGP	1500	7	–	–	–	–	–
	$(1 + \lambda)$-CGP	1500	7	–	–	16	–	–
	Canonical-CGP (Subgraph)	500	5	90	–	–	200	7
	Canonical-CGP (Block)	500	5	70	–	–	50	7
	$(\mu + \lambda)$-CGP (Subgraph)	500	8	75	25	125	–	–
	$(\mu + \lambda)$-CGP (Block)	500	5	75	4	16	–	–

Table 8. Results of the meta-optimization for the single output Boolean function problems. Table has been taken and revised from the work of Kalkreuth [9].

Problem	Algorithm	N	M	C	μ	λ	P	T
Parity-3	(1 + 4)-CGP	2000	1	–	–	–	–	–
	(1 + λ)-CGP	2000	1	–	–	1	–	–
	Canonical-CGP (Subgraph)	2000	1	70	–	–	10	2
	Canonical-CGP (Block)	2000	1	50	–	–	5	2
	(μ + λ)-CGP (Subgraph)	2000	1	90	4	1	–	–
	(μ + λ)-CGP (Block)	2000	1	50	4	1	–	–
Parity-4	(1 + 4)-CGP	1500	1	–	–	–	–	–
	(1 + λ)-CGP	1500	1	–	–	1	–	–
	Canonical-CGP (Subgraph)	1500	1	70	–	–	5	2
	Canonical-CGP (Block)	1500	1	50	–	–	5	2
	(μ + λ)-CGP (Subgraph)	1500	1	90	4	1	–	–
	(μ + λ)-CGP (Block)	1500	1	70	4	1	–	–
Parity-5	(1 + 4)-CGP	2000	1	–	–	–	–	–
	(1 + λ)-CGP	2000	1	–	–	1	–	–
	Canonical-CGP (Subgraph)	1000	1	70	–	–	5	2
	Canonical-CGP (Block)	2000	1	20	–	–	5	2
	(μ + λ)-CGP (Subgraph)	2000	1	25	4	1	–	–
	(μ + λ)-CGP (Block)	2000	1	25	4	1	–	–
Parity-6	(1 + 4)-CGP	2000	1	–	–	–	–	–
	(1 + λ)-CGP	2000	1	–	–	1	–	–
	Canonical-CGP (Subgraph)	2000	1	70	–	–	5	2
	Canonical-CGP (Block)	2000	1	70	–	–	5	2
	(μ + λ)-CGP (Subgraph)	2000	1	90	4	1	–	–
	(μ + λ)-CGP (Block)	2000	1	20	4	1	–	–
Parity-7	(1 + 4)-CGP	2500	1	–	–	–	–	–
	(1 + λ)-CGP	2500	1	–	–	1	–	–
	Canonical-CGP (Subgraph)	2500	1	50	–	–	5	2
	Canonical-CGP (Block)	2500	1	50	–	–	5	2
	(μ + λ)-CGP (Subgraph)	2500	1	50	4	1	–	–
	(μ + λ)-CGP (Block)	2500	1	50	4	1	–	–
Parity-8	(1 + 4)-CGP	2000	1	–	–	–	–	–
	(1 + λ)-CGP	2000	1	–	–	1	–	–
	Canonical-CGP (Subgraph)	2000	1	50	–	–	5	2
	Canonical-CGP (Block)	2000	1	50	–	–	5	2
	(μ + λ)-CGP (Subgraph)	2000	1	50	4	1	–	–
	(μ + λ)-CGP (Block)	2000	1	50	4	1	–	–

Table 9. Results of the meta-optimization for the multiple output Boolean function problems. Table has been taken and revised from the work of Kalkreuth [9].

Problem	Algorithm	N	M	C	μ	λ	P	T
Adder-1Bit	(1 + 4)-CGP	150	3	–	–	–	–	–
	(1 + λ)-CGP	150	3	–	–	1	–	–
	Canonical-CGP (Subgraph)	150	3	50	–	–	5	2
	Canonical-CGP (Block)	150	3	20	–	–	5	2
	(μ + λ)-CGP (Subgraph)	150	3	25	4	1	–	–
	(μ + λ)-CGP (Block)	150	3	25	4	1	–	–
Adder-2Bit	(1 + 4)-CGP	100	3	–	–	–	–	–
	(1 + λ)-CGP	150	1	–	–	1	–	–
	Canonical-CGP (Subgraph)	150	1	20	–	–	5	2
	Canonical-CGP (Block)	150	1	20	–	–	5	2
	(μ + λ)-CGP (Subgraph)	150	1	25	4	1	–	–
	(μ + λ)-CGP (Block)	150	1	25	4	1	–	–
Adder-3Bit	(1 + 4)-CGP	2000	1	–	–	–	–	–
	(1 + λ)-CGP	2000	1	–	–	1	–	–
	Canonical-CGP (Subgraph)	2000	1	50	–	–	5	2
	Canonical-CGP (Block)	2000	1	50	–	–	5	2
	(μ + λ)-CGP (Subgraph)	2000	1	50	8	1	–	–
	(μ + λ)-CGP (Block)	2000	1	25	8	1	–	–
Mult.-2Bit	(1 + 4)-CGP	1500	1	–	–	–	–	–
	(1 + λ)-CGP	1500	1	–	–	1	–	–
	Canonical-CGP (Subgraph)	1500	1	20	–	–	5	2
	Canonical-CGP (Block)	1500	1	20	–	–	5	2
	(μ + λ)-CGP (Subgraph)	1500	1	25	4	1	–	–
	(μ + λ)-CGP (Block)	1500	1	25	4	1	–	–
Mult.-3Bit	(1 + 4)-CGP	2000	1	–	–	–	–	–
	(1 + λ)-CGP	2000	1	–	–	1	–	–
	Canonical-CGP (Subgraph)	2000	1	20	–	–	5	2
	Canonical-CGP (Block)	2000	1	20	–	–	5	2
	(μ + λ)-CGP (Subgraph)	2000	1	25	4	1	–	–
	(μ + λ)-CGP (Block)	2000	1	25	4	1	–	–
Subtr.-2Bit	(1 + 4)-CGP	2000	1	–	–	–	–	–
	(1 + λ)-CGP	2000	1	–	–	1	–	–
	Canonical-CGP (Subgraph)	2000	1	20	–	–	5	2
	Canonical-CGP (Subgraph)	2000	1	20	–	–	5	2
	(μ + λ)-CGP (Subgraph)	2000	1	25	4	1	–	–
	(μ + λ)-CGP (Block)	2000	1	25	4	1	–	–

Table 10. Results of the meta-optimization for the image operator design problems.

Problem	Algorithm	N	M	C	μ	λ	P	T
Salt & Pepper noise	(1 + 4)-CGP	3000	1	–	–	–	–	–
	(1 + λ)-CGP	3000	1	–	–	1	–	–
	Canonical-CGP (Subgraph)	1000	1	50	–	–	10	2
	Canonical-CGP (Subgraph)	3000	1	50	–	–	5	2
	(μ + λ)-CGP (Subgraph)	3000	1	50	4	1	–	–
	(μ + λ)-CGP (Block)	3000	1	50	4	1	–	–
Gaussian noise	(1 + 4)-CGP	3000	1	–	–	–	–	–
	(1 + λ)-CGP	3000	1	–	–	1	–	–
	Canonical-CGP (Subgraph)	3000	1	50	–	–	5	2
	Canonical-CGP (Subgraph)	3000	1	50	–	–	5	2
	(μ + λ)-CGP (Subgraph)	3000	1	50	4	1	–	–
	(μ + λ)-CGP (Block)	3000	1	50	4	1	–	–

Table 11. Results of the algorithm comparison for the problems Koza 1, 2, 3 evaluated by the number of fitness evaluations (FE) to termination and the number of unfinished runs (#UR). Table has been taken and revised from the work of Kalkreuth [9].

Problem	Algorithm	Mean FE	SD	SEM	1Q	Median	3Q	#UR
Koza-1	(1 + 4)-CGP	8675635	16681422	±1668142	441477	1814344	7045961	2
	(1 + λ)-CGP	7370880‡	17384354	±1738435	204400	1050936	4294170	3
	Canonical-CGP (Subgraph)	663822‡	838546	±83854	135162	337950	710275	0
	Canonical-CGP (Block)	**243837‡**	**474790**	**±47479**	**13046**	**82124**	**261084**	0
	(μ + λ)-CGP (Subgraph)	7780751‡	15830735	±1583073	197284	1830312	6318740	3
	(μ + λ)-CGP (Block)	50188221	101041195	±10104119	1489364	16759132	61052140	8
Koza-2	(1 + 4)-CGP	8264426	19894512	±1989451	150140	888884	4378756	6
	(1 + λ)-CGP	8191549	20275790	±2027579	94290	559028	4710848	1
	Canonical-CGP (Subgraph)	444118‡	95000	±286700	627550	29650	78800	0
	Canonical-CGP (Block)	**114138‡**	**178463**	**±17846**	**12862**	**40523**	**141022**	0
	(μ + λ)-CGP (Subgraph)	5729778	11021660	±1102166.	238156	1320880	5878696	1
	(μ + λ)-CGP (Block)	22653191	54817816	±5481781	738020	3531700	19806332	3
Koza-3	(1 + 4)-CGP	600153	1214527	±121452	39076	177418	443038	0
	(1 + λ)-CGP	753551	2535215	±253521	29528	120368	431318	0
	Canonical-CGP (Subgraph)	**32870‡**	**57156**	**±10435**	**2488**	**6700**	**32713**	0
	Canonical-CGP (Block)	70795‡	216043	±21604	6161	15190	43414	0
	(μ + λ)-CGP (Subgraph)	926857	3473467	±347347	28548	121040	362180	0
	(μ + λ)-CGP (Block)	1167707	2287317	±228731	58832	220588	1150756	0

some of the tested problems the use of the block crossover also leads to better results when compared to the (1 + 4)-CGP.

Table 12. Results of the algorithm comparison for the symbolic regression problems evaluated with the *best-fitness-of-run* method. Table has been taken and revised from the work of Kalkreuth [9].

Problem	Algorithm Best Fitness	Mean	SD	SEM	1Q	Median	3Q
Nguyen-4	(1 + 4)-CGP	0.68	0.55	±0.05	0.34	0.58	0.77
	(1 + λ)-CGP	0.61	0.46	±0.04	0.35	0.54	0.74
	Canonical-CGP (Subgraph)	**0.50**†	**0.28**	**±0.04**	**0.31**	**0.47**	**0.60**
	Canonical-CGP (Block)	0.62	0.55	±0.05	0.30	0.54	0.63
	(μ + λ)-CGP (Subgraph)	0.60†	0.40	±0.04	0.36	0.54	0.76
	(μ + λ)-CGP (Block)	0.88	0.66	±0.07	0.53	0.67	1.01
Nguyen-5	(1 + 4)-CGP	0.45	0.42	±0.04	0.06	0.32	0.81
	(1 + λ)-CGP	0.39	0.33	±0.03	0.08	0.27	0.63
	Canonical-CGP (Subgraph)	0.29‡	0.27	±0.03	0.05	0.20	0.40
	Canonical-CGP (Block)	0.34	0.33	±0.03	0.05	0.26	0.59
	(μ + λ)**-CGP (Subgraph)**	**0.28**‡	**0.25**	**±0.02**	**0.06**	**0.19**	**0.45**
	(μ + λ)-CGP (Block)	0.48	0.40	±0.04	0.15	0.41	0.82
Nguyen-6	(1 + 4)-CGP	0.54	0.66	±0.06	0.16	0.29	0.61
	(1 + λ)-CGP	0.50	0.67	±0.06	0.15	0.22	0.50
	Canonical-CGP (Subgraph)	**0.31**‡	**0.31**	**±0.03**	**0.15**	**0.24**	**0.40**
	Canonical-CGP (Block)	0.44	0.63	±0.06	0.14	0.24	0.47
	(μ + λ)-CGP (Subgraph)	0.61	0.67	±0.06	0.16	0.35	0.67
	(μ + λ)-CGP (Block)	0.83	0.85	±0.08	0.23	0.38	1.28
Nguyen-7	(1 + 4)-CGP	0.79	0.48	±0.05	0.45	0.67	1.06
	(1 + λ)-CGP	0.71	0.45	±0.04	0.44	0.67	0.76
	Canonical-CGP (Subgraph)	**0.60**‡	**0.35**	**±0.03**	**0.36**	**0.60**	**0.68**
	Canonical-CGP (Block)	0.72	0.52	±0.05	0.47	0.65	0.70
	(μ + λ)-CGP (Subgraph)	0.62‡	0.40	±0.04	0.42	0.63	0.68
	(μ + λ)-CGP (Block)	1.39	1.13	±0.11	0.67	0.97	1.73
Keijzer-6	(1 + 4)-CGP	3.78	2.61	±0.26	2.16	3.24	4.59
	(1 + λ)-CGP	3.38	2.52	±0.25	2.41	3.03	3.158
	Canonical-CGP (Subgraph)	**2.81**†	**1.13**	**±0.11**	**1.78**	**2.90**	**3.75**
	Canonical-CGP (Block)	3.71	2.28	±0.22	2.37	3.15	4.01
	(μ + λ)-CGP (Subgraph)	2.88†	1.09	±0.10	2.25	3.14	3.15
	(μ + λ)-CGP (Block)	5.07	3.69	±0.36	3.16	3.94	5.46
Pagie-1	(1 + 4)-CGP	128.18	48.19	±4.81	87.81	119.09	161.08
	(1 + λ)-CGP	120.75	44.95	±4.49	86.14	120.91	155.06
	Canonical-CGP (Subgraph)	**98.52**‡	**50.57**	**±5.08**	**59.04**	**85.31**	**130.04**
	Canonical-CGP (Block)	119.97	45.44	±4.54	82.71	115.41	151.11
	(μ + λ)-CGP (Subgraph)	99.74‡	41.25	±4.12	65.32	95.79	131.76
	(μ + λ)-CGP (Block)	160.15	45.71	±4.57	133.84	157.33	187.67

Table 13. Results of the algorithm comparison for the single output Booleans function problem evaluated by the number of fitness evaluations (FE) to termination. Table has been taken and revised from the work of Kalkreuth [9].

Problem	Algorithm	Mean FE	SD	SEM	1Q	Median	3Q
Parity-3	(1 + 4)-CGP	3177	3417	±343	1246	2136	3760
	(1 + λ)-CGP	2182‡	1832	±183	885	1649	2868
	Canonical-CGP (Subgraph)	3107	3070	±307	1201	2104	3907
	Canonical-CGP (Block)	3013	2235	±223	1196	2672	3891
	(μ + λ)-CGP (Subgraph)	1808‡	1670	±167	645	1301	2119
	(μ + λ)-CGP Block)	**1781**‡	**1432**	**±143**	**739**	**1265**	**2509**
Parity-4	(1 + 4)-CGP	15420	14152	±1422	6292	10358	17726
	(1 + λ)-CGP	9615‡	6967	±696	4798	7703	13391
	Canonical-CGP (Subgraph)	54967	47042	±4727	24813	40612	71851
	Canonical-CGP (Block)	24463	23404	±2340	11158	17742	26914
	(μ + λ)-CGP (Subgraph)	**9211**‡	**7349**	**±735**	**4798**	**7392**	**11099**
	(μ + λ)-CGP (Block)	12202	8611	±861	6030	9593	15992
Parity-5	(1 + 4)-CGP	45542	33947	±3411	21524	36834	61222
	(1 + λ)-CGP	34375‡	28146	±2815	20685	27104	38941
	Canonical-CGP (Block)	102803	111238	±11123	47691	76170	122404
	Canonical-CGP (Subgraph)	**28413**‡	**25538**	**±2566**	**19640**	**23388**	**34876**
	(μ + λ)-CGP (Subgraph)	30801‡	23159	±2315	16315	24064	39778
	(μ + λ)-CGP (Block)	32298‡	18885	±1888	19514	289273	43804
Parity-6	(1 + 4)-CGP	199989	142915	±14291	107418	163234	242573
	(1 + λ)-CGP	92506‡	62113	±6211	45346	77795	117251
	Canonical-CGP (Subgraph)	242986	161762	±16257	134518	200196	309346
	Canonical-CGP (Block)	327097	218152	±21815	167538	286582	421521
	(μ + λ)-CGP (Subgraph)	**82428**‡	**50537**	**±5053**	**48087**	**68923**	**97539**
	(μ + λ)-CGP (Block)	82920‡	42766	±4276	51791	73296	102985
Parity-7	(1 + 4)-CGP	478055	301113	±30111	268210	393362	605372
	(1 + λ)-CGP	246241‡	196094	±19609	129927	204816	315353
	Canonical-CGP (Subgraph)	631568	548180	±54818	293613	453204	750792
	Canonical-CGP (Block)	850030	601124	±60112	452679	696466	1075600
	(μ + λ)-CGP (Subgraph)	**201532**‡	**131936**	**±13193**	**102515**	**172038**	**247155**
	(μ + λ)-CGP (Block)	263613‡	179168	±17916	128220	215078	339860*s*
Parity-8	(1 + 4)-CGP	1449319	961825	±175604	1027275	1257572	1883852
	(1 + λ)-CGP	512643‡	348361	±63601	299784	395583	604323
	Canonical-CGP (Subgraph)	1594115	1246555	±227588	935212	1152674	1675549
	Canonical-CGP (Block)	2809908	1942162	±354588	1486101	2270966	3622786
	(μ + λ)-CGP (Subgraph)	**436526**‡	**291573**	**±53233**	**239010**	**315086**	**523287**
	(μ + λ)-CGP (Block)	603235‡	274439	±50105	462495	548844	663047

Table 14. Results of the algorithm comparison for the multiple output Boolean function problems evaluated by the number of fitness evaluations (FE) to termination. Table has been taken and revised from the work of Kalkreuth [9].

Problem	Algorithm	Mean FE	SD	SEM	1Q	Median	3Q
Adder-1Bit	$(1+4)$-CGP	8373	9847	±984	2892	4826	10270
	$(1+\lambda)$-CGP	**4704**‡	**4526**	**±452**	**1996**	**3153**	**5440**
	Canonical-CGP (Subgraph)	12055	12744	±1274	4194	7710	15162
	Canonical-CGP (Block)	12915	15263	±1526	3904	7734	14993
	$(\mu+\lambda)$-CGP (Subgraph)	5193‡	4877	±487	2148	3699	6143
	$(\mu+\lambda)$-CGP (Block)	5564‡	4867	±486	1962	3900	7177
Adder-2Bit	$(1+4)$-CGP	146540	138275	±13827	68661	105746	176126
	$(1+\lambda)$-CGP	86168‡	84130	±8413	35658	58135	104502
	Canonical-CGP (Subgraph)	470193	380088	±38008	193837	329472	649081
	Canonical-CGP (Block)	232328	255735	±25573	90993	191414	286934
	$(\mu+\lambda)$-CGP (Subgraph)	**73312**‡	**62167**	**±6216**	**32654**	**50795**	**105657**
	$(\mu+\lambda)$-CGP (Block)	96382‡	94841	±9484	38355	63795	116310
Adder-3Bit	$(1+4)$-CGP	638856	506158	±50615	340366	474664	838796
	$(1+\lambda)$-CGP	293639‡	232145	±23214	144481	222136	417066
	Canonical-CGP	1034417	819093	±81909	504282	792500	1331676
	Canonical-CGP (Block)	838892	654062	±65406	432611	680524	1008478
	$(\mu+\lambda)$-CGP (Subgraph)	**266449**‡	**153200**	**±15320**	**139599**	**252469**	**352073**
	$(\mu+\lambda)$-CGP (Block)	349957‡	265435	±26543	204995	281461	390704
Multiplier-2Bit	$(1+4)$-CGP	13535	21488	±2148	4382	7590	13440
	$(1+\lambda)$-CGP	7381‡	7135	±713	3210	4988	8737
	Canonical-CGP (Subgraph)	14656	12720	±1272	7055	12276	18529
	Canonical-CGP (Block)	19215	32654	±3265	7055	12276	18529
	$(\mu+\lambda)$-CGP (Subgraph)	**6037**‡	**4069**	**±406**	**3590**	**4863**	**7160**
	$(\mu+\lambda)$-CGP (Block)	6246‡	4941	±494	3293	4812	7674
Multiplier-3Bit	$(1+4)$-CGP	118012	124585	±12458	49037	75972	131796
	$(1+\lambda)$-CGP	80884‡	104074	±10407	33267	54493	88733
	Canonical-CGP	205097	181375	±18137	85868	141202	253095
	Canonical-CGP (Block)	235011	215940	±21594	107160	163408	292391
	$(\mu+\lambda)$-CGP (Subgraph)	**70830**‡	**60434**	**±6043**	**33427**	**45230**	**90275**
	$(\mu+\lambda)$-CGP (Block)	74436‡	71610	±7161	31442	46240	99234
Subtractor-2Bit	$(1+4)$-CGP	22743	25293	±2529	8047	14404	27818
	$(1+\lambda)$-CGP	15942‡	17449	±1744	4609	10235	17203
	Canonical-CGP (Subgraph)	36874	42268	±4226	1241	19052	38932
	Canonical-CGP (Block)	35578	33030	±3303	16498	26910	40101
	$(\mu+\lambda)$-CGP (Subgraph)	**13525**‡	**15077**	**±1507**	**5589**	**8474**	**15262**
	$(\mu+\lambda)$-CGP (Block)	13198‡	7331	±733	7376	11916	18442

Table 15. Results of the algorithm comparison for the image operator design problems evaluated by the number of fitness evaluations (FE) to termination.

Problem	Algorithm	Mean FE	SD	SEM	1Q	Median	3Q
Salt & Pepper noise	(1 + 4)-CGP	48055	140673	±14067	3208	7256	21947
	(1 + λ)-CGP	16953	34010	±3401	3177	6220	16979
	Canonical-CGP (Subgraph)	57227	104822	±10482	7788	14124	44690
	Canonical-CGP (Block)	**10469**‡	**18697**	**±1869**	**2543**	**4672**	**9931**
	(μ + λ)-CGP (Subgraph)	51214	132168	±13216	4505	10556	34323
	(μ + λ)-CGP (Subgraph)	**12762**‡	**36587**	**±3658**	**1956**	**3479**	**8916**
Gaussian noise	(1 + 4)-CGP	62118	316015	±31601	4117	9202	21246
	(1 + λ)-CGP	28905	67476	±6747	3455	8475	20215
	Canonical-CGP (Subgraph)	83250	124696	±12469	12332	31270	105276
	Canonical-CGP (Block)	32998	111114	±11111	2992	5732	14185
	(μ + λ)-CGP (Subgraph)	35883	80896	±8089	4765	10243	30636
	(μ + λ)-CGP (Block)	**16992**‡	**46200**	**±4620**	**1586**	**4118**	**12664**

Table 13 and Table 14 show the results of the algorithm comparison in the Boolean domain. It is clearly visible that the choice of a $\lambda = 1$ results in a reduced number of fitness evaluations for all tested Boolean functions. This setting is also effective for the use of various setups of the ($\mu + \lambda$)-CGP. Overall, it can also be seen that the use of the phenotypic crossover operators lead to a significant decrease of the number of fitness evaluations to find the ideal solution when compared to the (1 + 4)-CGP.

Table 15 shows the results for the image operator design problems, and it can be seen that in this problem domain, the Canonical-CGP and the ($\mu + \lambda$)-CGP perform best.

5 Discussion and Analysis

The presented evaluation demonstrates that subgraph and block crossover can both contribute to the search performance by using a canonical or ($\mu + \lambda$) EA. Furthermore, the results of our experiments clearly demonstrate that the effectiveness of the (1 + λ)-CGP algorithms cannot be generalized. Our experiments in all three problem domains resulted in an improvement of the search performance of CGP when a crossover-based algorithm was used.

In the Boolean domain, our results regarding the parametrization are predominantly coherent with the findings of Miller and Smith [25]. Our experiments indicate that for the majority of our tested problems, the well-known CGP performance dogma of low population sizes and extremely high levels of redundancy is coherent with other studies [6, 12]. However, our experiments on the 1-Bit and 2-Bit Adder also indicate that this dogma cannot be generalized in this problem domain. Since it has been found that standard genotypic crossover operators do not contribute to the performance of CGP in this problem domain [24], our experiments demonstrate that advanced phenotypic operators can contribute significantly to the search performance by using a canonical and $(\mu + \lambda)$ EA. Our experiments also confirm findings of former work in the Boolean domain which have been presented in [12] and demonstrated that a setting of $\lambda = 1$ works more effective for the $(1 + \lambda)$-CGP as the common choice of $\lambda = 4$. Our results expand this finding to the recombination-based $(\mu + \lambda)$-CGP for which a setting of $\lambda = 1$ led to good results in the Boolean domain.

In the symbolic regression domain, the performance of the Canonical-CGP algorithm was superior to the (1 + 4)-CGP on all tested problems. Moreover, our evaluation of the whole evolutionary run for the more simple benchmark problems Koza 1, 2 & 3 revealed a big gap of the search performance between the (1 + 4)-CGP and the Canonical-CGP on these problems. This is underlined by the observation that the Canonical-CGP finished all runs successfully within the given budget of fitness evaluations. On the Salt & Pepper noise reduction problem, the search performance of the Canonical-CGP algorithm equipped with subgraph crossover was superior to the (1 + 4)-CGP. For the Gaussian noise reduction problem, we observed an improved search performance for the $(\mu + \lambda)$-CGP equipped with block crossover.

Since our meta optimization experiments demonstrate that the parameterization patterns vary for different problem domains it opens up the question, which conditions or types of problems require bigger or smaller population sizes and in which way subgraph and block crossover contribute to the evolutionary search. The same question arises for the length of the genotype. A preliminary assumption could be that the fitness landscape of certain problems requires more exploration abilities to overcome local optima or to explore a bigger and denser continuous fitness space. However, since the experiments of [25] mostly focused on Boolean function problems which are evaluated with discrete fitness more research is needed to study the behavior of CGP algorithms in continuous fitness spaces. Our results also demonstrate that the ideal parametrization of CGP depends on the problem itself. In this way, our results confirm the results of a former study which has been performed by Turner and Miller [35]. The study investigated different genotype lengths up to 100,000 nodes. In all cases, there was an ideal number of nodes. These vary depending on the problem. The ideal genotype lengths for the respective problems varied between 50 and 3000 function nodes.

5.1 Analysis of Hypotheses

H1. *Crossover Does Not Contribute to the Search Performance of Integer-based Standard CGP*

The hypothesis that crossover does not contribute to the search performance is rooted in the work of Miller [24] and Miller and Thompson [26]. The experiments of both works led to a general question of the benefits of crossover in CGP. However, further work, which has been contributed to standard CGP, rejected the use of crossover and focused on the well-known $(1 + \lambda)$-CGP. The results of our experiments clearly show that two algorithms that utilize subgraph and block crossover outperformed the (1 + 4)-CGP and $(1+\lambda)$-CGP on various benchmark problems in three different problem domains. Based on these findings we can disprove the hypothesis that crossover cannot contribute to the search performance of standard integer-based CGP.

H2. *Extremely Large Genotypes Perform Most Effective in CGP*
For the analysis of this hypothesis, we investigated the search performance of CGP after the meta-optimization. The claim that extremely high levels of redundancy, caused by extremely large genotypes, perform most efficient in CGP grew out of Miller and Smith's work [25]. Their experiments on Boolean function problems showed that CGP performed best with extremely high levels of redundancy on the Even-Parity-3 problem and the 2-Bit digital multiplier problem. The experiments were performed using the standard $(1 + \lambda)$-CGP algorithm and small population size of five individuals. The results of the experiments indicate that the efficiency of the combination of high levels of redundancy and small population sizes is valid for the majority of the tested Boolean function problems. However, on most tested symbolic regression problems, smaller genotypes and bigger population sizes performed best. The outcomes of our evaluation show that the dogma of high levels of redundancy cannot be generalized. Furthermore, it has been shown that also small genotypes can be efficiently used and recombined in CGP. The results of the experiments indicate that Miller and Smith's findings hold for discrete fitness spaces with fixed boundaries but can't be generalized for other types of fitness such as continuous fitness. Our results of the search performance experiments with continuous fitness indicate that the $(1 + 4)$-CGP algorithm is not an effective choice for the evolutionary search in this type of fitness spaces.

H3. *Small Populations Perform Most Effective in CGP*
Our results in the symbolic regression domain demonstrate that the claim that small population sizes generally perform most effectively in CGP cannot be generalized. In the Boolean domain, the results and findings of former studies seem to be coherent. However, it has been demonstrated that this claim does not hold for the symbolic regression and image operator design domain. Our results give clear evidence that the use of medium and high population sizes can lead to a significantly better search performance when compared to the traditional $(1 + 4)$-CGP in the symbolic regression domain.

6 Conclusion and Future Work

Our evaluation study demonstrates that the use of two phenotypic crossover operators is beneficial on various benchmark problems in three different problem domains. The results of the experiments also clearly show that popular performance dogmas of CGP are not coherent in other problem domains. However, our experiments demonstrate that former findings of the effective use of low population sizes and extremely large genotypes are valid in the Boolean domain. In the symbolic regression domain, our experiments revealed that this dogma cannot be generalized and showed that comparatively smaller genotypes and bigger populations can also perform effectively in CGP. On the image operator design problems, we observed that the crossover-based canonical and $(\mu + \lambda)$ EA contribute to the search performance of CGP. Overall, this study paves the way for further studies on the behavior of subgraph and block crossover based CGP algorithms. Therefore, our future work will focus on exploration analysis in fitness and phenotype space of CGP.

The following concluding remarks can be drawn from the results of our experiments:

- Subgraph and block crossover can improve the search performance of CGP
- Bigger populations can be effectively used with CGP on certain problems
- Well known parametrization patterns cannot be generalized

References

1. Atkinson, T., Plump, D., Stepney, S.: Evolving graphs by graph programming. In: Castelli, M., Sekanina, L., Zhang, M., Cagnoni, S., García-Sánchez, P. (eds.) EuroGP 2018. LNCS, vol. 10781, pp. 35–51. Springer, Cham (2018). https://doi.org/10.1007/978-3-319-77553-1_3. https://eprints.whiterose.ac.uk/126500/1/AtkinsonPlumpStepney.EuroGP.18.pdf
2. Clegg, J., Walker, J.A., Miller, J.F.: A new crossover technique for cartesian genetic programming. In: Thierens, D., et al. (eds.) GECCO 2007: Proceedings of the 9th Annual Conference on Genetic and Evolutionary Computation, vol. 2, pp. 1580–1587. ACM Press, London, 7–11 July 2007. https://doi.org/10.1145/1276958.1277276. https://www.cs.bham.ac.uk/~wbl/biblio/gecco2007/docs/p1580.pdf
3. Cramer, N.L.: A representation for the adaptive generation of simple sequential programs. In: Grefenstette, J.J. (ed.) Proceedings of an International Conference on Genetic Algorithms and the Applications, Carnegie-Mellon University, Pittsburgh, PA, USA, 24–26 July 1985, pp. 183–187 (1985). https://www.cs.ucl.ac.uk/staff/W.Langdon/ftp/papers/icga1985/icga85_cramer.pdf
4. Forsyth, R.: BEAGLE a Darwinian approach to pattern recognition. Kybernetes **10**(3), 159–166 (1981). https://doi.org/10.1108/eb005587. https://www.richardsandesforsyth.net/pubs/beagle81.pdf
5. Hicklin, J.: Application of the genetic algorithm to automatic program generation. Master's thesis, University of Idaho (1986)
6. Husa, J., Kalkreuth, R.: A comparative study on crossover in cartesian genetic programming. In: Castelli, M., Sekanina, L., Zhang, M., Cagnoni, S., García-Sánchez, P. (eds.) EuroGP 2018. LNCS, vol. 10781, pp. 203–219. Springer, Cham (2018). https://doi.org/10.1007/978-3-319-77553-1_13
7. Kalganova, T.: Evolutionary approach to design multiple-valued combinational circuits. In: Proceedings of the 4th International conference on Applications of Computer Systems (ACS 1997), Szczecin, Poland, pp. 333–339 (1997)
8. Kalkreuth, R.: Two new mutation techniques for cartesian genetic programming. In: Guervós, J.J.M., Garibaldi, J.M., Linares-Barranco, A., Madani, K., Warwick, K. (eds.) Proceedings of the 11th International Joint Conference on Computational Intelligence, IJCCI 2019, Vienna, Austria, 17–19 September 2019, pp. 82–92. ScitePress (2019). https://doi.org/10.5220/0008070100820092
9. Kalkreuth, R.: A comprehensive study on subgraph crossover in cartesian genetic programming. In: Guervós, J.J.M., Garibaldi, J.M., Wagner, C., Bäck, T., Madani, K., Warwick, K. (eds.) Proceedings of the 12th International Joint Conference on Computational Intelligence, IJCCI 2020, Budapest, Hungary, 2–4 November 2020, pp. 59–70. SCITEPRESS (2020). https://doi.org/10.5220/0010110700590070
10. Kalkreuth, R.: An empirical study on insertion and deletion mutation in cartesian genetic programming. In: Merelo, J.J., Garibaldi, J., Linares-Barranco, A., Warwick, K., Madani, K. (eds.) IJCCI 2019. SCI, vol. 922, pp. 85–114. Springer, Cham (2021). https://doi.org/10.1007/978-3-030-70594-7_4
11. Kalkreuth, R., Rudolph, G., Droschinsky, A.: A new subgraph crossover for cartesian genetic programming. In: McDermott, J., Castelli, M., Sekanina, L., Haasdijk, E., García-Sánchez, P. (eds.) EuroGP 2017. LNCS, vol. 10196, pp. 294–310. Springer, Cham (2017). https://doi.org/10.1007/978-3-319-55696-3_19

12. Kaufmann, P., Kalkreuth, R.: An empirical study on the parametrization of cartesian genetic programming. In: Proceedings of the Genetic and Evolutionary Computation Conference Companion, GECCO 2017, pp. 231–232. ACM, New York, NY, USA (2017). https://doi.org/10.1145/3067695.3075980
13. Kaufmann, P., Kalkreuth, R.: Parametrizing cartesian genetic programming: an empirical study. In: Kern-Isberner, G., Fürnkranz, J., Thimm, M. (eds.) KI 2017. LNCS (LNAI), vol. 10505, pp. 316–322. Springer, Cham (2017). https://doi.org/10.1007/978-3-319-67190-1_26
14. Koza, J.: Genetic programming: a paradigm for genetically breeding populations of computer programs to solve problems. Technical report STAN-CS-90-1314, Department of Computer Science, Stanford University, June 1990
15. Koza, J.R.: Genetic Programming: on the Programming of Computers by Means of Natural Selection. MIT Press, Cambridge, MA, USA (1992). https://mitpress.mit.edu/books/genetic-programming
16. Koza, J.R.: Genetic Programming II: Automatic Discovery of Reusable Programs. MIT Press, Cambridge Massachusetts, May 1994. https://www.genetic-programming.org/gpbook2toc.html
17. Langdon, W.B., Poli, R.: Fitness causes bloat. In: Chawdhry, P.K., Roy, R., Pant, R.K. (eds.) Soft Computing in Engineering Design and Manufacturing, pp. 13–22. Springer, London, 23–27 June 1997. https://doi.org/10.1007/978-1-4471-0427-8_2. https://www.cs.ucl.ac.uk/staff/W.Langdon/ftp/papers/WBL.bloat_wsc2.ps.gz
18. Langdon, W.B., Soule, T., Poli, R., Foster, J.A.: The evolution of size and shape. In: Spector, L., Langdon, W.B., O'Reilly, U.M., Angeline, P.J. (eds.) Advances in Genetic Programming, vol. 3, Chap. 8, pp. 163–190. MIT Press, Cambridge, MA, USA, June 1999. https://www.cs.ucl.ac.uk/staff/W.Langdon/aigp3/ch08.pdf
19. Luke, S.: ECJ then and now. In: Bosman, P.A.N. (ed.) Genetic and Evolutionary Computation Conference, Berlin, Germany, 15–19 July 2017, Companion Material Proceedings, pp. 1223–1230. ACM (2017). https://doi.org/10.1145/3067695.3082467
20. McDermott, J., et al.: Genetic programming needs better benchmarks. In: Soule, T., et al. (eds.) GECCO 2012: Proceedings of the Fourteenth International Conference on Genetic and Evolutionary Computation Conference, pp. 791–798. ACM, Philadelphia, Pennsylvania, USA, 7–11 July 2012. https://doi.org/10.1145/2330163.2330273
21. McPhee, N.F., Miller, J.D.: Accurate replication in genetic programming. In: Eshelman, L.J. (ed.) Proceedings of the 6th International Conference on Genetic Algorithms, Pittsburgh, PA, USA, 15–19 July 1995, pp. 303–309. Morgan Kaufmann (1995)
22. Mercer, R.E., Sampson, J.: Adaptive search using a reproductive meta-plan. Kybernetes (1978)
23. Miller, J.F., Thomson, P., Fogarty, T.: Designing electronic circuits using evolutionary algorithms. Arithmetic circuits: a case study. In: Genetic Algorithms and Evolution Strategies in Engineering and Computer Science, pp. 105–131. Wiley (1997)
24. Miller, J.F.: An empirical study of the efficiency of learning Boolean functions using a cartesian genetic programming approach. In: Banzhaf, W., et al. (eds.) Proceedings of the Genetic and Evolutionary Computation Conference, vol. 2, pp. 1135–1142. Morgan Kaufmann, Orlando, Florida, USA, 13–17 July 1999. https://citeseer.ist.psu.edu/153431.html
25. Miller, J.F., Smith, S.L.: Redundancy and computational efficiency in cartesian genetic programming. IEEE Trans. Evol. Comput. **10**(2), 167–174 (2006). https://doi.org/10.1109/TEVC.2006.871253
26. Miller, J.F., Thomson, P.: Cartesian genetic programming. In: Poli, R., et al. (eds.) EuroGP 2000. LNCS, vol. 1802, pp. 121–132. Springer, Heidelberg (2000). https://doi.org/10.1007/978-3-540-46239-2_9. https://www.elec.york.ac.uk/intsys/users/jfm7/cgp-eurogp2000.pdf
27. Miller, J.F.: Cartesian genetic programming: its status and future. Genet. Program Evolvable Mach. **21**(1), 129–168 (2020). https://doi.org/10.1007/s10710-019-09360-6

28. Scott, E.O., Luke, S.: ECJ at 20: toward a general metaheuristics toolkit. In: López-Ibáñez, M., Auger, A., Stützle, T. (eds.) Proceedings of the Genetic and Evolutionary Computation Conference Companion, GECCO 2019, Prague, Czech Republic, 13–17 July 2019, pp. 1391–1398. ACM (2019). https://doi.org/10.1145/3319619.3326865
29. Sekanina, L.: Image filter design with evolvable hardware. In: Applications of Evolutionary Computing, EvoWorkshops 2002: EvoCOP, EvoIASP, EvoSTIM/EvoPLAN, Kinsale, Ireland, 3–4 April 2002, Proceedings, pp. 255–266 (2002). https://doi.org/10.1007/3-540-46004-7_26
30. da Silva, J.E.H., Bernardino, H.: Cartesian genetic programming with crossover for designing combinational logic circuits. In: 7th Brazilian Conference on Intelligent Systems, BRACIS 2018, São Paulo, Brazil, 22–25 October 2018, pp. 145–150. IEEE Computer Society (2018). https://doi.org/10.1109/BRACIS.2018.00033
31. Soule, T., Foster, J.A.: Removal bias: a new cause of code growth in tree based evolutionary programming. In: 1998 IEEE International Conference on Evolutionary Computation, pp. 781–786. IEEE Press, Anchorage, Alaska, USA, 5–9 May 1998. https://doi.org/10.1109/ICEC.1998.700151. https://citeseer.ist.psu.edu/313655.html
32. Turner, A.J.: Improving crossover techniques in a genetic program. Master's thesis, Department of Electronics, University of York (2012)
33. Turner, A.J., Miller, J.F.: Cartesian genetic programming encoded artificial neural networks: a comparison using three benchmarks. In: Blum, C., et al. (eds.) GECCO 2013: Proceeding of the Fifteenth Annual Conference on Genetic and Evolutionary Computation Conference, pp. 1005–1012. ACM, Amsterdam, The Netherlands, 6–10 July 2013. https://doi.org/10.1145/2463372.2463484
34. Turner, A.J., Miller, J.F.: The importance of topology evolution in NeuroEvolution: a case study using cartesian genetic programming of artificial neural networks. In: Bramer, M., Petridis, M. (eds.) Research and Development in Intelligent Systems XXX, pp. 213–226. Springer, Cham (2013). https://doi.org/10.1007/978-3-319-02621-3_15
35. Turner, A.J., Miller, J.F.: Neutral genetic drift: an investigation using cartesian genetic programming. Genet. Program Evolvable Mach. **16**(4), 531–558 (2015). https://doi.org/10.1007/s10710-015-9244-6
36. White, D.R., et al.: Better GP benchmarks: community survey results and proposals. Genet. Program Evolvable Mach. **14**(1), 3–29 (2013). https://doi.org/10.1007/s10710-012-9177-2

An Information Granulation Approach Through m-Grams for Text Classification

Enrico De Santis, Antonino Capillo(✉), Emanuele Ferrandino, Fabio Massimo Frattale Mascioli, and Antonello Rizzi

Department of Information Engineering, Electronics and Telecommunications (DIET), University of Rome "La Sapienza", Via Eudossiana 18, 00184 Rome, Italy
{enrico.desantis,antonino.capillo,emanuele.ferrandino, fabiomassimo.frattalemascioli,antonello.rizzi}@uniroma1.it

Abstract. With the availability of a huge amount of electronic text sources within the World Wide Web and Social Network Sites, text mining and text classification are gaining more and more importance in AI related research fields. Nowadays, researchers and practitioners are focusing not only on high-performance systems in terms of classification capabilities, but also in generating gray-box interpretable models in which the relations belonging to the input-output mapping is also explainable, in line with the"explainable AI" paradigm. In this paper is proposed a text mining system capable of classifying text excerpts belonging to a suitable corpus through the human-centric Granular Computing approach. The system is grounded on two granulation levels of text, i.e. the set of extracted m-grams and a suitable hard partition of the latter obtained through both a well-suited dissimilarity measure between m-grams along with a clustering algorithm. The procedure allows embedding a text document – exploiting the Symbolic Histogram technique – in a real-valued vector space where standard Machine Learning algorithms, such as SVM, can safely operate. An evolutionary metaheuristic based on a Genetic Algorithm is, then, in charge of optimizing the system metaparameters, as well as performing a wrapper-like feature selection. This procedure allows carrying out knowledge discovery on the obtained models selecting relevant and interpretable information granules (i.e. m-grams) related to the classification task. The current study presents a new evaluation framework of the performance of the entire system in terms of generalization capabilities and the obtained results show the reliability of the overall procedure. Furthermore, a new conceptual framework is also presented in term of the "representation" problem in automated Pattern Recognition systems.

Keywords: Text mining · Text categorization · Granular computing · Explainable AI · Conceptual spaces

1 Introduction

It is well-known that we are living in the "information overloaded" Era – too much information, beyond the most relevant data [5] – thanks also to ICT advancements and the Internet, in which the World Wide Web and Social Networks can be conceived as

J. Garibaldi et al. (Eds.): IJCCI 2021, SCI 1119, pp. 73–89, 2023.
https://doi.org/10.1007/978-3-031-46221-4_4

a huge container of unstructured data, mostly in textual form. Textual data potentially contains a great value especially if there are available methodologies and techniques able to extract from text excerpts meaningful and semantically rich information. In fact, text documents in electronic format available on the Internet weave, together people's behavior, carry objective and subjective knowledge, such as the one in newspapers, blog posts, or comments and opinions on Social Network Sites [49]. Hence, extracting useful information from these electronic sources is a challenging problem faced in technical literature through the so-called text mining process, a specific branch of data mining also known as knowledge discovery in textual database (KDT) [19,23]. Particularly, text mining is defined as the process of extracting previously unknown, understandable, potential and practical patterns or knowledge from the collection of massive and unstructured text data or corpus [47]. Text mining is by now a really wide research area at the intersection with data mining, Artificial Intelligence, Natural Language Processing (NLP), Information Retrieval (IR), Knowledge Management and Discovery, corpus-based Computational Linguistics and other Computer Science related disciplines [7,20]. Similarly, Text Categorization or text classification (TC) is a subdiscipline that exploits Machine Learning for solving a supervised learning problem, where documents in a corpus are provided together with well-suited labels, reflecting the meaning of the category or topic in which the document falls [7].

On the other hand, the most relevant problem in automated Pattern Recognition approached through Machine Learning is the "representation", that is finding the right method to represent real-world objects, often heavily unstructured, in such a way that they can be elaborated by standard Machine Learning algorithms [30]. The representation problem is for some extent related to "meaning" that emerges in a not yet well-understood way from the underlying complex and hierarchical structure of text excerpts – produced by a really complex machinery, i.e. the brain. In fact, in textual data we can find a growing granulation starting from atomic objects, such as letters grouped in words and words suitably grouped in sentences that, in turn, are grouped in paragraphs, documents and corpora [30]. Moreover, as pointed out by Heeman [24], text documents possess a well-defined granular structure that can be divided in a high-level structure and a low-level one. While the high-level structure relies on the organization of a given document, the low-level structure reflects the hierarchical composition of the language including characters, words, phrases, concepts, clauses, sentences, paragraphs [25]. In this specific context, Granular Computing (GrC) can be conceived as a suitable toolbox convenient also for solving text mining problems, such as text classification as in [36]. The GrC approach grounds on the notion of "information granules", by which systems (and related data) can be perceived at different levels of specificity (detail), depending on the complexity of the problem and the objectives of the analysis [43]. In light of the so-called pyramid knowledge [1], "information granules", which arise in the process of data abstraction, allow discovering regularities in data, transforming raw data in knowledge and, hopefully, in wisdom. More precisely, "information granules" are complex information entities that arise in the process of data abstraction and knowledge formation. They are collections of entities that originate at the numeric level and are arranged together due to their similarity, functional or physical adjacency, indistinguishability, coherency, etc. [44]. In textual data, the m-gram decomposition can be seen as a granulation procedure of which the output consists of small pieces of text or phrases. Furthermore, grouping m-grams through some suitable similarity criterion within a clustering procedure is another granulation level. Anyway, in TC problems the GrC approach can help in finding regularities in text data, building a performing classification system able also to accomplish knowledge discovery tasks. This interesting property is in line with the "explainable AI" (XAI) paradigm [34], a corpus of practices and leading methodologies recently studied and developed by the scientific community, together with industry and government entities (such as DARPA in USA). In other words, the GrC approach allows building classifiers, as input-output processing systems, that are *interpretable* and *comprehensible*, where inputs are mathematically mapped to outputs (interpretability), while the adoption of information granules, as abstract but interpretable entities, also increases the comprehensibility of the model, since information granules can be semantically associated to well-defined concepts [16].

In the previous work [7], a novel text classifier, based on a GrC approach is presented. More precisely, documents to be classified are decomposed in their constituent m-grams or groups of tokens, with the aim of enhancing the text representation level and the classification performance. In addition, a novel dissimilarity measure between m-grams is proposed, which is a custom-based Edit distance [7]. Thus, the previous work is focused on assessing the feasibility of text classification with the above mentioned approach, showing it is suitable for achieving satisfying results. In this work, the purpose is to enhance the system efficacy by evaluating and adjusting the parameters configuration through several well-suited experiments. In addition, unlike [7], here we provide several hints about the conceptual background on which this work is based.

Specifically, the proposed TC system consists of several processing steps [7]. The procedure starts with a suitable granulation of text excerpts randomly extracting a number of m-grams – i.e. information granules – of a given size. The second step consists of a per-class clustering procedure, providing a proper hard partition of the space generated over a suitable numerical representation of the m-grams, obtained, in turn, through the *fastText* neural word embedding technique [3], performed on each token within the m-gram sequence. Hence, the information granule is represented as a variable-length ordered word-vector, lying in a rich semantic space. A suitable dissimilarity measure pertaining to the Edit family is then adopted to compare m-grams and to construct the partition through the clustering procedure. Then, the m-gram representatives belonging to the cluster model are interpreted as symbols of a proper alphabet used to estimate the statistical distribution of symbols in a given test document, through a well-suited membership rule. In other words, this approach leads to represent each document in a corpus as a Symbolic Histogram [14], i.e. a real-valued vector of fixed length useful for feeding standard Machine Learning algorithms. Finally, an evolutionary optimization procedure, through a Genetic Algorithm (GA), is in charge of both finding the (sub-)optimal hyper-parameters of the system and performing a wrapper-like features selection over symbols of the SH representation. The optimization meta-heuristic allows obtaining a simpler document representation that is computationally efficient and more interpretable.

In this work we propose an in-depth analysis of the system performance both varying the main meta-parameters and investigating two feature selection schemes within the optimization wrapper. In particular two binarization schemes are evaluated since the feature selection weights are encoded in the GA as real-valued genes. Moreover, the obtained symbols, that are small phrases, can be used for interpreting the decision rule generated by the learning system in assigning a class label, opening to the knowledge discovery paradigm in line with the XAI approach.

The paper is structured as follows. In Sect. 2 is provided a general background on text embedding methods in text mining, along with the proposed conceptual approach based on the GrC. Moreover, the section offers an overview of the main blocks that constitute the classification system. Section 3 describes the new criteria adopted to build the experimental framework for assessing the system performance. In Sect. 4, the setting of experiments and the main obtained results are shown and discussed. Finally, in Sect. 5 conclusions are drawn.

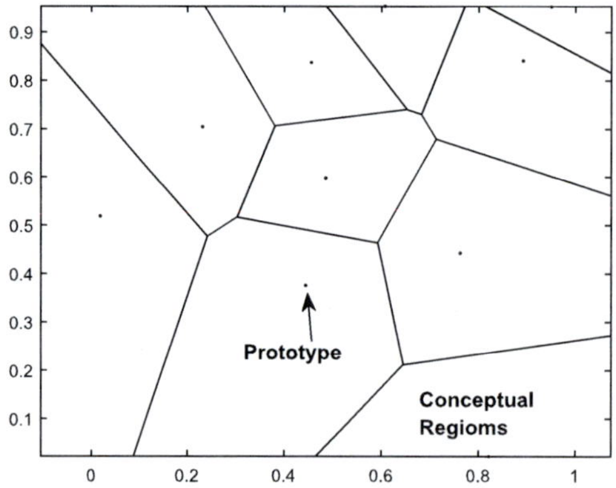

Fig. 1. Isomorphism between Voronoi regions and Conceptual Spaces [48].

2 The Text Categorization System

2.1 Background and Conceptual Framework

In the following work, natural language is conceived as a Complex System [11], being text a byproduct of a real-world powerful complex system that is the human brain. In fact, natural language arises from information processing and integration performed by different brain areas and can be considered as a communication system, defined and working at the interface between biology and social interactions [6,26]. In Complex Systems the representation problem is pivotal and this fact is true even in text mining applications.

One of the main steps in text mining is the representation of text excerpts in a well-suited algebraic space before feeding any automatic learning procedure. In general, the essence of such algebraic space, capturing some kind of co-occurrence between words and contexts, is built on the top of Distributional Semantics (DS) [4,18,28]. The main hypothesis of DS is that the similarity of meaning correlates with the similarity of distribution, a concept well-expressed by the American linguist Z.S. Harris, which sustained [22] that "words that are used and occur in the same contexts tend to purport similar meanings". This concept is traditionally captured by the Bag of Words (BoW) approach, where text documents are represented with the frequency of occurrence of some vocabulary words, that is the set of unique tokens belonging to the corpus. The BoW representation does not take into account the presentation order of terms in the text but it leads to define a dissimilarity measure between documents because the latter are represented as suitable vectors. This representation is traditionally adopted in Information Retrieval (IR) applications with the so called Vector Space Model [40], where the corpus is embedded in a term-document matrix and column vectors related to documents, generating a well-suited normed space endowed with a suitable distance measure. In many IR applications the cosine similarity is adopted, but with a suitable normalization procedure other distances – grounded on the scalar product operation – can be used. To improve the representation capabilities several weighting schemes can be introduced. The most adopted is the Term Frequency-Inverse Document Frequency (TF-IDF) scheme [39], where high -frequency words not specific for a document receive less weight compared to a low-frequency and specific terms. This technique captures the fact that words occurring roughly uniformly in the corpus tend to be less discriminatory. Even if the BOW embedding is highly adopted for its intrinsic simplicity, it has a number of drawbacks, such as the impossibility of capturing synonymy, the sparsity of the representation and the dimension of the feature space that is equivalent to the cardinality of the vocabulary words (often very large). A partial solution is offered by the Latent Semantic Indexing technique [12,27] grounded on the SVD decomposition of the term-document matrix, where documents are embedded in a low dimensional latent feature space. Hence, the numeric representation and the embedding of a text document in a suitable normed space able to capture some form of basic semantic is an interesting and challenging problem. Recently, the representation power of neural architectures has been fruitfully exploited, such as in the Continuous Bag of Words representation (CBOW) [33], where a shallow neural network is used to generate semantically rich word-vectors, and more advanced architectures capable of learning the syntactic-grammatical multilevel structure of a text excerpt, such as the "Transformers". The last one are able, through a smart multitasking learning scheme, to tackle the synonymy problems affecting the basic BoW scheme, providing also a very good degree of co-reference resolution [15]. In general, these architectures allow generating language models in an unsupervised fashion usable for other downstream tasks in NLP and text mining. *FastText* is a particular word embedding and text classification library – adopted in this study within the classification pipeline – created by Facebook's AI Research (FAIR) lab. *FastText* is capable to achieve really good performance for word representations and sentence classification, specially in the case of rare words by making use of character level information [3].

GrC has been fruitful adopted in text classifications in task such as concept formation and concept relationships identification and for inducing classification rules [29, 36,46]. In the approach proposed in the current work a two level granulation on text documents is performed. The first one, as already stated in the introduction, is at token level, where text is partitioned in several m-grams of a suitable length. Once defined a dissimilarity measure between m-grams, a document corpus (or a particular grouping dictated by a class label) can be partitioned, in turn, in a suitable number of regions, definable "Concepts Regions" in line with the Conceptual Spaces Theory outlined by P. Gärdenfors [21,45]. In other words, given a suitable dissimilarity measure between objects (here m-grams) the input space can be granulated in Voronoi regions – see Fig. 1. For example, if one adopts the well-known k-means clustering algorithm [17], centroidal Voronoi Regions are obtained. In this case the partition cardinality is a meta-parameter of the system regulating the granularity level. This fixed number of regions can be conceived as concepts for the given corpus or the given class of documents. Each region in this concept space is identified by a prototype and given a m-gram of a suitable length it can be associated, through a well-suited rule, to the region measuring the distance with the prototype. Hence, the region behaves as a receptive field [35] for the underlying concept. Moreover, a document, composed by a certain sequence of m-grams can be represented as a statistical distribution of prototypes (symbols of an alphabet), building the so-called Symbolic Histogram (SH) representation of a text excerpt. It is worth noting that GrC is a useful and general approach where, once defined a similarity measure between (numerical representations of) real-world objects, algorithms and procedures can be defined in a suitable high abstract level with the aim of modelling complex systems in different application areas, such as sequences modeling and prediction, graph embedding, image recognition in Computer Vision and the protein function prediction tasks, as in Bioinformatics [2,13,14,31,38]. In this setting the dissimilarity measure definition is very important, especially when the objects to be compared are unstructured and composed by different sub-patterns, deriving from heterogeneous measures taken on the real-world object to represent [9,10,41].

2.2 Overview of the Text Categorization System

In the following section it is provided a brief summary of the text categorization system grounded on the GrC approach and presented in details in [7]. The system flowchart is shown in Fig. 2.

The n documents which compose the textual dataset $\mathcal{D} = \{t_i\}_{i=1}^{n}$ t_i are tokenized in small word units named m-grams. The GrC model is synthesized by a per-class clustering process through the k-medoids algorithm [37], which leads to a suitable robust *embedding* procedure. In other words, we seek a suitable mapping $\mathcal{M}$ starting from the document (unstructured) space $\mathcal{D}$ to real-valued $\tilde{k}$-tuples, that is $\mathcal{M} : \mathcal{D} \rightarrow \mathbb{R}^{n \times \tilde{k}}$.

Thanks to the dissimilarity measure, which consists in a custom-based Edit distance, the clustering procedure is performed leading to a SH. The SH for a generic document, belonging to a class $\omega \in \Omega$, is an array of normalized counters $\mathbf{c} = [c_1, c_2, ..., c_k] \in \mathbb{R}^k$, each one related to an alphabet symbol $\lambda_s \in \Lambda^{\omega}$, $s = 1, 2, ..., k$, which represents the occurrence frequency of the m-grams for the given document t. In order to build the SH representation, once obtained a suitable partition, an association rule is needed for assigning an m-gram to a prototype (or alphabet symbol) and for increasing the associated counter. This rule is named "minDist" and is summarized in the pseudocode illustrated in Algorithm 1. The assignment is performed by a suitable method that relies on

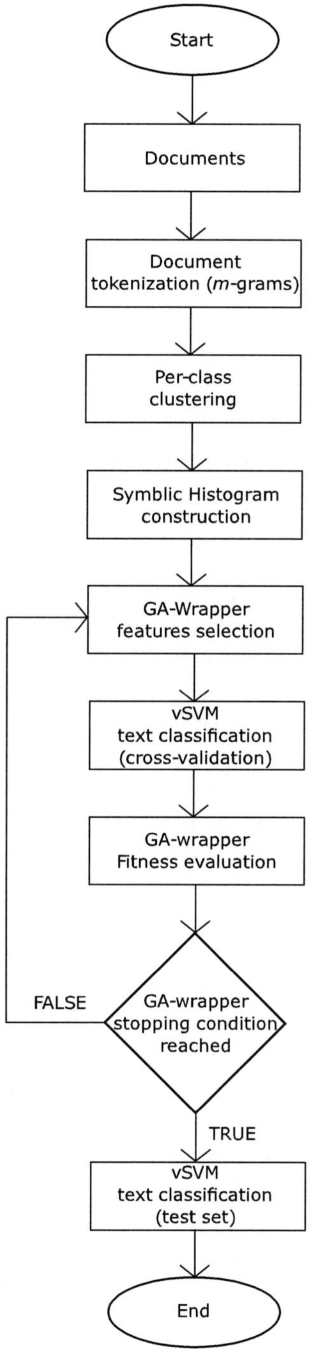

Fig. 2. The TC system flowchart.

Algorithm 1. The "minDist" algorithm, an important building block of the Symbolic Histogram approach.

Require: the dataset $\mathcal{D}$, the m-grams in each document t, the threshold τ, the alphabet Λ
Ensure: the SH matrix **C**

procedure MINDIST(DATASET,Λ,τ)
    for t-document in dataset **do**
        for m-gram i in t **do**
            for λ_s in Λ **do**
                $\mathbf{d}_{\phi(\eta)_{t,i}} \leftarrow edit(\lambda_s, \phi(\eta)_{t,i})$
            end for
            $d_{j,min} \leftarrow \min(\mathbf{d}_{\phi(\eta)_{t,i}})$
            if $d_{j,min} < \tau$ **then**
                $\mathbf{c}_t(j) \leftarrow \mathbf{c}_t(j) + 1$
            else
                $\mathbf{c}_t(j) \leftarrow \mathbf{c}_t(j)$
            end if
        end for
    end for
end procedure

selecting the nearest over k representatives and checking if the m-gram vector representation falls within a threshold τ – see Fig. 3. We indicate a generic m-gram (numerical) representation for the i-th m-gram η_i as a suitable mapping Φ, such as $\Phi(\eta_i)$. The mapping is constructed by training a *fastText* word embedding model [3], that is associating an embedding vector to each correspondent word within the m-gram.

In other words, the SH uniquely identifies the given document. Once completely defined, the SH is passed to the νSVM classifier with a Gaussian kernel paramtrized by γ. Specifically, the set of document-vectors obtained by the SH procedure is partitioned in two disjoint set, the training set $\mathcal{S}_{TR}$ and the test set $\mathcal{S}_{TS}$. The first is adopted for training the SVM classifier, the latter is used for measuring the generalization capabilities of the classification system.

As concerns the dissimilarity measure between two m-grams a custom-based Edit distance is adopted – see Algorithm 2, hence taking into account *substitutions*, *insertions* and *deletions*. Since the TC system computes m-grams and does not work at character level, the latter operators have been suitably redefined. An appealing characteristic of the proposed custom Edit distance consists in the possibility of comparing m-grams of different sizes. Specifically, the custom-based Edit distance is such that the substitution cost between the two given word-vectors within the m-grams representation, $\mathbf{x} = \phi_{\eta_1}[i_1]$ and $\mathbf{y} = \phi_{\eta_2}[i_2]$, belonging to different sequences, is the Euclidean distance $d^{cost}(\mathbf{x}, \mathbf{y})$, computed as follows:

$$d^{cost}(\mathbf{x}, \mathbf{y}) = \frac{||\mathbf{x} - \mathbf{y}||_2}{\sqrt{|u|}} = \frac{\sqrt{\sum_{i=1}^{u}(x_i - y_i)^2}}{\sqrt{|u|}}, \quad (1)$$

where $\sqrt{|u|}$ is a normalization factor.

In the provided pseudo-code – see Algorithm 2, d is the custom substitution cost, which in the classic Edit dissimilarity measure procedure is set to 1 like insertion and deletion operations.

Algorithm 2. The custom-based Edit dissimilarity measure algorithm.

Require: The two m-grams representations to compare ϕ_{η_1} and ϕ_{η_2}, the m-grams legths m_1 and m_2
Ensure: The Edit dissimilarity measure **d**

```
procedure CUSTEDITDIST(φη1, φη2, m1, m2)
    for i1 from 0 to m1 do
        d[i1, 0] := i1
    end for
    for i2 from 0 to m2 do
        d[0, i2] := i2
    end for
    for i1 from 1 to m1 do
        for i2 from 0 to m2 do
            if φη1[i1 − 1] = φη2[i2 − 1] then
                d ← 0
                d[i1, i2] ← d[i1 − 1, i2 − 1]
            else
                d ← d^cost (Euclidean distance)
                a ← d[i1 − 1, i2]
                b ← d[i1, i2 − 1]
                c ← d[i1 − 1, i2 − 1]
                d[0, i2] ← min(a, b, c)
            end if
        end for
    end for
    return d[m1, m2]
end procedure
```

The search space of the GA-Wrapper consists of the νSVM hyper-parameters, ν and γ, the threshold τ for the SH construction and the set of real-valued figures in the weight vector $\mathbf{v}$, which is used in the features selection procedure. It is worth to note that the dimension of the weights vector is $dim(\mathbf{v}) = \tilde{k}$, that is the number of alphabet symbols (i.e. Concept Regions).

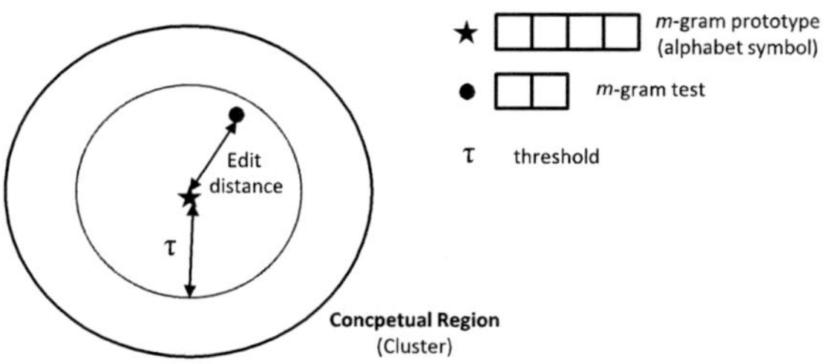

Fig. 3. The Concept Region and the decision threshold τ.

The optimization procedure is performed with a k-fold cross validation technique – see [7] – minimizing the following objective function:

$$J = \beta \cdot \varepsilon + (1 - \beta) \cdot \frac{\sum v_r}{\tilde{k}}, \tag{2}$$

that is a convex linear combination of the total (average) error rate ε and a structural complexity measure, which accounts for the number of selected features (i.e., the number of alphabet symbols).

In the features selection procedure, the generic alphabet symbol is selected or discarded according to the value of the related (binary) weight, 1 or 0, respectively. Since weights are real-valued numbers (as aforementioned) in the range $[0, 1]$, they have to be rounded to the nearest integer. Weights are encoded as GA genes such that rounding could be performed after each single generation or, alternatively, after the last one. Since it is not known a priori which approach could lead to better results, both of the above mentioned binarization approaches (A and B, respectively) are explored.

3 Enhancing the System Performance

3.1 Performance Exploration Strategy

The TC System has been previously tested on different datasets for assessing its effectiveness and also its usefulness as a grey-box model, without relying on the results on a single favorable dataset [7]. Since the reliability of the model is verified, it could be interesting to explore its performance more in deep. In other words, the best performance possible is sought for evaluating its potential.

The proposed strategy consists of running the model on a given dataset, changing its parameters configuration from an experiment to another by following a well defined criteria, which could led to a better performance. It is worth noting that in this study certain performance would be achieved only for the chosen dataset. Nevertheless, a rich dataset is adopted such that, even if the above-mentioned performance could not be considered valuable in general, results could be a good starting point to prove that. In fact, in this study the main aim is assessing the performance for a variety of configuration parameters and finding the right complexity level of the overall model. The chosen criteria are explained below while details about dataset and parameters configurations are reported in Sect. 4.

Number of Clusters. Better performance could be reached by finding the optimal number of clusters for the given dataset [8,32,42]. As shown in [8], the best number of clusters is achieved for different datasets and also a Particle Swarm Optimization procedure is applied to find it. Moreover, it seems that there is not a proportional relation between the number of clusters and performance. In other words, it is incorrect to state that more clusters lead to better results. Thus, it can be asserted that for a given dataset there is a number of clusters, (low or high), that leads to the best results and an optimization procedure is reasonably needed to find it. Thus, since an optimization procedure (suitable for not increasing the computational cost) has to be implemented yet, a more simple exploration of the performance has been though. Hence, the TC system performance has been tested setting firstly a lower and then an higher number of clusters, with a sensitivity analysis not present in the previous work [7].

Length of the m-grams. In the previous work [7], the set of the m-grams lengths to be extracted is fixed. Like the case related to number of clusters, it is reasonably to think that better results could be achieved considering the proper m-grams length for a given dataset. Specifically, if the number of clusters, thence the number of alphabet symbols-concepts, can be considered a degree of granulation at the coarse and high abstraction level, the m-grams length correspond to the degree of granulation at a finer level, within the information granulation hierarchy.

4 Simulation Settings and Results

4.1 Experimental Setup

The Dataset. The adopted dataset for the main experiments is called "Abstract" [7] and counts 575 scientific paper abstracts, which belong to the following five balanced classes: Anatomy, Information Theory, String Theory, Semiconductors and Smart Grid.

TC System Configuration and Experimental Setting. The GA-wrapper plays a relevant role in the system performance enhancement. The GA meta-parameters setting is shown in Table 1. As concerns the other settings regarding the text preprocessing or other configuration parameters, such as the ones related to the word embedding procedure, please see [7].

Table 1. GA meta-parameters setting.

Meta-parameter	Value
Population size	100
Crossover frac	0.8
Crossover type	Convex
Mutation type	Uniform
Selection type	Tournament
Stopping cond	Max stall
Max stall gen	50

According to the explanation in Sect. 3, the first step in the performance exploration strategy is to test the system with different number of clusters k. Since in the previous work k is set to 10 [7], new tests are performed with 5 and 15 clusters, respectively. Once k is fixed, five runs of the system are performed for estimating average results and the related variance, taking into consideration the stochastic nature of the GA-wrapper, such as in [7]. Following the same steps, k is then set to 15.

The configuration related to the most performing test is fixed for the next tests, which come from changing the m-grams lengths, extracted from the abstracts. In the previous work [7], m-grams of 1, 2 and 3 length values are chosen and also in the above-mentioned tests in this work. Thus, firstly m-grams from 2 to 4 length values are extracted, performing five runs of the system; secondly, m-grams from 3 to 5 in length are tested. In Table 4 the system output is reported.

Finally, the knowledge discovery for the best performing test is analyzed.

4.2 Performance Evaluation

The GrC classification model is tested on S_{TS} getting the overall classification performance through confusion matrix Π. The last is a square matrix of dimension $|\omega|$, where $|\omega|$ are the variable vector indicating the labels associated to each document. Rows represent true values while columns the predicted ones so that the generic element $\Pi(i, j)$ is the number of times the i-class has been classified as a j-class. We considered the following figures of merit, computed from confusion matrix:

(i) *True-Positive (TP)*: number of elements correctly classified for a class. The sum of the elements in the matrix diagonal;
ii) *False-Negative (FN)*: not correctly classified elements, in the row i that is the sum of all the elements in that row except $\Pi(i, i)$;
iii) *False-Positive (FP)*: the sum of all elements of the column j incorrectly assigned to the relative class because they belong to other classes;
iv) *True-Negative (TN)*: the sum of all elements of the row i incorrectly assigned to the relative class because they have been assigned to other classes;

The performance indicators rely on the fundamental figures of merit as follows. The Accuracy is given by:

$$\textit{Accuracy} = \mathrm{TP}/(\mathrm{TP} + \mathrm{FP} + \mathrm{FN} + \mathrm{TN}). \quad (3)$$

The Precision is the precentage of correctly classified patterns per-class:

$$\textit{Precision} = \mathrm{TP}/(\mathrm{TP} + \mathrm{FP}). \quad (4)$$

The Specificty gives information about how many patterns, not labeled for a given class, belong to other classes:

$$\textit{Specificity} = \mathrm{TN}/(\mathrm{TN} + \mathrm{FP}) \quad (5)$$

The Sensitivity gives information about how many patterns, labeled for a given class, belong to it:

$$\textit{Sensitivity} = \mathrm{TP}/(\mathrm{TP} + \mathrm{FN}) \quad (6)$$

The F1 Score is the harmonic mean of Precision and Sensitivity:

$$\textit{F1 Score} = \left(0.5 \cdot \textit{(Precision)}^{-1} + 0.5 \cdot \textit{(Sensitivity)}^{-1}\right)^{-1} \quad (7)$$

The Informedness is a tests that gives the same relevance to FP and FN:

$$\textit{Informedness} = \textit{Specificity} + \textit{Sensitivity} - 1 \quad (8)$$

4.3 Experimental Results

The first experiment aims at assessing the system performance with a number of clusters fixed to 15, both for the online binarization approach (Approach A) and for the "post-processing" binarization approach (Approach B) of the GA-wrapper Genes, which encode the Symbols in the Features Selection procedure. Results (reported in Table 2)

Table 2. Results of the first test ($k = 15$ clusters): performance metrics of the system for the two binarization approaches.

Perf. Metric	Approach A	Approach B
Accuracy	0.8975 (0.0025)	0.8802 (8.1034e-04)
Precision	0.9222 (0.0091)	0.9008(0.0083)
Specificity	0.9783 (7.6736e-04)	0.9587 (8.5031e-04)
Sensitivity	0.9478 (0.0023)	0.9826 (5.6767e-04)
F1Score	0.9334(7.6736e-04)	0.9170 (0.0016)
Informedness	0.9261 (0.0045)	0.9413 (9.4287e-05)

are averaged over five runs (with variance values in brackets), due to the stochastic nature of the GA algorithm.

It can be stated that the system performs better with Approach A, which is coherent with the results of the previous work [7]. The second experiment aims at assessing the system performance with a number of clusters fixed to 5 and with the same setting of the first test. Results (reported in Table 3) are averaged over five runs (with variance values in brackets).

Table 3. Results of the second test ($k = 5$ clusters): performance metrics of the system for the two binarization approaches.

Perf. Metric	Approach A	Approach B
Accuracy	**0.9304** (4.1456e-04)	0.9304 (0.0014)
Precision	0.9670 (0.0011)	0.9753(0.0013)
Specificity	0.9913 (8.2515e-05)	0.9935 (9.4287e-05)
Sensitivity	0.9688 (9.1957e-04)	0.9826 (5.6767e-04)
F1Score	0.9652(1.4469e-04)	0.9785 (4.5425e-04)
Informedness	0.9601 (5.5984e-04)	0.9761 (6.1389e-04)

Once again, the system returns better results with Approach A, like in the previous work [7]. Moreover, Accuracy is the best every returned by the system. Since the best results come from the second test (see Table 3), the number of clusters is fixed to 5. In addition, Approach A is also preferred to Approach B because, even though the Accuracy is the same for the two approaches, with the Approach A variance is lower than in Approach B. With the above mentioned configuration, a third test is performed by varying the length of the extracted m-grams, averaging the results over five runs as reported in Table 4. In the last column of Table 4 the best results of the second test are shown (with the original m-grams lengths set to 2 and 3) for an easier comparison.

Finally, the knowledge discovery related to the best performing test (with 15 clusters, Approach A and m-grams 1, 2 and 3 words long) is given in Table 5. More precisely: when the binary weight which comes from the evolutionary feature selection

Table 4. Results of the third experiment (approach A): performance metrics of the system with different m-grams lengths (L).

Perf. Metric	**L. 2 and 4**	**L. 3 and 5**	**L. 2 and 3**
$Accuracy$	**0.9304** (4.1456e-04)	0.9079 (5.4269e-04)	0.9217 (0.001)
$Precision$	0.9670 (0.0011)	0.9655(3.7934e-04)	0.9656 (3.7127e-04)
$Specificity$	0.9913 (8.2515e-05)	0.9913 (2.37627e-05)	0.9933 (3.2862e-05)
$Sensitivity$	0.9688 (9.1957e-04)	0.9739 (0.0034)	0.9739(5.6767e-04)
$F1Score$	0.9652(1.4469e-04)	0.9690 (0.0012)	0.9696(5.6767e-04)
$Informedness$	0.9601 (5.5984e-04)	0.9652 (0.0036)	0.9652 (7.3075e-04)

is 0, the corresponding alphabet symbol (in black in Table 5) is not considered during the classification; when the binary weight is 1, the corresponding alphabet symbol (in blue) is selected. Hence, for this dataset the TC system achieves its best classification performance for a higher model complexity in terms number of clusters and a lower number of n-grams, while the evolutionary metaheuristic seems retain a large number of m-grams to reach this results.

Table 5. Alphabet symbols (a.k.a. representative m-grams) extracted from the best performing experiment ("Abstract" dataset). In blue the selected symbols by the feature selection procedure.

Anatomy	*Information Theory*	*Smart Grid*	*String Theory*	*Semiconductor*
"just" "question"	"present" "first"	"place" "power" "process"	"second" "part" "turn"	"state" "first"
"present"	"analysis" "method" "information"	"system" "case"	"show" "process"	"state" "analysis" "show"
"case" "control" "study"	"present"	"smart" "grid" "present"	"development" "string" "theory"	"work"
"highrisk" "patient"	"way" "make" "full"	"kind"	"string" "theory" "pressureless"	"system" "electron" "semiconductor"
"young" "patient" "coronary"	"theory" "information" "entropy"	"grid" "smart"	"case"	"industry" "development" "current"
"show" "among" "patient"	"information" "information"	"technology" "yet" "come"	"time" "reparametrization"	"consequently" "new" "technology"
"death" "cause" "major"	"system" "hybrid" "system"	"grid" "propose" "smart"	"lightcone" "theory" "string"	"study" "effect"
"less" "coronary" "artery"	"local" "government" "make""	"power" "system"	"solution" "string"	"electron" "temperature" "latter"
"patient" "coronary"	"information" "entropy" "measure"	"system" "etc" "system"	"string" "theory"	"instance" "moreover" "specific"
"artery" "disease" "heart"	"system" "take"	"smart" "grid"	"theory" "action" "nevertheless"	"quantum" "device" "primary"
"case" "many"	"real" "system"	"present" "different" "way"	"present" "show" "interaction"	"heteroepitaxial"
"column" "common" "cause"	"traditional" "evaluation" "method"	"present" "different"	"theory" "background"	"current" "semiconductor"
"disease" "study"	"channel" "consider"	"case" "smart" "grid"	"dilatation"	"type" "semiconductor" "laser"
"coronary" "artery" "disease"	"call" "channel" "paper"	"distribution" "system" "consider"	"type" "semiconductor" "laser"	"semiconductor" "surface" "field"
"setting" "patient" "present"	"system"	"state" "power"	"semiconductor" "surface" "field"	"semiconductor" "device"

5 Conclusions

In this paper an enhancement of the text categorization system appeared in [7] and grounded on the Granular Computing approach is presented. The system is in charge of classifying text excerpts through a granulation procedure based on the extracted m-grams. A clustering algorithm is, then, adopted to partition m-grams trough a well-suited dissimilarity measure – a custom-based Edit distance – acting on sequences of word-vectors obtained encoding each word belonging to the m-gram with the fastText word embedding. In this way, sequences of word vectors (or phrases) of different length can be compared. The representative of each cluster – conceived as a Concept Region – is treated as an alphabed symbol. The set of alphabet symbols are, then, used for building the Symbolic Histogram representation of a text excerpts, i.e. measuring the statistical distribution of alphabet symbols for a given text document. An evolutionary metaheuristic with a well-suited gene encoding is adopted to perform a wrapper-like feature selection and to tune the system meta parameter. The feature selection lead to execute of Knowledge discovery on the obtained model. In the present study, the system effectiveness is explored through a different set of experiments. First of all, the system performance in term of generalization capabilities is evaluated by varying the number of clusters, i.e. the model complexity, by maintaining the comparison between two different approaches for the binarization of the evolutionary features selection weights [7]. Then, configurations with different values of the extracted m-grams lengths are proposed. The best (balanced) Accuracy is 0.9304, achieved for the configuration with 15 clusters, the online weights binarization setting and with a maximum length of the m-grams equal to 1, 2 and 3 words. That is also the best result ever on the proposed dataset and precisely about 2% better than the one reported in our previous study [7].

References

1. Ayed, S.B., Trichili, H., Alimi, A.M.: Data fusion architectures: a survey and comparison. In: 2015 15th International Conference on Intelligent Systems Design and Applications (ISDA), pp. 277–282. IEEE (2015)
2. Bianchi, F.M., Livi, L., Rizzi, A., Sadeghian, A.: A granular computing approach to the design of optimized graph classification systems. Soft. Comput. **18**(2), 393–412 (2014)
3. Bojanowski, P., Grave, E., Joulin, A., Mikolov, T.: Enriching word vectors with subword information. arXiv preprint arXiv:1607.04606 (2016)
4. Boleda, G.: Distributional semantics and linguistic theory. CoRR abs/1905.01896 (2019)
5. Buckland, M.: Information and society. MIT Press (2017)
6. Cameron, L., Larsen-Freeman, D.: Complex systems and applied linguistics. Int. J. Appl. Linguist. **17**(2), 226–239 (2007)
7. Capillo, A., De Santis, E., Mascioli, F.M.F., Rizzi, A.: Mining M-grams by a granular computing approach for text classification. In: IJCCI, pp. 350–360 (2020)
8. Chen, M., Ludwig, S.A.: Particle swarm optimization based fuzzy clustering approach to identify optimal number of clusters. J. Artif. Intell. Soft Comput. Res. **4**(1), 43–56 (2014)
9. De Santis, E., Martino, A., Rizzi, A., Mascioli, F.M.F.: Dissimilarity space representations and automatic feature selection for protein function prediction. In: 2018 International Joint Conference on Neural Networks (IJCNN), pp. 1–8. IEEE (2018)
10. De Santis, E., Rizzi, A., Sadeghian, A.: A cluster-based dissimilarity learning approach for localized fault classification in smart grids. Swarm Evol. Comput. **39**, 267–278 (2018)
11. De Santis, E., Sadeghian, A., Rizzi, A.: A smoothing technique for the multifractal analysis of a medium voltage feeders electric current. Int. J. Bifurcat. Chaos **27**(14), 1750211 (2017)
12. Deerwester, S., Dumais, S.T., Furnas, G.W., Landauer, T.K., Harshman, R.: Indexing by latent semantic analysis. J. Am. Soc. Inform. Sci. **41**(6), 391–407 (1990)
13. Del Vescovo, G., Rizzi, A.: Automatic classification of graphs by symbolic histograms. In: 2007 IEEE International Conference on Granular Computing (GRC 2007), pp. 410–410. IEEE (2007)
14. Del Vescovo, G., Rizzi, A.: Online handwriting recognition by the symbolic histograms approach. In: 2007 IEEE International Conference on Granular Computing (GRC 2007), pp. 686–686. IEEE (2007)

15. Devlin, J., Chang, M.W., Lee, K., Toutanova, K.: BERT: pre-training of deep bidirectional transformers for language understanding. arXiv preprint arXiv:1810.04805 (2018)
16. Doran, D., Schulz, S., Besold, T.R.: What does explainable AI really mean? a new conceptualization of perspectives. arXiv preprint arXiv:1710.00794 (2017)
17. Du, Q., Faber, V., Gunzburger, M.: Centroidal voronoi tessellations: applications and algorithms. SIAM Rev. **41**(4), 637–676 (1999)
18. Fabre, C., Lenci, A.: Distributional semantics today. ATALA, Sémantique distributionelle (2015)
19. Feldman, R., Dagan, I.: Knowledge discovery in textual databases (KDT). In: KDD, vol. 95, pp. 112–117 (1995)
20. Feldman, R., Sanger, J., et al.: The text mining handbook: advanced approaches in analyzing unstructured data. Cambridge University Press (2007)
21. Gärdenfors, P.: Conceptual spaces: The geometry of thought. MIT press (2004)
22. Harris, Z.S.: Distributional structure. Word **10**(2–3), 146–162 (1954)
23. Hearst, M.A.: Untangling text data mining. In: Proceedings of the 37th Annual meeting of the Association for Computational Linguistics, pp. 3–10 (1999)
24. Heeman, F.C.: Granularity in structured documents. Electron. publishing **5**(3), 143–155 (1992)
25. Jing, L., Lau, R.Y.K.: Granular computing for text mining: new research challenges and opportunities. In: Sakai, H., Chakraborty, M.K., Hassanien, A.E., Ślęzak, D., Zhu, W. (eds.) RSFDGrC 2009. LNCS (LNAI), vol. 5908, pp. 478–485. Springer, Heidelberg (2009). https://doi.org/10.1007/978-3-642-10646-0_58
26. Kwapień, J., Drożdż, S.: Physical approach to complex systems. Phys. Rep. **515**(3–4), 115–226 (2012)
27. Landauer, T.K., Foltz, P.W., Laham, D.: An introduction to latent semantic analysis. Discourse Process. **25**(2–3), 259–284 (1998)
28. Lenci, A.: Distributional semantics in linguistic and cognitive research. Ital. J Linguist. **20**(1), 1–31 (2008)
29. Liu, H., Cocea, M.: Fuzzy information granulation towards interpretable sentiment analysis. Granular Comput. **2**(4), 289–302 (2017). https://doi.org/10.1007/s41066-017-0043-8
30. Martino, A., De Santis, E., Rizzi, A.: An ecology-based index for text embedding and classification. In: 2020 International Joint Conference on Neural Networks (IJCNN), pp. 1–8. IEEE (2020)
31. Martino, A., Giuliani, A., Rizzi, A.: Granular computing techniques for bioinformatics pattern recognition problems in non-metric spaces. In: Pedrycz, W., Chen, S.-M. (eds.) Computational Intelligence for Pattern Recognition. SCI, vol. 777, pp. 53–81. Springer, Cham (2018). https://doi.org/10.1007/978-3-319-89629-8_3
32. Marutho, D., Handaka, S.H., Wijaya, E., et al.: The determination of cluster number at K-mean using elbow method and purity evaluation on headline news. In: 2018 International Seminar on Application for Technology of Information and Communication, pp. 533–538. IEEE (2018)
33. Mikolov, T., Chen, K., Corrado, G., Dean, J.: Efficient estimation of word representations in vector space. arXiv preprint arXiv:1301.3781 (2013)
34. Mueller, S.T., Hoffman, R.R., Clancey, W., Emrey, A., Klein, G.: Explanation in human-AI systems: a literature meta-review, synopsis of key ideas and publications, and bibliography for explainable AI. arXiv preprint arXiv:1902.01876 (2019)
35. Park, H.S., Pedrycz, W., Oh, S.K.: Granular neural networks and their development through context-based clustering and adjustable dimensionality of receptive fields. IEEE Trans. Neural Networks **20**(10), 1604–1616 (2009)

36. Possemato, F., Rizzi, A.: Automatic text categorization by a granular computing approach: facing unbalanced data sets. In: The 2013 International Joint Conference on Neural Networks (IJCNN), pp. 1–8. IEEE (2013)
37. Rdusseeun, L., Kaufman, P.: Clustering by means of medoids. In: Proceedings of the Statistical Data Analysis Based on the L1 Norm Conference, Neuchatel, Switzerland, pp. 405–416 (1987)
38. Rizzi, A., Del Vescovo, G., Livi, L., Mascioli, F.M.F.: A new granular computing approach for sequences representation and classification. In: The 2012 International Joint Conference on Neural Networks (IJCNN), pp. 1–8. IEEE (2012)
39. Robertson, S.: Understanding inverse document frequency: on theoretical arguments for IDF. J. documentation **60**(5), 503–520 (2004)
40. Salton, G., Wong, A., Yang, C.S.: A vector space model for automatic indexing. Commun. ACM **18**(11), 613–620 (1975)
41. Santis, E.D., Livi, L., Sadeghian, A., Rizzi, A.: Modeling and recognition of smart grid faults by a combined approach of dissimilarity learning and one-class classification. Neurocomputing **170**, 368–383 (2015). https://doi.org/10.1016/j.neucom.2015.05.112, Advances on Biological Rhythmic Pattern Generation: Experiments, Algorithms and Applications Selected Papers from the 2013 International Conference on Intelligence Science and Big Data Engineering (IScIDE 2013) Computational Energy Management in Smart Grids
42. Xu, Z., Yin, Y., Chen, H., Xu, H., Li, P.: Algorithm for determining number of clusters based on dichotomy. In: 2020 International Conferences on Internet of Things (iThings) and IEEE Green Computing and Communications (GreenCom) and IEEE Cyber, Physical and Social Computing (CPSCom) and IEEE Smart Data (SmartData) and IEEE Congress on Cybermatics (Cybermatics), pp. 180–185. IEEE (2020)
43. Yao, J.T., Vasilakos, A.V., Pedrycz, W.: Granular computing: perspectives and challenges. IEEE Trans. Cybern. **43**(6), 1977–1989 (2013)
44. Yao, Y.: Granular computing for data mining. In: Data Mining, Intrusion Detection, Information Assurance, and Data Networks Security 2006, vol. 6241, p. 624105. International Society for Optics and Photonics (2006)
45. Zenker, F., Gärdenfors, P. (eds.): Applications of Conceptual Spaces. SL, vol. 359. Springer, Cham (2015). https://doi.org/10.1007/978-3-319-15021-5
46. Zhang, X., Yin, Y., Yu, H.: An application on text classification based on granular computing. Commun. IIMA **7**(2), 1 (2007)
47. Zhang, Y., Chen, M., Liu, L.: A review on text mining. In: 2015 6th IEEE International Conference on Software Engineering and Service Science (ICSESS), pp. 681–685. IEEE (2015)
48. De Santis, E., Rizzi, A.: Prototype theory meets word embedding: a novel approach for text categorization via granular computing. Cogn. Comput. **15**(3), 976–997 (2023). https://doi.org/10.1007/s12559-023-10132-9
49. De Santis, E., Martino, A., Rizzi, A.: An infoveillance system for detecting and tracking relevant topics from Italian tweets during the COVID-19 event. IEEE Access **8**, 132527–132538 (2020)

Recent Research Topics in Evolutionary Multiobjective Optimization: A Personal Perspective

Carlos A. Coello Coello(✉)

CINVESTAV-IPN, Evolutionary Computation Group, Departamento de Computación, Av. Instituto Politécnico Nacional No. 2508, Col. San Pedro Zacatenco, México D.F. 07300, Mexico
carlos.coellocoello@cs.cinvestav.mx

Abstract. The field which is now known as evolutionary multi-objective optimization (EMOO) started in the mid-1980s and since then, it has experienced a significant growth. In this chapter, we discuss some of the (several) research topics that remain open and which, from the author's perspective, are worth exploring in the next few years.

Keywords: Capacitated vehicle routing problem with time windows · Strength pareto evolutionary algorithm · Hill-climbing · Multi objective optimization

1 Introduction

Many of the real-world problems that we aim to solve have two or more (often conflicting) objectives that we aim to optimize simultaneously. These are the so-called multiobjective optimization problems (MOPs) and their solution has attracted a significant attention, particularly in the last 15 years. Because of the conflicting nature of the objectives, solving an MOP generates not one, but a set of solutions representing the best possible trade-offs among the objectives (i.e., solutions in which one objective cannot be improved without worsening at least another one). Such solutions constitute the *Pareto optimal set* and the image of this set (i.e., the corresponding objective function values) form the so-called *Pareto front*.

Research on multi-objective optimization started in the 1970s within Operations Research, and a wide variety of algorithms have been developed since then [135]. However, such mathematical programming techniques present several limitations, from which the most relevant are two: (1) they are very susceptible to the shape and/or continuity of the Pareto front and (2) they generate one element of the Pareto optimal set per execution. Additionally, departing from different starting points, there is no guarantee that the algorithm will generate different solution from those generated before. Furthermore, some mathematical programming techniques require that both the objective functions and the constraints of the MOP are provided in algebraic form and cannot properly operate when these values come from a simulator. All of these limitations have motivated the use of evolutionary algorithms for solving MOPs. The so-called multi-objective evolutionary algorithms (MOEAs) have several advantages with respect to

J. Garibaldi et al. (Eds.): IJCCI 2021, SCI 1119, pp. 90–120, 2023.
https://doi.org/10.1007/978-3-031-46221-4_5

mathematical programming techniques, from which the main ones are their flexibility (i.e., they require little domain specific information to operate) and their ease of use. Additionally, MOEAs are less susceptible to the shape and continuity of the Pareto front and operate at all times with a set of solutions (the so-called population). This allows (with a proper manipulation) to generate several elements of the Pareto optimal set in a single algorithmic execution [34]. This research area is now called Evolutionary Multi-Objective Optimization (EMOO).

The first MOEA was the *Vector Evaluated Genetic Algorithm* (VEGA) which was proposed by J. David Schaffer in the mid-1980s [170–172]. Remarkably, there was not much interest in EMOO for almost a decade. It was until the mid-1990s when research in this area started to increase and this trend has been kept since then.[1]

The remainder of this paper is organized as follows. In Sect. 2, we provide some basic mathematical concepts related to multiobjective optimization, with the aim of making of this a self-contained document. Section 3 provides the list of research topics on which this chapter is focused. Each of these topics is briefly described and then, some of the research opportunities available around them are also briefly discussed. In Sect. 4 we provide the main challenges associated to each of the research areas previously discussed. Finally, Sect. 5 provides the conclusions of the chapter.

2 Basic Concepts

In multiobjective optimization, the aim is to solve problems of the type[2]:

$$\text{minimize } \vec{f}(\vec{x}) := [f_1(\vec{x}), f_2(\vec{x}), \ldots, f_k(\vec{x})] \tag{1}$$

subject to:

$$g_i(\vec{x}) \leq 0 \quad i = 1, 2, \ldots, m \tag{2}$$

$$h_i(\vec{x}) = 0 \quad i = 1, 2, \ldots, p \tag{3}$$

where $\vec{x} = [x_1, x_2, \ldots, x_n]^T$ is the vector of decision variables, $f_i : \mathbb{R}^n \rightarrow \mathbb{R}$, $i = 1, ..., k$ are the objective functions and $g_i, h_j : \mathbb{R}^n \rightarrow \mathbb{R}$, $i = 1, ..., m$, $j = 1, ..., p$ are the constraint functions of the problem.

A few additional definitions are required to introduce the notion of optimality used in multiobjective optimization:

Definition 1. Given two vectors $\vec{x}, \vec{y} \in \mathbb{R}^k$, we say that $\vec{x} \leq \vec{y}$ if $x_i \leq y_i$ for $i = 1, ..., k$, and that $\vec{x}$ **dominates** $\vec{y}$ (denoted by $\vec{x} \prec \vec{y}$) if $\vec{x} \leq \vec{y}$ and $\vec{x} \neq \vec{y}$.

Definition 2. We say that a vector of decision variables $\vec{x} \in \mathcal{X} \subset \mathbb{R}^n$ is **nondominated** with respect to $\mathcal{X}$, if there does not exist another $\vec{x}' \in \mathcal{X}$ such that $\vec{f}(\vec{x}') \prec \vec{f}(\vec{x})$.

[1] The author maintains the EMOO repository [32] which currently contains over 13,150 bibliographic references on evolutionary multiobjective optimization. The EMOO repository is located at: https://emoo.cs.cinvestav.mx.

[2] Without loss of generality, we will assume only minimization problems.

Definition 3. We say that a vector of decision variables $\vec{x}^* \in \mathcal{F} \subset \mathbb{R}^n$ ($\mathcal{F}$ is the feasible region) is **Pareto-optimal** if it is nondominated with respect to $\mathcal{F}$.

Definition 4. The **Pareto Optimal Set** $\mathcal{P}^*$ is defined by:

$$\mathcal{P}^* = \{\vec{x} \in \mathcal{F} | \vec{x} \text{ is Pareto-optimal}\}$$

Definition 5. The **Pareto Front** $\mathcal{PF}^*$ is defined by:

$$\mathcal{PF}^* = \{\vec{f}(\vec{x}) \in \mathbb{R}^k | \vec{x} \in \mathcal{P}^*\}$$

Therefore, our aim is to obtain the Pareto optimal set from the set $\mathcal{F}$ of all the decision variable vectors that satisfy (2) and (3). Note however that in practice, not all the Pareto optimal set is normally desirable or achievable, and decision makers tend to prefer certain types of solutions or regions of the Pareto front [18].

3 Recent Research Topics

This chapter will focus on a handful of topics:

- Algorithms
- Scalability
- Computationally Expensive MOPs
- Hyper-heuristics

Each of these topics will be briefly discussed and some of the research opportunities available around them will also be mentioned in the following subsections.

3.1 Algorithms

As one would expect, in their early days (i.e., the period going from 1984 to the mid-1990s) MOEAs were rather simple and naïve. Let's consider, for example, the previously mentioned *Vector Evaluated Genetic Algorithm (VEGA)* [171]. In this MOEA, the population of a simple genetic algorithm was subdivided into as many sub-populations as objectives defined in the MOP to be solved. It is worth emphasizing that at that time, only problems having two objectives were considered. Once these sub-populations were established, solutions were selected in each of them based on a single objective. In other words, in sub-population one, solutions that were the best with respect to objective one were favored by the selection process. In sub-population two, solutions that were the best with respect to objective two were favored by the selection process. In a further step, individuals from all sub-populations were shuffled with the aim of recombining solutions that were the best with respect to the first objective with those that were the best with respect to the second objective. When using proportional selection (e.g., the roulette-wheel selection strategy), VEGA behaved similarly to the use of a linear aggregating function that combines all the objective functions into a single scalar value [36]. It is worth noting, however, that strictly speaking, the selection mechanism of VEGA, really opposes Pareto optimality, because solutions that represent the best possible trade-offs between the objectives won't be selected, because they are not the best with respect to any of them. Nevertheless, and in spite of its limitations, Schaffer was able to successfully solve classification problems with VEGA [172], and some other researchers eventually found applications in which this sort of scheme could be useful (see for example [31]).

Linear aggregating functions were among the most popular approaches adopted in the early days of MOEAs [78], but their inability for dealing with non-convex Pareto fronts was soon pointed out by some researchers [41]) and this made them impopular within EMOO. It is worth indicating, however, that linear aggregating functions have shown to be a good option when dealing with multi-objective combinatorial optimization problems (see for example [95]).

Other approaches that were popular in the early days include the use of the ϵ-constraint method (this approach transform a MOP into several single-objective optimization problems that are solved sequentially) [109], lexicographic ordering (in this approach, objectives are sorted by their importance and are solved sequentially, one at a time) [7] and target vector approaches (these are nonlinear aggregating functions in which the aim is to minimize the differences between the objective function values obtained by a MOEA and the target values defined by the user) [33]. These approaches have several limitations (e.g., the ϵ-constraint method is computationally expensive and requires the Nadir point [135]), but they are still occasionally used by some authors, but these publications normally occur outside the EMOO literature (e.g., in engineering optimization journals [143]).

After VEGA, there was practically no research activity around EMOO during several years. In 1989, David Goldberg reviewed in his famous book on genetic algorithms [68] the work that had been done on EMOO (which basically consisted of VEGA) and proposed a mechanism called *Pareto ranking*, which is informally described in the book. The idea of Pareto ranking is to rank the individuals of the population of a genetic algorithm using Pareto optimality in such a way that all the nondominated solutions obtain the best possible rank and are sampled at the same rate (i.e., all nondominated solutions are considered equally good and, therefore, have the same probability of survival). Dominated solutions would obviously obtained lower ranks and their survival probability would be lower. It is worth noting that Goldberg did not provide a full pseudo-code of this procedure and over the years, different implementations were proposed.

In 1993, Fonseca and Fleming [63] proposed a Pareto ranking method within an algorithm called *Multi-Objective Genetic Algorithm* (MOGA) in which, for each individual, a counter (r) of the solutions that dominate it is kept. Then, fitness can be assigned, for example, using the following expression:

$$\text{fitness} = \frac{1}{1+r} \tag{4}$$

This way, all nondominated solutions will have a fitness of 1.0 (because $r = 0$ in this case, since these solutions are not dominated by any other solution in the population). Dominated solutions obtain a fitness which decreases proportionally to the number of individuals that dominate them.

In 1994, Srinivas and Deb [177] proposed another ranking procedure (which was called *nondominated sorting*) within an algorithm called *Nondominated Sorting Genetic Algorithm* (NSGA) which operated in a different way as MOGA's ranking procedure. In the case of NSGA, the nondominated solutions were identified in the original population and they were removed and assigned a "dummy" fitness value (e.g., 1,000,000). Using the remainder of the population, the nondominated solutions were identified again (this would be the second layer), would be removed and would be assigned a dummy fitness value which had to be inferior to the value used for the first layer (e.g., 800,000). This procedure had to be repeated until the entire population had a fitness value. It's clear that NSGA was more inefficient than MOGA, since the ranking procedure had to be repeated several times.

In his book, Goldberg [68] also realized that in MOEAs, diversity would be lost because of stochastic noise (digital computers use pseudo-random numbers) and this went agains the goal of being able to generate several elements of the Pareto optimal set in a single execution of a genetic algorithm, since this would be a clear advantage of the population-based nature of MOEAs. Thus, in order to block the loss of diversity that normally occurs in a genetic algorithm, Goldberg proposed to use a mechanism which was eventually called *density estimator*, and which has the task of maintaining diversity in the population of a MOEA (in other words, it must retain different nondominated solutions in the population). Goldberg's original proposal was to adopt a mechanism called *fitness sharing* which he proposed in 1987 [69]. In fact, both MOGA and NSGA adopted fitness sharing but the former applied in objective function space and the latter applied in decision variable space. Over the years, a wide variety of density estimators have been proposed, including: clustering [212], adaptive grids [100], crowding [46], entropy [151] and parallel coordinates [85].

In the late 1990s, another mechanism became standard in MOEAs: *elitism*. In single-objective genetic algorithms, elitism consists in retaining the best individual in the population, passing it intact (without recombining nor mutating it) to the next generation. In MOEAs, since all nondominated solutions are considered equally good, elitism should preserve them all. When Zitzler developed the *Strength Pareto Evolutionary Algorithm* (SPEA) in the late 1990s [212] he proposed the use of an *external archive* to retain all the nondominated solutions generated by his algorithm and such solutions also participated in the selection procedure. However, in his experiments, Zitzler realized that the number of nondominated solutions would grow very quickly and this diluted selection pressure. Because of this, he proposed the use of a bound on the size of the external archive and used clustering to remove redundant solutions (i.e., solutions that were too similar, for example, in objective function space). It is important to indicate that elitism is important not only for practical reasons (e.g., it is easier and more fair to compare the performance of two MOEAs that adopt external archives of the same size), but also for theoretical reasons since it has been (mathematically) proved that elitism is required in a MOEA in order to guarantee convergence [164].

During the second half of the 1990s, a wide variety of MOEAs were proposed [30]. However, few of these approaches were used by people from a group different to the one in which the algorithm was proposed (besides MOGA and NSGA, the *Niched-Pareto Genetic Algorithm* (NPGA) [86] was another relatively popular non-elitist MOEA). Regarding elitist MOEAs, besides SPEA, the *Pareto Archived Evolution Strategy* (PAES) [103] was also relatively popular. However, with no doubt, the most popular of the elitist Pareto-based MOEAs has been the *Nondominated Sorting Genetic Algorithm-II* (NSGA-II) [43, 46] which adopts a more efficient nondominated sorting algorithm (which generates different layers in a single pass) than the original NSGA, and incorporates a very clever density estimator called *crowded comparison operator*, which does not require any user-defined parameters. NSGA-II is still used today by an important number of researchers, in spite of the well-known limitations of its crowded comparison operator in high dimensionality in objective function space[3] (the so-called *many-objective optimization problems* [34]).

For over 15 years, Pareto-based MOEAs were, by far, the most popular approaches in the specialized literature. However, in 2004, a new type of algorithmic design was proposed: indicator-based selection. Although some researchers had already proposed the use of a performance indicator to prune the external archive of a MOEA (see for example [102]), the *Indicator-Based Evolutionary Algorithm* [211] was the first indicator-based MOEA. IBEA consists of an algorithmic framework that allows the incorporation of any performance indicator into the selection mechanism of a MOEA. IBEA was originally tested with the hypervolume [210] and the binary ϵ indicator [211].

At the beginning, indicator-based MOEAs were seen as more of a curiosity, since it was unclear if they could offer any clear advantages over Pareto-based MOEAs. However, as many-objective optimization started to gain popularity, it became very evident that scalability was a serious limitation of Pareto-based MOEAs and researchers started to get interested in indicator-based MOEAs which did not seem to have the same scalability limitations as Pareto-based MOEAs. One of the main responsibles for this interest was the *S Metric Selection Evolutionary Multiobjective Algorithm* (SMS-EMOA) originally introduced in 2005 [53]. SMS-EMOA randomly generates an initial population and then produces a single solution per iteration (i.e., it uses the so-called steady state selection) using the crossover and mutation operators from NSGA-II. Then, it applies nondominated sorting (as in NSGA-II). When the last nondominated front has more than one solution, SMS-EMOA uses the hypervolume indicator [210] to decide which solution should be removed. Beume et al. [15] proposed an improved version of SMS-EMOA in which the hypervolume contribution is not used when, in the nondominated sorting process, more than one front is obtained (i.e., the hypervolume is used as a density estimator). In this case, they use the number of solutions that dominate to a certain individual (in other words, the solution that is dominated by the largest number of solutions is removed). This makes SMS-EMOA a bit more efficient than its original version. However, since this MOEA relies on the use of exact hypervolume contributions, it becomes too computationally expensive as we increase the number of objectives [14].

SMS-EMOA was the first of a series of indicator-based MOEAs (several of which rely on the use of the hypervolume indicator) that were developed during several years. It is worth indicating, however, that in most of these approaches, the performance indicator is used as a density estimator (see for example [90]) as is the case of SMS-EMOA. The use of an indicator in the selection mechanism of a MOEA remains as something

[3] In fact, there is empirical evidence indicating that the crowded comparison operator has difficulties even in MOPs with only 3 objectives (see for example [106]).

relatively rare (see for example [134]) mainly because of the high computational cost of the hypervolume in high dimensionality (in objective function space).

An obvious question at this point is: why is that researchers preferred to use the *hypervolume* for indicator-based MOEAs? Does it offer any advantages over other performance indicators? The hypervolume (which is also known as the $\mathcal{S}$ metric or the Lebesgue Measure) of a set of solutions measures the size of the portion of objective space that is dominated by those solutions collectively. One of its main advantages of this performance indicator are its mathematical properties, since it has been proved that the maximization of this performance measure is equivalent to finding the Pareto optimal set [62]. Additionally, empirical studies have shown that (for a certain number of points previously defined by the user) maximizing the hypervolume indeed produces subsets of the true Pareto front [53, 100]. Also, the hypervolume assesses both convergence and, to a certain extent, also the spread of solutions along the Pareto front (although without necessarily enforcing a uniform distribution of solutions).

However, and in spite of its advantages, there are several issues regarding the use of the hypervolume. First, the computation of this performance indicator depends of a reference point, which can influence the results in a significant manner [92]. Some people have proposed to use the worst objective function values in the current population, but this requires scaling the objectives. Nevertheless, the most serious limitation of the hypervolume is its high computational cost. The best algorithms known to compute hypervolume have a polynomial complexity on the number of points used, but such complexity grows exponentially on the number of objectives [14]. This has triggered a significant amount of research regarding algorithms that can reduce the computational cost of computing the hypervolume and the hypervolume contributions, which is what we need for a hypervolume-based MOEA[4] (see for example [38,73,96,112,165,175,184,198]).

Evidently, the use of the hypervolume raises some interesting questions, since when used to select solutions, we only care about the so-called hypervolume contribution [19]. So, the obvious question is: can we estimate the hypervolume contribution using some approximation method? In this case, having a certain error in the estimation is acceptable, for as long as this error is consistent so that the selection mechanism remains reliable. This is precisely the sort of approach adopted in the *Hypervolume Estimation Algorithm for Multi-Objective Optimization* (HyPE) [6] in which Monte Carlo simulations are used to approximate the hypervolume contribution. In spite of the fact that HyPE can efficiently solve MOPs having a very large number of objectives (up to 50), its results are not as competitive as when using exact hypervolume contributions. Other approximation methods have been proposed (see for example [47] in which polar coordinates are adopted), but they are not very popular within indicator-based MOEAs [56].

Another obvious possibility is to use a different performance indicator whose computation is relatively inexpensive. Unfortunately, the hypervolume is the only unary indicator which is known to be Pareto compliant [213], which makes less attractive the

[4] See also:
http://ls11-www.cs.uni-dortmund.de/rudolph/hypervolume/start
http://people.mpi-inf.mpg.de/~tfried/HYP/
http://iridia.ulb.ac.be/~manuel/hypervolume.

use of other performance indicators. Nevertheless, there are some other performance indicators which are weakly Pareto compliant, such as $R2$ [20] and the Inverted Generational Distance plus ($IGD+$) [93]. Although several efficient and effective indicator-based MOEAs have been proposed around these performance indicators (see for example [21, 83, 117, 125, 126, 186]), their use has remained relatively rare in the specialized literature.

In 2007, a different sort of algorithmic design was proposed with the *Multi-Objective Evolutionary Algorithm based on Decomposition* (MOEA/D) [206]. The idea of using *decomposition* (or scalarization) methods was originally proposed in mathematical programming more than 20 years ago [40] and it consists in transforming an MOP into several single-objective optimization problems which are then solved to generate the nondominated solutions of the original problem. Unlike linear aggregating functions, the use of scalarization (or decomposition) methods allows the generation of non-convex portions of the Pareto front and works even in disconnected Pareto fronts. MOEA/D presents an important advantage with respect to methods proposed in the mathematical programming literature (such as Normal Boundary Intersection (NBI) [40]): it uses neighborhood search to solve simultaneously all the single-objective optimization problems generated from the transformation. Additionally, MOEA/D is not only effective and efficient, but can also be used for solving problems with more than 3 objectives although in such cases it may require higher population sizes.

Decomposition-based MOEAs became fashionable at around 2010 and have remained as an active research area since then [167]. In fact, this sort of approach influenced the development of the Nondominated Sorting Genetic Algorithm-III (NSGA-III) [44] which adopts both decomposition and reference points to deal with many-objective problems. However, it was recently found that decomposition-based MOEAs do not work properly with certain Pareto front geometries [94]. This will certainly trigger a lot of research in the next few years, given the popularity of decomposition-based MOEAs [187, 194].

In recent years, most of the research on algorithms has focused on solving specific classes of problems (e.g., the so-called multimodal MOPs [183], problems having irregular Pareto fronts [87], many-objective problems [116], constrained MOPs [64, 139], dynamic MOPs [156] and large-scale MOPs [185] among others) using variants of existing MOEAs.

3.2 Scalability

During the early days of EMOO (from the mid-1980s to the late 1990s), MOEAs were adopted, in most cases, to solve MOPs having only two or three objectives. However, over the years, the need for tackling many-objective problems became more evident. When Pareto-based MOEAs were applied to such problems, it was soon identified that scalability in objective function space is a serious limitation for them [88].

Nevertheless, an interesting question is if scalability is, per se, a feature that makes any MOP more difficult. The answer is, indeed, counterintuitive. When Schütze et al. [173] were studying the actual source of difficulty in many-objective problems in 2011, they concluded that adding more objectives to an MOP does not necessarily make it harder. According to this study, the difficulty is really associated to the intersection of the descent cones of the objectives (these descent cones are obtained with the combination of the gradients of each objective). This was somehow corroborated by an empirical study conducted in 2015 by Ishibuchi et al. [91] in which it was shown that NSGA-II could properly solve many-objective knapsack problems in which the objectives were highly correlated.

So, the question remains: why is that many-objective problems turn out to be difficult to solve in practice when using Pareto-based MOEAs? A series of experimental [153, 191] and analytical studies [37, 91, 101] have identified the following limitations of Pareto-based MOEAs in many-objective problems:

1. *Deterioration of the Search Ability:* The proportion of nondominated solutions in a population increases rapidly with the number of objectives [59]. According to Bentley *et al.* [10] the number of nondominated k-dimensional vectors on a set of size n is $O(\ln^{k-1} n)$. This implies that in problems with a large number objectives, the selection of solutions is carried out almost at random or guided by the density estimator. In fact, Mostaghim and Schmeck [140] experimentally showed that a random search optimizer could achieve better results than NSGA-II [46] in a problem with 10 objectives. This problem is the reason why most indicator-based MOEAs are able to tackle many-objective problems simply by using a more powerful density estimator that properly guides the search (i.e., that breaks ties among the many nondominated solutions generated when using Pareto optimality).
2. *Dimensionality of the Pareto Front:* A topic that is rarely discussed in EMOO is the so-called "curse of dimensionality": the number of points required to represent accurately a Pareto front increases exponentially with the number of objectives. The number of points necessary to represent a k-dimensional Pareto front with resolution r is given by $O(kr^{k-1})$ [174]. This raises several interesting issues [91]. For example, what sort of data structures can be used to efficiently manage the large number of points required for many-objective problems? Also, what sort of density estimators would be required in very high-dimensional problems? Does it make sense to enforce a uniform distribution of solutions in problems having a very large number of objectives? What sort of performance indicators can be adopted in such problems? Today, most MOEAs normally use relatively small population sizes (less than 300 individuals) even when tackling MOPs with more than six objectives in spite of the fact that such population sizes are clearly inappropriate for sampling such high-dimensional Pareto fronts.
3. *Visualization of the Pareto Front:* How can we visualize Pareto fronts when we have more than three objectives? Clearly, a regular plot is not possible in such cases. This is a serious problem since visualization plays a key role for a proper decision making process. Over the years, a variety of visualization techniques have been proposed for many-objective problems (see for example [188]), and this is still an active research area (see for example [16, 81, 180]).

In order to properly deal with many-objective optimization problems, three main approaches have been normally adopted [9, 116, 121]:

1. In the early days of EMOO, it was popular to use relaxed forms of Pareto dominance, although this approach is rarely used today. Some of the relaxed forms of

Pareto dominance that have been proposed are, for example: k-optimality [60], preference order ranking [49], the *favour* relation [179], and a method that controls the dominance area [168], among others. These approaches are the *mathematically correct* way of dealing with many-objective problems, because these relaxed forms of Pareto dominance provide a richer ordering of the solutions while obtaining an optimal set that is usually a subset of the Pareto optimal set.
2. Indicator-based MOEAs have also been used to deal with many-objective optimization problems, in spite of the (previously discussed) limitations of some performance indicators such as the hypervolume (see for example [97]).
3. Another interesting approach that was popular some years ago is the so-called *dimensionality reduction*. In this case, the idea is to reduce the number of objectives of the MOP either during the search process or in an *a posteriori* manner, during the decision making process [22, 119, 169]. The main goal, in this case, is to identify redundant (i.e., non-conflicting) objectives in order to discard them. A *redundant* objective is one that can be removed without changing the dominance relation induced by the original objective set. Evidently, if each objective is conflicting with respect to every other objective, no reduction is possible. However, this is rarely the case in practice.

Evidently, many other approaches can be (and have been) used for solving many-objective problems. Some examples are the following: the use of alternative ranking schemes (different from nondominated sorting) (see for example [65]), the use of machine learning techniques (as in MONEDA [128]), the use of ϵ-dominance[5] [104], and the two-archive MOEA, which uses one archive for convergence and another for diversity [153], among others.

3.3 Computationally Expensive MOPs

Although MOEAs offer a number of advantages when solving complex real-world MOPs (e.g., they don't need the objective functions to be provided in algebraic form), their most important limitation is that, because of their stochastic nature, they normally require a large number of objective function evaluations to produce a good sample of the search space that allows the algorithm to define a proper search direction. This is, indeed, a serious limitation that can certainly affect the applicability of MOEAs in certain real-world MOPs having objective functions which are very costly to evaluate (e.g., MOPs from aeronautical and aerospace engineering [144]).

In general, MOEAs can become computationally unaffordable for an application when:

- The evaluation of the fitness functions is computationally expensive (i.e., it takes from minutes to hours, depending on the quality or granularity of the model and the available computational resources).

[5] This is a relaxed form of Pareto dominance which was proposed in 2002 [114] and is adopted as an archiving technique.

- The fitness functions cannot be defined in an algebraic form (e.g., when the fitness functions are generated by a computational simulation of the physics of the real system).
- The total number of evaluations of the fitness functions is limited by some financial constraints (i.e., there is a financial cost involved in computing the fitness functions and we cannot exceed a certain pre-defined budget).

As we get access to more computational power each year at more affordable prices, the interest in using MOEAs for solving computationally expensive MOPs has significantly increased in the last few years. The main approaches that have been developed in this area can be roughly divided into three main groups [166]:

1. **Parallelism.** Clearly, the use of parallel processing is the most obvious choice for tackling computationally expensive MOPs, particularly with the decrease in the cost of high-speed multi-core processors (see for example [42,98,130,146,162]). It is worth noticing however, that in spite of the existence of an important number of papers on parallel MOEAs [57,182], basic research in this area has remained scarce since the origins of MOEAs [34]. Most papers on parallel MOEAs focus on applications [200] or relatively straightforward parallel extensions of well-established MOEAs such as NSGA-II [152] or SMS-EMOA [82]. For many years, emphasis has been placed on developing synchronous parallel MOEAs. Nevertheless, in recent years, the development of asynchronous implementations of MOEAs (which are more appropriate for the heterogeneous computer architectures that are more common nowadays) has gained popularity [48,79,197,201,205].
2. **Surrogates.** When using this approach, an empirical model that approximates the real problem is built through the use of information gathered from actual objective function evaluations [50,142,181]. Then, the empirical model (on which evaluating the fitness function is computationally inexpensive) is used to predict promising new solutions [181]. Current functional approximation models include polynomials (response surface methodologies [54,199]), artificial neural networks (e.g., multilayer perceptrons (MLPs) [5], radial-basis function (RBF) networks [2,204], Gaussian processes [17,207], support vector machines [3,4,203] and Kriging [131,148], among others. Although frequently used in engineering applications, surrogate methods can normally be adopted only in problems of low dimensionality, which is an important limitation when dealing with real-world MOPs. Additionally, surrogate models tend to lack robustness which is also an important issue in optimization problems. Nevertheless, there has been a lot of research oriented towards overcoming the scalability and robustness limitations of surrogate methods (see for example [122,147,157,202]).
3. **Fitness Inheritance.** This is an approach originally proposed by Smith et al. [176] with the goal of reducing the total number of fitness function evaluations performed by a (single-objective) evolutionary algorithm. The core idea of this mechanism is the following: when assigning fitness to an individual, in some cases we evaluate the objective function as usual, but the rest of the time, we assign as the fitness value of an individual the average of the fitness of its parents. This allows saving one fitness function evaluation. This idea is based on the assumption of similarity of

an offspring to its parents. Evidently, fitness inheritance must not be always applied, since the algorithm needs to use the true fitness function value from time to time, in order to obtain enough information to properly guide the search. The percentage of time in which fitness inheritance is applied is called *inheritance proportion*. If this inheritance proportion is one (i.e., 100%), the algorithm is most likely to prematurely converge [27]. Extending fitness inheritance to multi-objective optimization involves several issues, mainly related to its apparent limitation for dealing with non-convex Pareto fronts [52]. However, some researchers have managed to successfully adapt fitness inheritance to MOEAs [66, 150, 158, 195], reporting important savings on the total number of objective function evaluations performed.

3.4 Hyper-Heuristics

A hyper-heuristic is a search method or learning mechanism for selecting or generating heuristics to solve (complex) search problems [23]. Hyper-heuristics are high-level approaches that operate on a search space of heuristics or on a pool of heuristics instead of the search space of the problem at hand. Hyper-heuristics aim to provide more general search methodologies. Typically, simple heuristics tend to work well on a particular type of problem. However, when facing a new problem or even slightly modified instances of the same problem, such heuristics tend to perform poorly. Additionally, identifying which heuristic works efficiently on a certain problem is a very tedious and time-consuming task whose computational cost may become prohitive in some applications.

Burke et al. [24] proposed a taxonomy of hyper-heuristics considering two dimensions:

1. The nature of the heuristics' search space, and
2. the different sources of feedback information.

Regarding the nature of the search space, there are two options:

1. Heuristic selection, which are methodologies for choosing existing heuristics,
2. heuristic generation, which are methodologies for generating new heuristics from the components of existing ones.

Regarding the source of feedback information obtained during the search process, there are three options:

1. **No-learning**, in which there is no learning mechanism and the heuristic selection is based on either a random or an exhaustive process,
2. **Offline learning**, in which knowledge is gathered in the form of rules from a set of training instances, that will hopefully generalize to solve unseen instances, and
3. **Online learning**, in which the learning takes place while the algorithm is solving an instance of a problem.

The use of collaborative approaches which work as hyper-heuristics can be found across Operations Research, Computer Science and Artificial Intelligence. Although the ideas behind hyper-heuristics can be traced back to the early 1960s in single-objective optimization, until relatively recently, their potential hadn't been explored in multi-objective optimization. Early attempts in this field date back to 2005, when hyper-heuristics started to be used to solve multi-objective combinatorial optimization problems, such as space allocation and timetabling [25], decision-tree induction algorithms [8], bin packing and cutting stock problems [70], integration and test order problems [74,75,127], spanning trees [107], job shop scheduling [189], knapsack problems [108] and software module clustering [110,111], among others.

Multi-objective hyper-heuristics for continuous search spaces are still rare in the specialized literature. For example Vrugt et al. [190], proposed the Multi-ALgorithm Genetically Adaptive Method (AMALGAM), which is an online selection hyper-heuristic that operates similarly to NSGA-II [46]. However, in this case, the offspring are also created using the variation operators of other stochastic search methods, such as differential evolution [178], particle swarm optimization [99] and adaptive Metropolis search [76]. Although all these methods participate during the optimization process, those showing the highest reproductive success are favored. Additionally, the initialization of the population adopts Latin hypercubes sampling.

León et al. [115] proposed the Metaheuristics-based Extensible Tool for Cooperative Optimization (METCO) which is based on the island model and the cooperation of a set of MOEAs, which grants more computational resources to those algorithms that show a more promising behavior. A coordinator node is in charge of maintaining the global solution and selecting the configurations that are executed on the islands. A configuration consists of a MOEA plus the variation operators and the set of parameters which define them (population size, mutation and crossover rates, etc.). These parameters are defined by the user. The global solution set is obtained by merging local results achieved by each of the islands and its size is limited using the crowding distance operator. Besides the global stopping criterion, a local stopping criterion is defined for the execution of the MOEAs on the islands. When the local stopping criterion is reached, the configuration is scored using a performance indicator. Then, the coordinator applies the hyper-heuristic, selecting the configuration that will continue executing on the idle island. If the configuration has changed, the subpopulation is replaced by a random subset of the currently global solution.

Wang and Li [196] proposed the Multi-Strategy Ensemble Dynamic Multi-Objective Evolutionary Algorithm (MS-MOEA), which is an offline selection hyper-heuristic that adopts the fundamental principle of AMALGAM of combining different variation operators. This approach works as the ϵ-MOEA [45] with an external archive that is pruned to a limit size using the hypervolume indicator. The heuristics for generating new individuals consists of two re-initialization techniques, which are based on random sampling and Gaussian distribution with mean around the previous optimal solutions; the genetic operators SBX (Simulated Binary Crossover) and polynomial-based mutation; the Differential Evolution strategies DE/rand/1 and DE/current to best/1; as well as Gaussian mutation. This approach was designed to solve dynamic multi-objective optimization problems (i.e., problems in which either the Pareto optimal set or the Pareto optimal front change over time). If there is an environmental change, then one re-initialization technique is applied under certain probability. Genetic operators are used at early stages of the evolutionary process. Once convergence is detected, Differential Evolution is adopted to enhance diversity. Under this scheme, each strategy creates an offspring. After a fixed number of solutions is created, Gaussian mutation is launched for escaping from local optima. It is worth noticing that convergence is detected when the external archive has been full during a certain number of generations.

McClymont et al. [132] proposed the Markov Chain Hyper-Heuristic (MCHH), which is a selection heuristic with online learning working as a $(\mu + \lambda)$-Evolution Strategy with an unbounded external archive. The pool of heuristics is composed of four variation operators: mutation, replication, transposition, and cloning. All of them operate on the decision variables of a given solution. The heuristic selection mechanism uses a Markov chain, which can be seen as a directed graph where every vertex is connected to each other vertex and to itself. A vertex represents a state (heuristic), and the weight of an edge out represents the probability of moving from the current state to the destination state. All edges out of a state must sum to one. The Markov chain stochastically selects the next heuristic biased by these probabilities. The selected heuristic is employed during a certain number of generations. Then, its performance is measured by counting the number of parents that were dominated by each offspring produced by such heuristic. This performance is used to update the corresponding probability in the Markov chain using reinforcement learning.

Maashi [124] proposed an online learning selection choice function based hyper-heuristic framework for multiobjective optimization. Her proposed approach controls and combines the strengths of three well-known MOEAs (NSGA-II, SPEA2, and MOGA), which are adopted as her low-level heuristics.

Gonçalves et al. [71] proposed the MOEA/D Hyper-Heuristic (MOEA/D-HH), which is an online selection hyper-heuristic that is coupled to a MOEA/D variant [206]. In this approach, an adaptive choice function is used to determine the Differential Evolution (DE) strategy that should be applied to generate individuals at each iteration.

Walker and Keedwell [192] proposed the indicator-based multiobjective sequence-based hyper-heuristic (MOSSHH) algorithm. This seems to be the first attempt to use a hyper-heuristic in many-objective problems. This online selection hyper-heuristic is based on a hidden Markov model to determine the mutation strategy to be applied for generating a single child from the current parent. Thus, this approach works as a $(1 + 1)$-Evolution Strategy complemented with an external archive, which keeps all the non-dominated solutions discovered so far. The pool of seven mutation heuristics consists primarily of 1) adding noise to the current solution using three different continuous probability distributions, and 2) replacing the parent (or only a variable) with another one, whether randomly created or taken from the archive. At each iteration, the child replaces the parent if the former dominates the second. However, in a further paper [193], this comparison rule was changed by strategies based on the hypervolume indicator [210], the *favour* relation [51] and the average rank [11]. Moreover, the hidden Markov model is updated if the child is added to the archive and if it was better than the parent.

Hernández Gómez and Coello Coello [84] proposed a hyper-heuristic which combines the strengths and compensates for the weaknesses of different scalarizing functions. In this case, the selection process is conducted through an indicator called s-energy [80], which measures the even distribution of a set of points in k-dimensional manifolds.

4 Challenges

After reviewing some of the most relevant research done on the topics selected in Sect. 3, we provide here some of the (many) challenges that lie ahead in each of them.

- **Algorithmic Design.** Although there has been some debate regarding the paradigm that will become more common in the next few years, it seems quite obvious at this point that decomposition-based MOEAs are the clear winner so far. For indicator-based MOEAs to become more popular, other performance indicators need to be proposed. Although there has been some research activity in this regard (see for example [20,93,133,163]) none of these other performance indicators has become as popular as the hypervolume. Another interesting idea is the combination of performance indicators in order to take advantage of their strenghts and compensate for their limitations (see for example [55]). Recently, Falcón et al. [58] showed that it is possible to produce new Pareto-compliant indicators from the combination of weakly Pareto-compliant indicators and at least one that is Pareto-compliant. Although these results are limited by the fact that the hypervolume remains as the only known Pareto-compliant indicator, an interesting outcome of this work was that the new Pareto-compliant indicators generated with this procedure have a different search bias than the hypervolume. This opens a new research path within indicator-based MOEAs that is worth exploring in the near future.
  Nevertheless, a more relevant question in this area is the following: can we design MOEAs in a different way? This is a very relevant question, since algorithmic development has been at the heart of research on EMOO and it is quite important that new algorithmic proposals are made in the next few years in order to keep this research area alive. Evidently, it is not trivial to produce a selection mechanism that is not Pareto-based, decomposition-based or indicator-based, but this is indeed possible. For example, Molinet Berenguer and Coello Coello [12,138], proposed an approach that transforms an MOP into a linear assignment problem using a set of weight vectors uniformly scattered. Uniform design is adopted to obtain the set of weights, and the Kuhn-Munkres (Hungarian) algorithm [105] is used to solve the resulting assignment problem. This approach was found to perform quite well (and at a low computational cost) even in many-objective optimization problems. As such, this approach does not belong to any of the three algorithmic families previously discussed and it constitutes an intriguing new family of MOEAs.
  So, it should be clear that a challenge is to develop selection mechanisms for MOEAs that are different from those that have been developed so far. Additionally, popularizing the use of such MOEAs is certainly another (perhaps more difficult) challenge.
  Another challenge in this area is to gain a deeper understanding of the limitations of current MOEAs. For example, knowing that some scalarizing functions offer

advantages over others is very useful to design good decomposition-based and even indicator-based MOEAs (see for example [149]).

Another interesting idea is to combine components of MOEAs under a single framework that allows to exploit their advantages. This is the basic idea of Borg [77], which adopts ϵ-dominance, a measure of convergence speed called ϵ progress, an adaptive population size, multiple recombination operators and a steady-state selection mechanism.

A relevant question in this regard is if this sort of scheme could lead us to the automated design of MOEAs as has been suggested by researchers from automated parameter tuning for single-objective evolutionary algorithms [89].

- **Scalability.** This area presents several challenges. For example, what sort of performance indicator should we use to assess diversity in many-objective problems. In low-dimensional Pareto fronts, the aim is normally to achieve a uniform distribution of solutions along the Pareto front. However, what would be a desirable distribution in an MOP having, for example, 10 objectives, if we are using only 300 solutions to sample it? In recent years, the use of some performance indicators such as the Riesz s-energy [80] have shown promise in this regard, but more research in this area is required.

  In contrast with the significant interest that researchers have had on many-objective optimization in recent years, scalability in decision variable space (i.e., the solution of the so-called large-scale MOPs) has been only recently studied (see for example [123, 136, 209]). This is remarkable if we consider that large-scale multiobjective optimization problems (i.e., problems having more than 100 decision variables) are not rare in real-world applications (see for example [120]). In this area, the use of cooperative coevolutionary approaches (which have been very successful in single-objective large-scale optimization) is the most common research trend. However, new test suites are required for large-scale multiobjective optimization and some work has already been done in this direction (see for example [28]).

  Another challenge in this area is the solution of large-scale many-objective problems which is a relatively recent research topic in which some results are already available (see for example [26, 208]).
- **Dealing with Expensive Objective Functions.** Other approaches for dealing with expensive objective functions are also possible. For example, some researchers have adopted cultural algorithms [35, 39, 155, 159], which gather knowledge during the evolutionary process and use it to perform a more efficient search at the expense of a significantly larger memory usage. Cultural algorithms were proposed by Reynolds [160, 161], as an approach that tries to add domain knowledge to an evolutionary algorithm during the search process, avoiding the need to add it *a priori*. This approach uses, in addition to the population space commonly adopted in evolutionary algorithms, a belief space, which encodes the knowledge obtained from the search points and their evaluation, in order to influence the evolutionary operators that guide the search. However, the belief space is commonly designed based on the group of problems that is to be solved. At each generation, the cultural algorithm selects some exemplar individuals from the population, in order to extract information from them that can be useful during the search. Such an information is used to update the belief space. The belief space will then influence the operators of the evolutionary

algorithm, to transform them into informed operators that can enhance the search process. Cultural algorithms can be an effective means of saving objective function evaluations, but since a map of decision variable space must be kept at all times, their cost soon becomes prohibitive even for problems of moderate dimensionality (in decision variable space).

- **Hyper-heuristics.** We certainly need theoretical studies on the use of hyper-heuristics in multiobjective optimization and some work in that direction has been already done. Qian et al. [154] provided a theoretical study on the effectiveness of selection hyperheuristics for multiobjective optimization. This paper concluded that applying selection hyper-heuristics to any of the three major components of a MOEA (selection, mutation and acceptance), can exponentially speed up the optimization process.
  Another interesting idea is to combine different performance indicators within an indicator-based MOEA as proposed by Falcón-Cardona and Coello Coello [55]. In this case, IGD^+, $\epsilon+$, Δ_p and $R2$ are adopted as possible density estimators (i.e., the low-level heuristics).
  One more interesting area of research would be the use of genetic programming to generate components of MOEAs (e.g., evolutionary operators or even scalarizing functions) that can improve their performance when adopted within a multiobjective hyper-heuristic. Some work in this direction has been already reported (see for example [13]).

5 Conclusions

It should be clear that EMOO is still a research area that has a wide variety of topics worth exploring in the next few years. However, research on EMOO requires nowadays, more creativity than 25 years ago. Additionally, we also need to generate more profound ideas and to understand more complex interactions within an ever changing field which has evolved in different directions.

It is important, for example, to design new mechanisms (e.g., operators, encodings, etc.) for MOEAs based on specific features of real-world problems (e.g., variable length encodings, expensive objective functions, uncertainty, etc.). See for example [118]. The way in which coevolutionary approaches can help us to solve complex multiobjective optimization problems is another interesting venue for future research. Besides large scale problems, coevolution can help us in other domains (e.g., dynamic multiobjective optimization problems [67]), but its potential has been scarcely studied in this area (see [137]).

Nevertheless, it's important to keep in mind that a great source of diversity regarding research ideas is the knowledge coming from other disciplines. For example, EMOO has adopted advanced data structures (e.g., red-black trees [61]), concepts from computational geometry (e.g., convex hulls [29, 129], quadtrees [141] and Voronoi maps [145]), and from economics (e.g., game theory [72]) to design novel MOEAs and operators.

We also need to explore more ways of bridging the gap between Operations Research and EMOO. An example is the development of hybrid approaches that combine a MOEA with a mathematical programming technique (see for example [113]).

Another one is the use of the Karush-Kuhn-Tucker optimality conditions to estimate proximity of a solution to the Pareto optimal set (see [1]).

Hopefully, EMOO will continue to be a very active area of research for many more years, but this depends on the newcomers who can produce new interesting research paths. The work by analogy is normal in any field, but should not become the only work done to avoid stagnation of a research topic.

Acknowledgements. The author acknowledges support from CONACyT project no. 2016-01-1920 (*Investigación en Fronteras de la Ciencia 2016*) and from a project from the 2018 SEP-Cinvestav Fund (application no. 4).

References

1. Abouhawwash, M., Deb, K.: Karush-Kuhn-Tucker proximity measure for multi-objective optimization based on numerical gradients. In: 2016 Genetic and Evolutionary Computation Conference (GECCO'2016), pp. 525–532. ACM Press, Denver, Colorado, USA (20–24 July 2016). ISBN: 978-1-4503-4206-3
2. Akhtar, T., Shoemaker, C.A.: Multi objective optimization of computationally expensive multi-modal functions with RBF surrogates and multi-rule selection. J. Global Optim. **64**(1), 17–32 (2016)
3. Alves Ribeiro, V.H., Reynoso-Meza, G.: Multi-objective support vector machines ensemble generation for water quality monitoring. In: 2018 IEEE Congress on Evolutionary Computation (CEC'2018), pp. 608–613. IEEE Press, Rio de Janeiro, Brazil (8–13 July 2018). ISBN: 978-1-5090-6017-7
4. Aytug, H., Sayin, S.: using support vector machines to learn the efficient set in multiple objective discrete optimization. Eur. J. Oper. Res. **193**(2), 510–519 (1 March 2009)
5. Azzouz, N., Bechikh, S., Said, L.B.: Steady state IBEA assisted by MLP neural networks for expensive multi-objective optimization problems. In: 2014 Genetic and Evolutionary Computation Conference (GECCO 2014), pp. 581–588. ACM Press, Vancouver, Canada (12–16 July 2014). ISBN: 978-1-4503-2662-9
6. Bader, J., Zitzler, E.: HypE: an algorithm for fast hypervolume-based many-objective optimization. Evolut. Comput. **19**(1), 45–76 (2011)
7. Basgalupp, M.P., Barros, R.C., de Carvalho, A.C., Freitas, A.A., Ruiz, D.D.: LEGAL-Tree: a lexicographic multi-objective genetic algorithm for decision tree induction. In: Proceedings of the 2009 ACM symposium on Applied Computing (SAC2009), pp. 1085–1090. ACM Press, New York, USA (2009)
8. Basgalupp, M.P., Barros, R.C., Pogdorelec, V.: Evolving decision-tree induction algorithms with a multi-objective hyper-heuristic. In: Proceedings of the 30th Annual ACM Symposium on Applied Computing (SAC2015), pp. 110–117. ACM Press, New York, USA (2015). ISBN: 978-1-4503-3196-8
9. Bechikh, S., Elarbi, M., Ben Said, L.: Many-objective optimization using evolutionary algorithms: a survey. In: Bechikh, S., Datta, R., Gupta, A. (eds.) Recent Advances in Evolutionary Multi-objective Optimization. ALO, vol. 20, pp. 105–137. Springer, Cham (2017). https://doi.org/10.1007/978-3-319-42978-6_4
10. Bentley, J., Kung, H., Schkolnick, M., Thompson, C.: On the average number of maxima in a set of vectors and applications. J. Assoc. Comput. Mach. **25**(4), 536–543 (1978)
11. Bentley, P.J., Wakefield, J.P.: Finding acceptable solutions in the pareto-optimal range using multiobjective genetic algorithms. In: Chawdhry, P.K., Roy, R., Pant, R.K. (eds.) Soft Computing in Engineering Design and Manufacturing. Springer, London (1997). https://doi.org/10.1007/978-1-4471-0427-8_25

12. Molinet Berenguer, J.A., Coello Coello, C.A.: Evolutionary many-objective optimization based on Kuhn-Munkres' algorithm. In: Gaspar-Cunha, A., Henggeler Antunes, C., Coello, C.C. (eds.) EMO 2015. LNCS, vol. 9019, pp. 3–17. Springer, Cham (2015). https://doi.org/10.1007/978-3-319-15892-1_1
13. Bernabé Rodríguez, A.V., Coello Coello, C.A.: Generation of new scalarizing functions using genetic programming. In: Bäck, T., et al. (eds.) PPSN 2020. LNCS, vol. 12270, pp. 3–17. Springer, Cham (2020). https://doi.org/10.1007/978-3-030-58115-2_1
14. Beume, N., Fonseca, C.M., Lopez-Ibanez, M., Paquete, L., Vahrenhold, J.: On the complexity of computing the hypervolume indicator. IEEE Trans. Evol. Comput. **13**(5), 1075–1082 (2009)
15. Beume, N., Naujoks, B., Emmerich, M.: SMS-EMOA: multiobjective selection based on dominated hypervolume. Eur. J. Oper. Res. **181**(3), 1653–1669 (16 September 2007)
16. Blasco, X., Herrero, J.M., Sanchis, J., Martinez, M.: A new graphical visualization of n-dimensional Pareto front for decision-making in multiobjective optimization. Inf. Sci. **178**(20), 3908–3924 (15 October 2008)
17. Bradford, E., Schweidtmann, A.M., Lapkin, A.: Efficient multiobjective optimization employing Gaussian processes, spectral sampling and a genetic algorithm. J. Global Optim. **71**(2), 407–438 (2018)
18. Branke, J., Deb, K.: Integrating user preferences into evolutionary multi-objective optimization. In: Jin, Y. (ed.) Knowledge Incorporation in Evolutionary Computation. Studies in Fuzziness and Soft Computing, vol. 167. Springer, Berlin, Heidelberg (2005). https://doi.org/10.1007/978-3-540-44511-1_21
19. Bringmann, K., Friedrich, T.: An efficient algorithm for computing hypervolume contributions. Evol. Comput. **18**(3), 383–402 (2010)
20. Brockhoff, D., Wagner, T., Trautmann, H.: On the properties of the $R2$ indicator. In: 2012 Genetic and Evolutionary Computation Conference (GECCO'2012), pp. 465–472. ACM Press, Philadelphia, USA (July 2012). ISBN: 978-1-4503-1177-9
21. Brockhoff, D., Wagner, T., Trautmann, H.: R2 indicator-based multiobjective search. Evolut. Comput. **23**(3), 369–395 (2015)
22. Brockhoff, D., Zitzler, E.: Objective reduction in evolutionary multiobjective optimization: theory and applications. Evolut. Comput. **17**(2), 135–166 (2009)
23. Burke, E.K., et al.: Hyper-heuristics: a survey of the state of the art. J. Operat. Res. Soc. **64**, 1695–1724 (2013)
24. Burke, E.K., Hyde, M.R., Kendall, G., Ochoa, G., Özcan, E., Woodward, J.R.: A classification of hyper-heuristic approaches: revisited. In: Gendreau, M., Potvin, J.-Y. (eds.) Handbook of Metaheuristics. ISORMS, vol. 272, pp. 453–477. Springer, Cham (2019). https://doi.org/10.1007/978-3-319-91086-4_14
25. Burke, E.K., Landa Silva, J.D., Soubeiga, E.: Multi-objective hyper-heuristic approaches for space allocation and timetabling. In: Ibaraki, T., Nonobe, K., Yagiura, M. (eds.) Metaheuristics: Progress as Real Problem Solvers. Operations Research/Computer Science Interfaces Series, vol. 32. Springer, Boston (2005). https://doi.org/10.1007/0-387-25383-1_6
26. Cao, B., et al.: Distributed parallel particle swarm optimization for multi-objective and many-objective large-scale optimization. IEEE Access **5**, 8214–8221 (2017)
27. Chen, J.H., Goldberg, D.E., Ho, S.Y., Sastry, K.: Fitness inheritance in multi-objective optimization. In: Langdon, W., et al. (eds.) Proceedings of the Genetic and Evolutionary Computation Conference (GECCO'2002), pp. 319–326. Morgan Kaufmann Publishers, San Francisco, California (2002)
28. Cheng, R., Jin, Y., Olhofer, M., Sendhoff, B.: Test problems for large-scale multiobjective and many-objective optimization. IEEE Trans. Cybern. **47**(12), 4108–4121 (2017)

29. Cococcioni, M., Ducange, P., Lazzerini, B., Marcelloni, F.: A new multi-objective evolutionary algorithm based on convex hull for binary classifier optimization. In: 2007 IEEE Congress on Evolutionary Computation (CEC'2007), pp. 3150–3156. IEEE Press, Singapore (2007)
30. Coello Coello, C.A.: An updated survey of GA-based multiobjective optimization techniques. ACM Comput. Surv. **32**(2), 109–143 (2000)
31. Coello Coello, C.A.: Treating constraints as objectives for single-objective evolutionary optimization. Eng. Optim. **32**(3), 275–308 (2000)
32. Coello Coello, C.A.: The EMOO repository: a resource for doing research in evolutionary multiobjective optimization. IEEE Comput. Intell. Mag. **1**(1), 37–45 (2006)
33. Coello Coello, C.A., Christiansen, A.D.: Two New GA-based methods for multiobjective optimization. Civ. Eng. Syst. **15**(3), 207–243 (1998)
34. Coello Coello, C.A., Lamont, G.B., Van Veldhuizen, D.A.: Evolutionary algorithms for solving multi-objective problems. Springer, New York, 2nd edn. (2007). https://doi.org/10.1007/978-0-387-36797-2
35. Coello Coello, C.A., Landa Becerra, R.: Evolutionary multiobjective optimization using a cultural algorithm. In: 2003 IEEE Swarm Intelligence Symposium Proceedings, pp. 6–13. IEEE Service Center, Indianapolis, Indiana, USA (2003)
36. Coello Coello, C.A.: An empirical study of evolutionary techniques for multiobjective optimization in engineering design, Ph. D. thesis, Department of Computer Science, Tulane University, New Orleans, Louisiana, USA (1996)
37. Corne, D., Knowles, J.: Techniques for highly multiobjective optimisation: some nondominated points are better than others. In: Thierens, D. (ed.) 2007 Genetic and Evolutionary Computation Conference (GECCO'2007), vol. 1, pp. 773–780. ACM Press, London, UK (2007)
38. Cox, W., While, L.: Improving the IWFG algorithm for calculating incremental hypervolume. In: 2016 IEEE Congress on Evolutionary Computation (CEC'2016), pp. 3969–3976. IEEE Press, Vancouver, Canada (24–29 July 2016). iSBN 978-1-5090-0623-6
39. Daneshyari, M., Yen, G.G.: Cultural-based multiobjective particle swarm optimization. IEEE Trans. Syst. Man Cybern. Part B-Cybern. **41**(2), 553–567 (2011)
40. Das, I., Dennis, J.: Normal-boundary intersection: a new method for generating the pareto surface in nonlinear multicriteria optimization problems. SIAM J. Optim. **8**(3), 631–657 (1998)
41. Das, I., Dennis, J.: A closer look at drawbacks of minimizing weighted sums of objectives for pareto set generation in multicriteria optimization problems. Struct. Optimiz. **14**(1), 63–69 (1997)
42. de Oliveira, L.B., Marcelino, C.G., Milanés, A., Almeida, P.E.M., Carvalho, L.M.: A successful parallel implementation of NSGA-II on GPU for the energy dispatch problem on hydroelectric power plants. In: 2016 IEEE Congress on Evolutionary Computation (CEC'2016), pp. 4305–4312. IEEE Press, Vancouver, Canada (24–29 July 2016). ISBN 978-1-5090-0623-6
43. Deb, K., Agrawal, S., Pratap, A., Meyarivan, T.: A fast elitist non-dominated sorting genetic algorithm for multi-objective optimization: NSGA-II. In: Schoenauer, M., et al. (eds.) PPSN 2000. LNCS, vol. 1917, pp. 849–858. Springer, Heidelberg (2000). https://doi.org/10.1007/3-540-45356-3_83
44. Deb, K., Jain, H.: An evolutionary many-objective optimization algorithm using reference-point-based nondominated sorting approach, part i: solving problems with box constraints. IEEE Trans. Evol. Comput. **18**(4), 577–601 (2014)
45. Deb, K., Mohan, M., Mishra, S.: Evaluating the ϵ-domination based multi-objective evolutionary algorithm for a quick computation of pareto-optimal solutions. Evolut. Comput. **13**(4), 501–525 (2005)

46. Deb, K., Pratap, A., Agarwal, S., Meyarivan, T.: A fast and elitist multiobjective genetic algorithm: NSGA-II. IEEE Trans. Evol. Comput. **6**(2), 182–197 (2002)
47. Deng, J., Zhang, Q.: Approximating hypervolume and hypervolume contributions using polar coordinate. IEEE Trans. Evol. Comput. **23**(5), 318–913 (2019)
48. Depolli, M., Trobec, R., Filipic, B.: asynchronous master-slave parallelization of differential evolution for multi-objective optimization. Evolut. Comput. **21**(2), 261–291 (2013)
49. di Pierro, F., Khu, S.T., Savić, D.A.: An investigation on preference order ranking scheme for multiobjective evolutionary optimization. IEEE Trans. Evol. Comput. **11**(1), 17–45 (2007)
50. Diaz-Manriquez, A., Toscano, G., Hugo Barron-Zambrano, J., Tello-Leal, E.: A review of surrogate assisted multiobjective evolutionary algorithms. Computat. Intell. Neurosci. **2016**, 9420460 (2016)
51. Drechsler, N., Drechsler, R., Becker, B.: Multi-objective optimisation based on relation *Favour*. In: Zitzler, E., Thiele, L., Deb, K., Coello Coello, C.A., Corne, D. (eds.) EMO 2001. LNCS, vol. 1993, pp. 154–166. Springer, Heidelberg (2001). https://doi.org/10.1007/3-540-44719-9_11
52. Ducheyne, E., De Baets, B., De Wulf, R.: Is fitness inheritance useful for real-world applications? In: Fonseca, C.M., Fleming, P.J., Zitzler, E., Thiele, L., Deb, K. (eds.) EMO 2003. LNCS, vol. 2632, pp. 31–42. Springer, Heidelberg (2003). https://doi.org/10.1007/3-540-36970-8_3
53. Emmerich, M., Beume, N., Naujoks, B.: An EMO algorithm using the hypervolume measure as selection criterion. In: Coello Coello, C.A., Hernández Aguirre, A., Zitzler, E. (eds.) EMO 2005. LNCS, vol. 3410, pp. 62–76. Springer, Heidelberg (2005). https://doi.org/10.1007/978-3-540-31880-4_5
54. Esfe, M.H., Hajmohammad, M.H., Wongwises, S.: Pareto optimal design of thermal conductivity and viscosity of ND-Co3O4 Nanofluids by MOPSO and NSGA II using Response surface methodology. Curr. Nanosci. **14**(1), 62–70 (2018)
55. Falcón-Cardona, J.G., Coello Coello, C.A.: A multi-objective evolutionary hyper-heuristic based on multiple indicator-based density estimators. In: 2018 Genetic and Evolutionary Computation Conference (GECCO'2018). pp. 633–640. ACM Press, Kyoto, Japan (15–19 July 2018). ISBN: 978-1-4503-5618-3
56. Falcón-Cardona, J.G., Coello Coello, C.A.: Indicator-based multi-objective evolutionary algorithms: a comprehensive survey. ACM Comput. Surv. **53**(2), 29 (2020)
57. Falcon-Cardona, J.G., Gomez, R.H., Coello, C.A.C., Castillo Tapia, M.G.: Parallel multi-objective evolutionary algorithms: a comprehensive survey. Swarm Evolut. Comput. **67**, 100960 (2021)
58. Falcón-Cardona, J., Emmerich, M., Coello Coello, C.: On the construction of pareto-compliant combined indicators. Evolut. Comput. **30**(3), 381–408 (2022)
59. Farina, M., Amato, P.: On the optimal solution definition for many-criteria optimization problems. In: Proceedings of the NAFIPS-FLINT International Conference'2002, pp. 233–238. IEEE Service Center, Piscataway, New Jersey (2002)
60. Farina, M., Amato, P.: A fuzzy definition of "optimality" for many-criteria optimization problems. IEEE Trans. Syst. Man Cybern. Part A-Syst. Humans **34**(3), 315–326 (2004)
61. Fieldsend, J.E., Everson, R.M., Singh, S.: Using unconstrained elite archives for multiobjective optimization. IEEE Trans. Evol. Comput. **7**(3), 305–323 (2003)
62. Fleischer, M.: The measure of pareto optima applications to multi-objective metaheuristics. In: Fonseca, C.M., Fleming, P.J., Zitzler, E., Thiele, L., Deb, K. (eds.) EMO 2003. LNCS, vol. 2632, pp. 519–533. Springer, Heidelberg (2003). https://doi.org/10.1007/3-540-36970-8_37

63. Fonseca, C.M., Fleming, P.J.: Genetic algorithms for multiobjective optimization: formulation, discussion and generalization. In: Forrest, S. (ed.) Proceedings of the Fifth International Conference on Genetic Algorithms, pp. 416–423. Morgan Kauffman Publishers, San Mateo, California, USA (1993)
64. Garcia, J.L.L., Monroy, R., Hernandez, V.A.S., Coello, C.A.C.: COARSE-EMOA: an indicator-based evolutionary algorithm for solving equality constrained multi-objective optimization problems. Swarm Evolut. Comput. **67**, 100983 (2021)
65. Garza-Fabre, M., Pulido, G.T., Coello, C.A.C.: Ranking methods for many-objective optimization. In: Aguirre, A.H., Borja, R.M., Garciá, C.A.R. (eds.) MICAI 2009. LNCS (LNAI), vol. 5845, pp. 633–645. Springer, Heidelberg (2009). https://doi.org/10.1007/978-3-642-05258-3_56
66. Giannakoglou, K.C., Kampolis, I.C.: Multilevel optimization algorithms based on metamodel- and fitness inheritance-assisted evolutionary algorithms. In: Tenne, Y., Goh, C.-K. (eds.) Computational Intelligence in Expensive Optimization Problems. ALO, vol. 2, pp. 61–84. Springer, Heidelberg (2010). https://doi.org/10.1007/978-3-642-10701-6_3
67. Goh, C.K., Tan, K.C.: A competitive-cooperative coevolutionary paradigm for dynamic multiobjective optimization. IEEE Trans. Evol. Comput. **13**(1), 103–127 (2009)
68. Goldberg, D.E.: Genetic algorithms in search. optimization and machine learning. Addison-Wesley Publishing Company, Reading, Massachusetts (1989)
69. Goldberg, D.E., Richardson, J.: Genetic algorithm with sharing for multimodal function optimization. In: Grefenstette, J.J. (ed.) Genetic Algorithms and Their Applications: Proceedings of the Second International Conference on Genetic Algorithms, pp. 41–49. Lawrence Erlbaum, Hillsdale, New Jersey (1987)
70. Gomez, J.C., Terashima-Marín, H.: Approximating multi-objective hyper-heuristics for solving 2D irregular cutting stock problems. In: Sidorov, G., Hernández Aguirre, A., Reyes García, C.A. (eds.) MICAI 2010. LNCS (LNAI), vol. 6438, pp. 349–360. Springer, Heidelberg (2010). https://doi.org/10.1007/978-3-642-16773-7_30
71. Gonçalves, R.A., Kuk, J.N., Almeida, C.P., Venske, S.M.: MOEA/D-HH: a hyper-heuristic for multi-objective problems. In: Gaspar-Cunha, A., Henggeler Antunes, C., Coello, C.C. (eds.) EMO 2015. LNCS, vol. 9018, pp. 94–108. Springer, Cham (2015). https://doi.org/10.1007/978-3-319-15934-8_7
72. Greiner, D., Periaux, J., Emperador, J.M., Galvan, B., Winter, G.: Game theory based evolutionary algorithms: a review with Nash applications in structural engineering optimization problems. Archives of Computational Methods in Engineering **24**(4), 703–750 (2017)
73. Guerreiro, A.P., Fonseca, C.M.: Computing and updating hypervolume contributions in up to four dimensions. IEEE Trans. Evol. Comput. **22**(3), 449–463 (2018)
74. Guizzo, G., Fritsche, G.M., Vergilio, S.R., Pozo, A.T.R.: A hyper-heuristic for the multi-objective integration and test order problem. In: 2015 Genetic and Evolutionary Computation Conference (GECCO 2015), pp. 1343–1350. ACM Press, Madrid, Spain (11–15 July 2015). ISBN 978-1-4503-3472-3
75. Guizzo, G., Vergilio, S.R., T.R.Pozo, A., Fritsche, G.M.: a multi-objective and evolutionary hyper-heuristic applied to the integration and test order problem. Appl. Soft Comput. **56**, 331–344 (2017)
76. Haario, H., Saksman, E., Taminen, J.: An adaptive metropolis algorithm. Bernoulli **7**(2), 223–242 (2001)
77. Hadka, D., Reed, P.: Borg: An auto-adaptive many-objective evolutionary computing framework. Evolut. Comput. **21**(2), 231–259 (2013)
78. Hajela, P., Lin, C.Y.: Genetic search strategies in multicriterion optimal design. Struct. Optim. **4**, 99–107 (1992)

79. Harada, T., Takadama, K.: Performance comparison of parallel asynchronous multi-objective evolutionary algorithm with different asynchrony. In: 2017 IEEE Congress on Evolutionary Computation (CEC'2017), pp. 1215–1222. IEEE Press, San Sebastián, Spain (5–8 June 2017). ISBN 978-1-5090-4601-0
80. Hardin, D.P., Saff, E.B.: Discretizing manifolds via minimum energy points. Notices AMS **51**(10), 1186–1194 (2004)
81. He, Z., Yen, G.G.: Visualization and performance metric in many-objective optimization. IEEE Trans. Evol. Comput. **20**(3), 386–402 (2016)
82. Hernández Gómez, R., Coello Coello, C.A., Alba, E.: A parallel version of SMS-EMOA for many-objective optimization problems. In: Handl, J., Hart, E., Lewis, P.R., López-Ibáñez, M., Ochoa, G., Paechter, B. (eds.) PPSN 2016. LNCS, vol. 9921, pp. 568–577. Springer, Cham (2016). https://doi.org/10.1007/978-3-319-45823-6_53
83. Hernández Gómez, R., Coello Coello, C.A.: Improved metaheuristic based on the $R2$ indicator for many-objective optimization. In: 2015 Genetic and Evolutionary Computation Conference (GECCO 2015), pp. 679–686. ACM Press, Madrid, Spain (11–15 July 2015). ISBN 978-1-4503-3472-3
84. Hernández Gómez, R., Coello Coello, C.A.: a hyper-heuristic of scalarizing functions. In: 2017 Genetic and Evolutionary Computation Conference (GECCO'2017), pp. 577–584. ACM Press, Berlin, Germany (15–19 July 2017). ISBN 978-1-4503-4920-8
85. Hernández Gómez, R., Coello Coello, C.A., Alba Torres, E.: A multi-objective evolutionary algorithm based on parallel coordinates. In: 2016 Genetic and Evolutionary Computation Conference (GECCO'2016), pp. 565–572. ACM Press, Denver, Colorado, USA (20–24 July 2016). ISBN 978-1-4503-4206-3
86. Horn, J., Nafpliotis, N., Goldberg, D.E.: A niched pareto genetic algorithm for multiobjective optimization. In: Proceedings of the First IEEE Conference on Evolutionary Computation, IEEE World Congress on Computational Intelligence, vol. 1, pp. 82–87. IEEE Service Center, Piscataway, New Jersey (1994)
87. Hua, Y., Liu, Q., Hao, K., Jin, Y.: A survey of evolutionary algorithm for multi-objective optimization problems with irregular pareto front. IEEE-CAA J. Autom. Sinica **8**(2), 303–318 (2021)
88. Hughes, E.J.: Evolutionary many-objective optimisation: many once or one many? In: 2005 IEEE Congress on Evolutionary Computation (CEC'2005), vol. 1, pp. 222–227. IEEE Service Center, Edinburgh, Scotland (2005)
89. Hutter, F., Hoos, H.H., Leyton-Brown, K., Stützle, T.: ParamILS: an automatic algorithm configuration framework. J. Artif. Intell. Res. **36**, 267–306 (2009)
90. Igel, C., Hansen, N., Roth, S.: covariance matrix adaptation for multi-objective optimization. Evolut. Comput. **15**(1), 1–28 (2007)
91. Ishibuchi, H., Akedo, N., Nojima, Y.: Behavior of multiobjective evolutionary algorithms on many-objective Knapsack problems. IEEE Trans. Evol. Comput. **19**(2), 264–283 (2015)
92. Ishibuchi, H., Imada, R., Setoguchi, Y., Nojima, Y.: How to specify a reference point in hypervolume calculation for fair performance comparison. Evol. Comput. **26**(3), 411–440 (2018)
93. Ishibuchi, H., Masuda, H., Tanigaki, Y., Nojima, Y.: Modified distance calculation in generational distance and inverted generational distance. In: Gaspar-Cunha, A., Henggeler Antunes, C., Coello, C.C. (eds.) EMO 2015. LNCS, vol. 9019, pp. 110–125. Springer, Cham (2015). https://doi.org/10.1007/978-3-319-15892-1_8
94. Ishibuchi, H., Setoguchi, Y., Masuda, H., Nojima, Y.: Performance of decomposition-based many-objective algorithms strongly depends on pareto front shapes. IEEE Trans. Evol. Comput. **21**(2), 169–190 (2017)

95. Jaszkiewicz, A.: On the performance of multiple-objective genetic local search on the 0/1 Knapsack problem-a comparative experiment. IEEE Trans. Evol. Comput. **6**(4), 402–412 (2002)
96. Jaszkiewicz, A.: Improved quick hypervolume algorithm. Comput. Oper. Res. **90**, 72–83 (2018)
97. Jiang, S., Zhang, J., Ong, Y.S., Zhang, A.N., Tan, P.S.: A simple and fast hypervolume indicator-based multiobjective evolutionary algorithm. IEEE Trans. Cybern. **45**(10), 2202–2213 (2015)
98. Kato, T., Shimoyama, K.: Evolutionary algorithm with parallel evaluation strategy using constrained penalty-based boundary intersection. In: 2016 IEEE Congress on Evolutionary Computation (CEC'2016), pp. 3702–3709. IEEE Press, Vancouver, Canada (24–29 July 2016). ISBN 978-1-5090-0623-9
99. Kennedy, J., Eberhart, R.C.: swam intelligence. Morgan Kaufmann Publishers, San Francisco, California, USA (2001). ISBN: 978-1-55860-595-4
100. Knowles, J., Corne, D.: Properties of an adaptive archiving algorithm for storing nondominated vectors. IEEE Trans. Evol. Comput. **7**(2), 100–116 (2003)
101. Knowles, J., Corne, D.: Quantifying the effects of objective space dimension in evolutionary multiobjective optimization. In: Obayashi, S., Deb, K., Poloni, C., Hiroyasu, T., Murata, T. (eds.) EMO 2007. LNCS, vol. 4403, pp. 757–771. Springer, Heidelberg (2007). https://doi.org/10.1007/978-3-540-70928-2_57
102. Knowles, J.D.: Local-search and hybrid evolutionary algorithms for pareto optimization, Ph. D. thesis, The University of Reading, Department of Computer Science, Reading, UK (2002)
103. Knowles, J.D., Corne, D.W.: Approximating the nondominated front using the pareto archived evolution strategy. Evol. Comput. **8**(2), 149–172 (2000)
104. Kowatari, N., Oyama, A., Aguirre, H.E., Tanaka, K.: A study on large population MOEA using adaptive ϵ-box dominance and neighborhood recombination for many-objective optimization. In: Hamadi, Y., Schoenauer, M. (eds.) LION 2012. LNCS, pp. 86–100. Springer, Heidelberg (2012). https://doi.org/10.1007/978-3-642-34413-8_7
105. Kuhn, H.W.: The hungarian method for the assignment problem. Naval Res. Logist. Quart.**2**(1–2), 83–97 (1955)
106. Kukkonen, S., Deb, K.: Improved pruning of non-dominated solutions based on crowding distance for bi-objective optimization problems. In: 2006 IEEE Congress on Evolutionary Computation (CEC'2006), pp. 1164–1171. IEEE, Vancouver, BC, Canada (2006)
107. Kumar, R., Bal, B.K., Rockett, P.: Multiobjective genetic programming approach to evolving heuristics for the bounded diameter minimum spanning tree problem. In: 2009 Genetic and Evolutionary Computation Conference (GECCO'2009), pp. 309–316. ACM Press, Montreal, Canada (8–12 July 2009). ISBN: 978-1-60558-325-9
108. Kumar, R., Joshi, A.H., Banka, K.K., Rockett, P.I.: Evolution of Hyperheuristics for the Biobjective 0/1 Knapsack Problem by Multiobjective Genetic Programming. In: 2008 Genetic and Evolutionary Computation Conference (GECCO'2008), pp. 1227–1234. ACM Press, Atlanta, USA (2008). ISBN: 978-1-60558-131-6
109. Kumar, S.V., Ranjithan, S.R.: Evaluation of the constraint method-based evolutionary algorithm (cmea) for a three-objective optimization problem. In: Langdon, W., et al. (eds.) Proceedings of the Genetic and Evolutionary Computation Conference (GECCO'2002), pp. 431–438. Morgan Kaufmann Publishers, San Francisco, California (2002)
110. Kumari, A.C., Srinivas, K.: Hyper-heuristic approach for multi-objective software module clustering. J. Syst. Softw. **117**, 384–401 (2016)

111. Kumari, A.C., Srinivas, K., Gupta, M.: Software module clustering using a hyper-heuristic based multi-objective genetic algorithm. In: Proceedings of the 2013 3rd IEEE International Advance Computing Conference, pp. 813–818. IEEE Press, Ghaziabad, India (22–23 February 2013). ISBN: 978-1-4673-4528-6
112. Lacour, R., Klamroth, K., Fonseca, C.M.: A box decomposition algorithm to compute the hypervolume indicator. Comput. Operat. Res. **79**, 347–360 (2017)
113. Lara, A., Sanchez, G., Coello Coello, C.A., Schütze, O.: HCS: a new local search strategy for memetic multi-objective evolutionary algorithms. IEEE Trans. Evol. Comput. **14**(1), 112–132 (2010)
114. Laumanns, M., Thiele, L., Deb, K., Zitzler, E.: Combining convergence and diversity in evolutionary multi-objective optimization. Evolut. Comput. **10**(3), 263–282 (2002)
115. León, C., Miranda, G., Segura, C.: Parallel hyperheuristic: a self-adaptive island-based model for multi-objective optimization. In: 2008 Genetic and Evolutionary Computation Conference (GECCO'2008), pp. 757–758. ACM Press, Atlanta, USA (2008). ISBN: 978-1-60558-131-6
116. Li, B., Li, J., Tang, K., Yao, X.: Many-objective evolutionary algorithms: a survey. ACM Comput. Surv. **48**(1), 1–35 (2015)
117. Li, F., Cheng, R., Liu, J., Jin, Y.: A two-stage R2 indicator based evolutionary algorithm for many-objective optimization. Appl. Soft Comput. **67**, 245–260 (2018)
118. Li, H., Deb, K.: Challenges for evolutionary multiobjective optimization algorithms in solving variable-length problems. In: 2017 IEEE Congress on Evolutionary Computation (CEC'2017), pp. 2217–2224. IEEE Press, San Sebastián, Spain (5–8 June 2017). ISBN: 978-1-5090-4601-0
119. López Jaimes, A., Coello Coello, C.A., Chakraborty, D.: Objective reduction using a feature selection technique. In: 2008 Genetic and Evolutionary Computation Conference (GECCO'2008), pp. 674–680. ACM Press, Atlanta, USA (2008). ISBN: 978-1-60558-131-6
120. Lu, R., Guan, X., Li, X., Hwang, I.: A large-scale flight multi-objective assignment approach based on multi-island parallel evolution algorithm with cooperative coevolutionary. Sci. Chin.-Inf. Sci. **59**(7), 823876 (2016)
121. von Lücken, C., Baran, B., Brizuela, C.: A survey on multi-objective evolutionary algorithms for many-objective problems. Comput. Optim. Appl. **58**(3), 707–756 (2014)
122. Luo, C., Shimoyama, K., Obayashi, S.: Effects of the number of design variables on performances in Kriging-model-based many-objective optimization. In: 2015 IEEE Congress on Evolutionary Computation (CEC'2015), pp. 1901–1908. IEEE Press, Sendai, Japan (25–28 May 2015). ISBN: 978-1-4799-7492-4
123. Ma, X., et al.: A multiobjective evolutionary algorithm based on decision variable analyses for multiobjective optimization problems with large-scale variables. IEEE Trans. Evol. Comput. **20**(2), 275–298 (2016)
124. Maashi, M.: An investigation of multi-objective hyper-heuristics for multi-objective optimisation, Ph. D. thesis, The University of Nottingham, UK (2014)
125. Manoatl Lopez, E., Coello Coello, C.A.: IGD^+-EMOA: a multi-objective evolutionary algorithm based on IGD^+. In: 2016 IEEE Congress on Evolutionary Computation (CEC'2016), pp. 999–1006. IEEE Press, Vancouver, Canada (24–29 July 2016). ISBN: 978-1-5090-0623-9
126. Manoatl Lopez, E., Coello Coello, C.A.: An improved version of a reference-based multi-objective evolutionary algorithm based on IGD+. In: 2018 Genetic and Evolutionary Computation Conference (GECCO'2018), pp. 713–720. ACM Press, Kyoto, Japan (15–19 July 2018). ISBN: 978-1-4503-5618-3

127. Mariani, T., Guizzo, G., Vergilio, S.R., Pozo, A.T.: Grammatical evolution for the multi-objective integration and test order problem. In: 2016 Genetic and Evolutionary Computation Conference (GECCO'2016), pp. 1069–1076. ACM Press, Denver, Colorado, USA (20–24 July 2016). ISBN: 978-1-4503-4206-3
128. Martí, L., García, J., Berlanga, A., Molina, J.M.: Introducing MONEDA: scalable multiobjective optimization with a neural estimation of distribution algorithm. In: 2008 Genetic and Evolutionary Computation Conference (GECCO'2008), pp. 689–696. ACM Press, Atlanta, USA (2008). ISBN: 978-1-60558-131-6
129. Martínez, S.Z., Coello, C.A.C.: An archiving strategy based on the convex hull of individual minima for MOEAs. In: 2010 IEEE Congress on Evolutionary Computation (CEC'2010), pp. 912–919. IEEE Press, Barcelona, Spain (18–23 July 2010)
130. Maruf Hussain, M., Fujimoto, N.: Parallel multi-objective particle swarm optimization for large swarm and high dimensional problems. In: 2018 IEEE Congress on Evolutionary Computation (CEC'2018), pp. 1546–1555. IEEE Press, Rio de Janeiro, Brazil (8–13 July 2018). ISBN: 978-1-5090-6017-7
131. Mazumdar, A., Chugh, T., Miettinen, K., López-Ibáñez, M.: On dealing with uncertainties from kriging models in offline data-driven evolutionary multiobjective optimization. In: Deb, K., et al. (eds.) EMO 2019. LNCS, vol. 11411, pp. 463–474. Springer, Cham (2019). https://doi.org/10.1007/978-3-030-12598-1_37
132. McClymont, K., Keedwell, E.C.: Markov chain hyper-heuristic (mchh): an online selective hyper-heuristic for multi-objective continuous problems. In: 2011 Genetic and Evolutionary Computation Conference (GECCO'2011), pp. 2003–2010. ACM Press, Dublin, Ireland (12–16 July 2011)
133. Menchaca-Mendez, A., Coello Coello, C.A.: Selection mechanisms based on the maximin fitness function to solve multi-objective optimization problems. Inf. Sci. **332**, 131–152 (2016)
134. Menchaca-Mendez, A., Coello Coello, C.A.: An alternative hypervolume-based selection mechanism for multi-objective evolutionary algorithms. Soft. Comput. **21**(4), 861–884 (2017)
135. Miettinen, K.M.: Nonlinear Multiobjective Optimization. Kluwer Academic Publishers, Boston, Massachusetts (1999)
136. Miguel Antonio, L., Coello Coello, C.A.: Decomposition-based approach for solving large scale multi-objective problems. In: Handl, J., Hart, E., Lewis, P.R., López-Ibáñez, M., Ochoa, G., Paechter, B. (eds.) PPSN 2016. LNCS, vol. 9921, pp. 525–534. Springer, Cham (2016). https://doi.org/10.1007/978-3-319-45823-6_49
137. Miguel Antonio, L., Coello Coello, C.A.: Coevolutionary multiobjective evolutionary algorithms: survey of the state-of-the-art. IEEE Trans. Evol. Comput. **22**(6), 851–865 (2018)
138. Miguel Antonio, L., Molinet Berenguer, J.A., Coello Coello, C.A.: Evolutionary many-objective optimization based on linear assignment problem transformations. Soft. Comput. **22**(16), 5491–5512 (2018)
139. Ming, F., Gong, W., Wang, L., Lu, C.: A tri-population based co-evolutionary framework for constrained multi-objective optimization problems. Swarm Evol. Comput. **70**, 101055 (2022)
140. Mostaghim, S., Schmeck, H.: Distance based ranking in many-objective particle swarm optimization. In: Rudolph, G., Jansen, T., Beume, N., Lucas, S., Poloni, C. (eds.) PPSN 2008. LNCS, vol. 5199, pp. 753–762. Springer, Heidelberg (2008). https://doi.org/10.1007/978-3-540-87700-4_75
141. Mostaghim, S., Teich, J.: Quad-trees: a data structure for storing pareto sets in multiobjective evolutionary algorithms with elitism. In: Abraham, A., Jain, L., Goldberg, R. (eds.) Evolutionary Multiobjective Optimization. Advanced Information and Knowledge Processing. Springer, London (2005)

142. Muller, J.: SOCEMO: surrogate optimization of computationally expensive multiobjective problems. INFORMS J. Comput. **29**(4), 581–596 (2017)
143. Murugan, M.S., Suresh, S., Ganguli, R., Mani, V.: Target vector optimization of composite box beam using real-coded genetic algorithm: a decomposition approach. Struct. Multidiscip. Optim. **33**(2), 131–146 (2007)
144. no, A.A.M., Coello Coello, C.A., Mezura-Montes, E.: Multi-objective evolutionary algorithms in aeronautical and aerospace engineering. IEEE Trans. Evolut. Comput. **16**(5), 662–694 (2012)
145. Okabe, T., Jin, Y., Sendhoff, B., Olhofer, M.: Voronoi-based estimation of distribution algorithm for multi-objective optimization. In: 2004 Congress on Evolutionary Computation (CEC'2004), vol. 2, pp. 1594–1601. IEEE Service Center, Portland, Oregon, USA (2004)
146. Ortega, G., Filatovas, E., Garzon, E.M., Casado, L.G.: Non-dominated sorting procedure for pareto dominance ranking on multicore CPU and/or GPU. J. Global Optim. **69**(3), 607–627 (2017)
147. Palar, P.S., Shimoyama, K.: Multiple metamodels for robustness estimation in multi-objective robust optimization. In: Trautmann, H., et al. (eds.) EMO 2017. LNCS, vol. 10173, pp. 469–483. Springer, Cham (2017). https://doi.org/10.1007/978-3-319-54157-0_32
148. Palar, P.S., Shimoyama, K.: On multi-objective efficient global optimization via universal Kriging surrogate model. In: 2017 IEEE Congress on Evolutionary Computation (CEC'2017), pp. 621–628. IEEE Press, San Sebastián, Spain (5–8 June 2017). ISBN: 978-1-5090-4601-0
149. Pescador-Rojas, M., Hernández Gómez, R., Montero, E., Rojas-Morales, N., Riff, M.-C., Coello Coello, C.A.: An overview of weighted and unconstrained scalarizing functions. In: Trautmann, H., et al. (eds.) EMO 2017. LNCS, vol. 10173, pp. 499–513. Springer, Cham (2017). https://doi.org/10.1007/978-3-319-54157-0_34
150. Pilato, C., Palermo, G., Tumeo, A., Ferrandi, F., Sciuto, D., Lanzi, P.L.: Fitness inheritance in evolutionary and multi-objective high-level synthesis. In: 2007 IEEE Congress on Evolutionary Computation (CEC'2007), pp. 3459–3466. IEEE Press, Singapore (2007)
151. Pires, E.J.S., Machado, J.A.T., de Moura Oliveira, P.B.: Entropy diversity in multi-objective particle swarm optimization. Entropy **15**(12), 5475–5491 (2013)
152. Powell, D., Hollingsworth, J.: A NSGA-II, web-enabled, parallel optimization framework for NLP and MINLP. In: 2007 Genetic and Evolutionary Computation Conference (GECCO'2007), vol. 2, pp. 2145–2150. ACM Press, London, UK (2007)
153. Praditwong, K., Yao, X.: How well do multi-objective evolutionary algorithms scale to large problems. In: 2007 IEEE Congress on Evolutionary Computation (CEC'2007), pp. 3959–3966. IEEE Press, Singapore (2007)
154. Qian, C., Tang, K., Zhou, Z.-H.: Selection hyper-heuristics can provably be helpful in evolutionary multi-objective optimization. In: Handl, J., Hart, E., Lewis, P.R., López-Ibáñez, M., Ochoa, G., Paechter, B. (eds.) PPSN 2016. LNCS, vol. 9921, pp. 835–846. Springer, Cham (2016). https://doi.org/10.1007/978-3-319-45823-6_78
155. Qin, H., Zhou, J., Lu, Y., Li, Y., Zhang, Y.: Multi-objective cultured differential evolution for generating optimal trade-offs in reservoir flood control operation. Water Resour. Manage **24**(11), 2611–2632 (2010)
156. Raquel, C., Yao, X.: Dynamic multi-objective optimization: a survey of the state-of-the-art. In: Yang, S., Yao, X. (eds.) Evolutionary Computation for Dynamic Optimization Problems. Studies in Computational Intelligence, vol. 490. Springer, Berlin, Heidelberg (2013). https://doi.org/10.1007/978-3-642-38416-5_4
157. Ray, T., Smith, W.: A surrogate assisted parallel multiobjective evolutionary algorithm for robust engineering design. Eng. Optim. **38**(8), 997–1011 (2006)

158. Reyes Sierra, M., Coello Coello, C.A.: Fitness inheritance in multi-objective particle swarm optimization. In: 2005 IEEE Swarm Intelligence Symposium (SIS'05), pp. 116–123. IEEE Press, Pasadena, California, USA (2005)
159. Reynolds, R., Liu, D.: Multi-objective cultural algorithms. In: 2011 IEEE Congress on Evolutionary Computation (CEC'2011), pp. 1233–1241. IEEE Service Center, New Orleans, Louisiana, USA (5–8 June 2011)
160. Reynolds, R.G.: An introduction to cultural algorithms. In: Sebald, A.V., Fogel, L.J. (eds.) Proceedings of the Third Annual Conference on Evolutionary Programming, pp. 131–139. World Scientific, River Edge, New Jersey (1994)
161. Reynolds, R.G., Michalewicz, Z., Cavaretta, M.: Using cultural algorithms for constraint handling in GENOCOP. In: McDonnell, J.R., Reynolds, R.G., Fogel, D.B. (eds.) Proceedings of the Fourth Annual Conference on Evolutionary Programming, pp. 298–305. MIT Press, Cambridge, Massachusetts (1995)
162. Rocha, H., Peretta, I.S., Lima, G.F.M., Marques, L.G., Yamanaka, K.: Exterior lighting computer-automated design based on multi-criteria parallel evolutionary algorithm: optimized designs for illumination quality and energy efficiency. Expert Syst. Appl. **45**, 208–222 (2016)
163. Rodríguez Villalobos, C.A., Coello Coello, C.A.: A new multi-objective evolutionary algorithm based on a performance assessment indicator. In: 2012 Genetic and Evolutionary Computation Conference (GECCO'2012), pp. 505–512. ACM Press, Philadelphia, USA (2012). ISBN: 978-1-4503-1177-9
164. Rudolph, G., Agapie, A.: Convergence properties of some multi-objective evolutionary algorithms. In: Proceedings of the 2000 Conference on Evolutionary Computation, vol. 2, pp. 1010–1016. IEEE Press, Piscataway, New Jersey (2000)
165. Russo, L.M.S., Francisco, A.P.: Extending quick hypervolume. J. Heuristics **22**(3), 245–271 (2016)
166. Santana-Quintero, L.V., Montaño, A.A., Coello, C.A.C.: A review of techniques for handling expensive functions in evolutionary multi-objective optimization. In: Tenne, Y., Goh, C.-K. (eds.) Computational Intelligence in Expensive Optimization Problems. ALO, vol. 2, pp. 29–59. Springer, Heidelberg (2010). https://doi.org/10.1007/978-3-642-10701-6_2
167. Santiago, A., et al.: A survey of decomposition methods for multi-objective optimization. In: Castillo, O., Melin, P., Pedrycz, W., Kacprzyk, J. (eds.) Recent Advances on Hybrid Approaches for Designing Intelligent Systems. SCI, vol. 547, pp. 453–465. Springer, Cham (2014). https://doi.org/10.1007/978-3-319-05170-3_31
168. Sato, H., Aguirre, H.E., Tanaka, K.: Controlling dominance area of solutions and its impact on the performance of MOEAs. In: Obayashi, S., Deb, K., Poloni, C., Hiroyasu, T., Murata, T. (eds.) EMO 2007. LNCS, vol. 4403, pp. 5–20. Springer, Heidelberg (2007). https://doi.org/10.1007/978-3-540-70928-2_5
169. Saxena, D.K., Duro, J.A., Tiwari, A., Deb, K., Zhang, Q.: Objective reduction in many-objective optimization: linear and nonlinear algorithms. IEEE Trans. Evol. Comput. **17**(1), 77–99 (2013)
170. Schaffer, J.D.: Multiple objective optimization with vector evaluated genetic algorithms, Ph. D. thesis, Vanderbilt University (1984)
171. Schaffer, J.D.: Multiple objective optimization with vector evaluated genetic algorithms. In: Genetic Algorithms and their Applications: Proceedings of the First International Conference on Genetic Algorithms, pp. 93–100. Lawrence Erlbaum (1985)
172. Schaffer, J.D., Grefenstette, J.J.: Multiobjective learning via genetic algorithms. In: Proceedings of the 9th International Joint Conference on Artificial Intelligence (IJCAI-85), pp. 593–595. AAAI, Los Angeles, California (1985)

173. Schütze, O., Lara, A., Coello Coello, C.A.: On the influence of the number of objectives on the hardness of a multiobjective optimization problem. IEEE Trans. Evol. Comput. **15**(4), 444–455 (2011)
174. Sen, P., Yang, J.B.: Multiple Criteria Decision Support in Engineering Design, 1st edn. Springer-Verlag, London (1998). https://doi.org/10.1007/978-1-4471-3020-8
175. Shang, K., Ishibuchi, H., He, L., Meng Pang, L.: A survey on the hypervolume indicator in evolutionary multiobjective optimization. IEEE Trans. Evol. Comput. **25**(1), 1–20 (2021)
176. Smith, R.E., Dike, B.A., Stegmann, S.A.: Fitness inheritance in genetic algorithms. In: SAC 1995: Proceedings of the 1995 ACM Symposium on Applied Computing, pp. 345–350. ACM Press, New York, NY, USA (1995)
177. Srinivas, N., Deb, K.: Multiobjective Optimization Using Nondominated Sorting in Genetic Algorithms. Evolut. Comput. **2**(3), 221–248 (1994)
178. Storn, R., Price, K.: Differential evolution - a simple and efficient heuristic for global optimization over continuous spaces. J. Global Optim. **11**(4), 341–359 (1997)
179. Sülflow, A., Drechsler, N., Drechsler, R.: Robust multi-objective optimization in high dimensional spaces. In: Obayashi, S., Deb, K., Poloni, C., Hiroyasu, T., Murata, T. (eds.) EMO 2007. LNCS, vol. 4403, pp. 715–726. Springer, Heidelberg (2007). https://doi.org/10.1007/978-3-540-70928-2_54
180. Suzuki, N., Okamoto, T., Koakutsu, S.: Visualization of pareto optimal solutions using MIGSOM. In: 2015 IEEE Congress on Evolutionary Computation (CEC'2015), pp. 2556–2564. IEEE Press, Sendai, Japan (25–28 May 2015). ISBN: 978-1-4799-7492-4
181. Tabatabaei, M., Hakanen, J., Hartikainen, M., Miettinen, K., Sindhya, K.: A survey on handling computationally expensive multiobjective optimization problems using surrogates: non-nature inspired methods. Struct. Multidiscip. Optim. **52**(1), 1–25 (2015)
182. Talbi, E.-G., Mostaghim, S., Okabe, T., Ishibuchi, H., Rudolph, G., Coello Coello, C.A.: Parallel approaches for multiobjective optimization. In: Branke, J., Deb, K., Miettinen, K., Słowiński, R. (eds.) Multiobjective Optimization. LNCS, vol. 5252, pp. 349–372. Springer, Heidelberg (2008). https://doi.org/10.1007/978-3-540-88908-3_13
183. Tanabe, R., Ishibuchi, H.: A review of evolutionary multimodal multiobjective optimization. IEEE Trans. Evol. Comput. **24**(1), 193–200 (2020)
184. Tang, W., Liu, H.L., Chen, L., Tan, K.C., ming Cheung, Y.: Fast hypervolume approximation scheme based on a segmentation strategy. Inf. Sci. **509**, 320–342 (2020)
185. Tian, Y., Si, L., Zhang, X., Cheng, R., He, C., Tan, K.C., Jin, Y.: Evolutionary large-scale multi-objective optimization: a survey. ACM Comput. Surv. **54**, 147 (2021)
186. Tian, Y., Zhang, X., Cheng, R., Jin, Y.: A multi-objective evolutionary algorithm based on an enhanced inverted generational distance metric. In: 2016 IEEE Congress on Evolutionary Computation (CEC'2016), pp. 5222–5229. IEEE Press, Vancouver, Canada (24–29 July 2016). ISBN: 978-1-5090-0623-9
187. Trivedi, A., Srinivasan, D., Sanyal, K., Ghosh, A.: A survey of multiobjective evolutionary algorithms based on decomposition. IEEE Trans. Evol. Comput. **21**(3), 440–462 (2017)
188. Tušar, T., Filipič, B.: Visualization of pareto front approximations in evolutionary multiobjective optimization: a critical review and the prosection method. IEEE Trans. Evol. Comput. **19**(2), 225–245 (2015)
189. Vazquez-Rodriguez, J.A., Petrovic, S.: A new dispatching rule based genetic algorithm for the multi-objective job shop problem. J. Heuristics **16**(6), 771–793 (2010)
190. Vrugt, J.A., Robinson, B.A.: Improved evolutionary optimization from genetically adaptive multimethod search. Proceed. Nat. Acad. Sci. United States Am. **104**(3), 708–711 (2007)
191. Wagner, T., Beume, N., Naujoks, B.: Pareto-, aggregation-, and indicator-based methods in many-objective optimization. In: Obayashi, S., Deb, K., Poloni, C., Hiroyasu, T., Murata, T. (eds.) EMO 2007. LNCS, vol. 4403, pp. 742–756. Springer, Heidelberg (2007). https://doi.org/10.1007/978-3-540-70928-2_56

192. Walker, D.J., Keedwell, E.: Multi-objective optimisation with a sequence-based selection hyper-heuristic. In: Proceedings of the 2016 on Genetic and Evolutionary Computation Conference Companion, pp. 81–82. ACM Press, New York, USA (20–24 July 2016)
193. Walker, D.J., Keedwell, E.: Towards many-objective optimisation with hyper-heuristics: identifying good heuristics with indicators. In: Handl, J., Hart, E., Lewis, P.R., López-Ibáñez, M., Ochoa, G., Paechter, B. (eds.) PPSN 2016. LNCS, vol. 9921, pp. 493–502. Springer, Cham (2016). https://doi.org/10.1007/978-3-319-45823-6_46
194. Wang, J., et al.: A survey of decomposition approaches in multiobjective evolutionary algorithms. Neurocomputing **408**, 308–330 (2020)
195. Wang, T.C., Ting, C.K.: Fitness inheritance assisted MOEA/D-CMAES For complex multi-objective optimization problems. In: 2018 IEEE Congress on Evolutionary Computation (CEC'2018), pp. 1013–1020. IEEE Press, Rio de Janeiro, Brazil (8–13 July 2018). ISBN: 978-1-5090-6017-7
196. Wang, Y., Li, B.: Multi-strategy ensemble evolutionary algorithm for dynamic multi-objective optimization. Memetic Comput. **2**(1), 3–24 (2010)
197. Wessing, S., Rudolph, G., Menges, D.A.: Comparing asynchronous and synchronous parallelization of the SMS-EMOA. In: Handl, J., Hart, E., Lewis, P.R., López-Ibáñez, M., Ochoa, G., Paechter, B. (eds.) PPSN 2016. LNCS, vol. 9921, pp. 558–567. Springer, Cham (2016). https://doi.org/10.1007/978-3-319-45823-6_52
198. While, L., Bradstreet, L., Barone, L.: A fast way of calculating exact hypervolumes. IEEE Trans. Evol. Comput. **16**(1), 86–95 (2012)
199. Wismans, L., Berkum, E.V., Bliemer, M.: Acceleration of solving the dynamic multi-objective network design problem using response surface methods. J. Intell. Transp. Syst. **18**(1), 17–29 (2014)
200. Wong, M.L., Cui, G.: Data mining using parallel multi-objective evolutionary algorithms on graphics processing units. In: Tsutsui, S., Collet, P. (eds.) Massively Parallel Evolutionary Computation on GPGPUs. NCS, pp. 287–307. Springer, Heidelberg (2013). https://doi.org/10.1007/978-3-642-37959-8_14
201. Yagoubi, M., Schoenauer, M.: Asynchronous master/slave moeas and heterogeneous evaluation costs. In: 2012 Genetic and Evolutionary Computation Conference (GECCO'2012), pp. 1007–1014. ACM Press, Philadelphia, USA (2012). ISBN: 978-1-4503-1177-9
202. Yang, D., Sun, Y., di Stefano, D., Turrin, M., Sariyildiz, S.: Impacts of problem scale and sampling strategy on surrogate model accuracy. An application of surrogate-based optimization in building design. In: 2016 IEEE Congress on Evolutionary Computation (CEC'2016), pp. 4199–4207. IEEE Press, Vancouver, Canada (24–29 July 2016). ISBN: 978-1-5090-0623-6
203. Yun, Y., Yoon, M., Nakayama, H.: Multi-objective optimization based on meta-modeling by using support vector regression. Optim. Eng. **10**(2), 167–181 (2009)
204. Zapotecas Martínez, S., Coello Coello, C.A.: MOEA/D assisted by RBF networks for expensive multi-objective optimization problems. In: 2013 Genetic and Evolutionary Computation Conference (GECCO'2013), pp. 1405–1412. ACM Press, New York, USA (6–10 July 2013). ISBN: 978-1-4503-1963-8
205. Zavoianu, A.C., Lughofer, E., Koppelstaetter, W., Weidenholzer, G., Amrhein, W., Klement, E.P.: Performance comparison of generational and steady-state asynchronous multi-objective evolutionary algorithms for computationally-intensive problems. Knowl.-Based Syst. **87**, 47–60 (2015)
206. Zhang, Q., Li, H.: MOEA/D: a multiobjective evolutionary algorithm based on decomposition. IEEE Trans. Evol. Comput. **11**(6), 712–731 (2007)
207. Zhang, Q., Liu, W., Tsang, E., Virginas, B.: Expensive multiobjective optimization by MOEA/D with Gaussian process model. IEEE Trans. Evol. Comput. **14**(3), 456–474 (2010)

208. Zhang, X., Tian, Y., Cheng, R., Jin, Y.: A decision variable clustering-based evolutionary algorithm for large-scale many-objective optimization. IEEE Trans. Evol. Comput. **22**(1), 97–112 (2018)
209. Zille, H., Ishibuchi, H., Mostaghim, S., Nojima, Y.: A framework for large-scale multiobjective optimization based on problem transformation. IEEE Trans. Evol. Comput. **22**(2), 260–275 (2018)
210. Zitzler, E.: Evolutionary algorithms for multiobjective optimization: methods and applications, Ph. D. thesis, Swiss Federal Institute of Technology (ETH), Zurich, Switzerland (1999)
211. Zitzler, E., Künzli, S.: Indicator-based selection in multiobjective search. In: Yao, X., et al. (eds.) PPSN 2004. LNCS, vol. 3242, pp. 832–842. Springer, Heidelberg (2004). https://doi.org/10.1007/978-3-540-30217-9_84
212. Zitzler, E., Thiele, L.: Multiobjective evolutionary algorithms: a comparative case study and the strength pareto approach. IEEE Trans. Evol. Comput. **3**(4), 257–271 (1999)
213. Zitzler, E., Thiele, L., Laumanns, M., Fonseca, C.M., da Fonseca, V.G.: Performance assessment of multiobjective optimizers: an analysis and review. IEEE Trans. Evol. Comput. **7**(2), 117–132 (2003)

A Multi-objective Optimization Approach for the Capacitated Vehicle Routing Problem with Time Windows (CVRPTW)

Wissam Marrouche[1(✉)], Haidar M. Harmanani[2], and Janka Chlebíková[1]

[1] School of Computing, University of Portsmouth, Portsmouth, UK
{wissam.marrouche,janka.chlebikova}@port.ac.uk
[2] Computer Science Department, Lebanese American University, Byblos, Lebanon
haidar@lau.edu.lb

Abstract. Investigating complex combinatorial optimization problems such as the capacitated vehicle routing problem with time windows (CVRPTW) commonly benefits the advancement of logistic systems and telecommunications fields. CVRPTW is a elaborate strain of the well studied capacitated vehicle routing problem (CVRP) with a time window framework as an added constraint. As the problems are known to be NP-hard, meta-heuristics proficiently explore the search space to obtain near optimal solutions. In this manuscript, we explore the strength Pareto evolutionary algorithm SPEA2 hybridized with a hill climbing method as a local search, which yields a very sophisticated yet promising population based meta-heuristic algorithm. The novelty of our approach is in the selection of multi objective evolutionary algorithm namely SPEA2, its proposed evolutionary operators, and the optimization for three targets: the total distance traveled, the number of routes, and the average time per route - a newly introduced objective -, which offers an interesting trade-off opportunity scenarios for practitioners or supply chain managers. Based on our experimental results, such multiple objective optimization can guide the search toward favorable reduction of the total distance as well as offering a margin of compromise in the selection of the objectives of the final solution in general. The presented algorithm is a Python implementation tested extensively on the `Solomon's instances benchmark`. Encouraging results are recorded (sometimes reaching the best known in literature) and hints for further refinement are proposed for future research.

Keywords: Capacitated Vehicle routing problem with time windows · Strength pareto evolutionary algorithm · Hill climbing · Multi objective optimization · Evolutionary operators · Meta-heuristics

1 Introduction

The vehicle routing problem (VRP) is one of the most studied combinatorial optimization problem that arises in application logistic systems such as organizing delivery and/or collection of payloads in a geographical region. The advancement brought to

J. Garibaldi et al. (Eds.): IJCCI 2021, SCI 1119, pp. 121–143, 2023.
https://doi.org/10.1007/978-3-031-46221-4_6

the problem can not only benefit to applied operational research or management science [27], but it has a major economic (and environmental) impact [1]. For example, in food and drink industry "added cost" of distribution can contribute up to 70% of intrinsic good cost [7].

The VRP problem [30] can be *stretched* out by imposing constraints on one or more variables of the standard problem. The capacitated vehicle routing problem (CVRP) is the standard VRP, where a limited vehicle capacity is taken into account along with routes optimization. If we relax the multiple route and vehicle capacity constraints, hence all customers must be served by a single route, the CVRP reduces into the *Travelling Salesman Problem*, an NP-hard problem [27].

In this paper we study the CVRP problem with time windows constraints, CVRPTW, similarly to [30]. In this paper, we expand on the work in [22] and consider multi-identical vehicles and single depot while optimizing for some specific objectives that we will describe next. However this work can be extended to multi-depot by partitioning the customers to zones and running the CVRPTW algorithm version in parallel [9].

Exact methods like integer linear programming (ILP) are very computationally demanding to be even considered to solve for the NP-hard CVRPTW problem [6]. Moreover, the time windows introduces a series of precedence on visits, which turns the problem asymmetric, despite the fact that the distance and time matrices were initially symmetric. In practice, heuristics and meta-heuristic methods can be used in order to achieve near-optimal solutions within reasonable amount of computation time. Such approaches provide good solutions for industrial scale problems [27]. We approach the CVRPTW problem utilizing the population-based strength Pareto evolutionary algorithm (SPEA2) as conceived by Zitzler et al. [36] but augmented - or hybridized - - with a hill climbing local search to guide the search toward favorable solutions. More precisely, this work experiments with different genetic operators types such as uniform/fixed point recombination operators and their hybrid combination with other operators, where the genetic operators are followed by hill climbing operator in order to improve the final quality of the solution. Therefore, we also tuned the parameter setting of the proposed operators to reach satisfactory solutions. The advantage of our approach is that it ameliorates the search space toward good solutions by optimizing for three main objectives namely the total distance, the number of routes (equal to the number of vehicles deployed) and the average time per route which is an objective primarily needed to counterbalance the distinct vehicles operation time and route lengths. As a consequence, the proposed algorithmic approach applied to well entrenched CVRPTW benchmarks – and compared to a relatively close existing one – offers a myriad of Pareto optimal/sub-optimal diverse solutions and thus allowing a trade-off analysis in the selection of a desired solution that matches the implementation needs of the domain expert practitioners or supply chain managers.

The remainder of the paper unfolds as follows: Sect. 2 provides a formal introduction of the CVRPTW problem, Sect. 3 lists concisely the VRP related work found in the literature and Sect. 4 gives a comprehensive synopsis of the methodology of our brought forward approach. Section 5 presents and analyzes the simulation results. Lastly, Sect. 6 summarizes the findings and provides suggestions for further research directions.

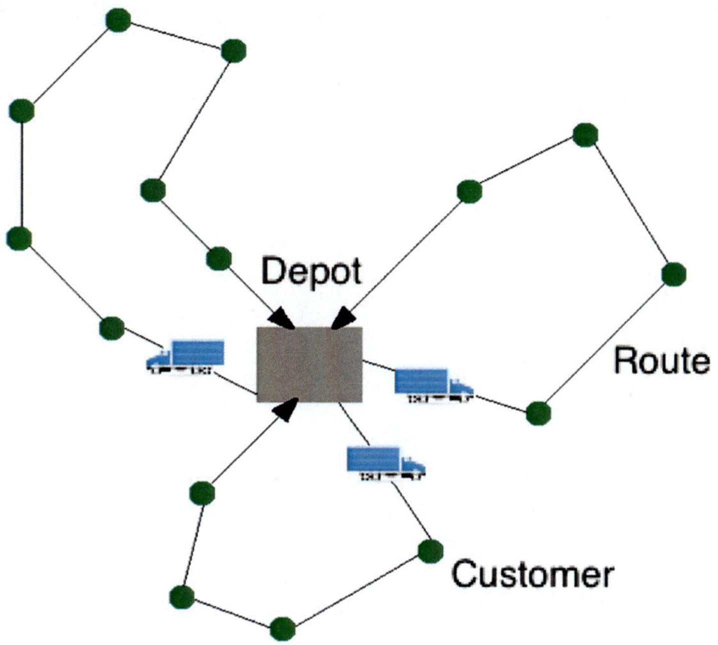

Fig. 1. Typical representative solution of VRP [22].

2 Formal Model

Our formal definition of the CVRTPW problem is based on the one from [16]. The problem formulation presupposes $N + 1$ customers and K vehicles. Customer 0 is positioned at the depot node, and each arc in the network constitutes a link between two customers i, j and determine the direction of travel with travel time t_{ij} and cost c_{ij}. Every route starts from the depot, navigates through a set of customers, and finally goes back to the depot as in Fig. 1. Each out of K vehicles is related to one route in the network and has the same capacity q_k and the demand m_i of every customer must be serviced only one time by one vehicle. The aggregation of the capacity of all the demands on a given route can't exceed q_k capacity of vehicle k embarking on this route.

The time window restrictions for each customer i are inferred by a predefined time interval, an earliest arrival time (e_i), and a latest arrival time (l_i). The vehicle must visit the customer before the l_i; if it reaches before the e_i, then it must stand by. Moreover, a service time f_i for each customer for a given loading and unloading time is taken into consideration. Each vehicle is expected to complete its route within the total route time r_k.

There are further simplifications in the model, as the speed of the vehicle is considered to be unitary; which means that just one unit of time is needed to traverse one unit of distance. Also loading and unloading time to be unique regardless of the amount of the demands.

This is formalized by the following subsequent parameters and variables:

- $x_{ijk} = \begin{cases} 1 \text{ if vehicle } k \text{ travels from customer } i \text{to customer } j \\ 0 \text{ otherwise} \end{cases}$
  $i \neq j; i, j \in \{0, 1, 2, \ldots N\}; k \in \{1, \ldots, K\}$
- K number of vehicles
- N number of customers (0 denotes the central depot)
- c_{ij} cost incurred in arc from customer i to customer j
- t_{ij} travel time between customer i and customer j
- m_i demand at customer i
- q_k capacity of vehicle k
- e_i earliest arrival time at customer i
- l_i latest arrival time at customer i
- f_i service time at customer i
- t_i arrival time at customer i
- w_i waiting time at customer i for the service operation to be ready to start
- r_k maximum route time for vehicle k

Based on these parameters, the goal of the CVRPTW problem is to minimise the objective function

$$\min \sum_{i=0}^{N} \sum_{j=0, j \neq i}^{N} \sum_{k=1}^{K} c_{ij} x_{ijk} \tag{1}$$

with subject to the following constraints:

$$\sum_{k=1}^{K} \sum_{j=1}^{N} x_{0jk} \leq K \quad \text{(maximum } K \text{ vehicles leave the depot)} \tag{2}$$

$$\sum_{j=1}^{N} x_{0jk} = \sum_{j=1}^{N} x_{j0k} \leq 1 \quad \text{for } k \in \{1, 2, \ldots, K\} \tag{3}$$

(every route starts and terminates at the depot)

$$\sum_{k=1}^{K} \sum_{j=0, j \neq i}^{N} x_{ijk} = 1 \quad \text{for } i \in \{1, 2, \ldots, N\} \tag{4}$$

$$\sum_{k=1}^{K} \sum_{i=0, j \neq i}^{N} x_{ijk} = 1 \quad \text{for } j \in \{1, 2, \ldots, N\} \tag{5}$$

(4 and 5 ensure that each customer is served by exactly one vehicle)

$$\sum_{i=1}^{N} m_i \sum_{j=0, j \neq i}^{N} x_{ijk} \leq q_k \quad \text{for } k \in \{1, 2, \ldots, K\} \tag{6}$$

(capacity of each vehicle can't be exceeded)

$$\sum_{i=1}^{N} \sum_{j=0, j \neq i}^{N} x_{ijk}(t_{ij} + f_i + w_i) \leq r_k \quad \text{for } k \in \{1, 2, \ldots, K\} \tag{7}$$

(travel time constraints for each vehicle can't be exceeded)

$$t_0 = w_0 = f_0 = 0 \tag{8}$$

$$\sum_{k=1}^{K} \sum_{i=0, j \neq i}^{N} x_{ijk}(t_i + t_{ij} + f_i + w_i) \leq t_j \quad \text{for } j \in \{1, 2, \ldots, N\} \tag{9}$$

$$e_i \leq (t_i + w_i) \leq l_i \quad \text{for } i \in \{1, 2, \ldots, N\} \tag{10}$$

(constraints (8), (9), and (10) are related to the times window)

$$c_{ij} = \sqrt{(i_x - j_x)^2 + (i_y - j_y)^2} \tag{11}$$

Equation 11 computes the travel cost, where the Euclidean i_x is the x coordinate and i_y is the y coordinate of customer i.

All these parameters are considered in `Solomon`'s 1987 model, and `Solomon`'s 56 100-customers instances of CVRPTW model that we'll be using for comparison of our results.

In this paper we look on the correlation of three objectives: total distance, number of routes, and average time per route concurrently in attempt to find the best solution for the single depot CVRPTW in a reasonable time. Reducing the average time per route is an important aspect to consider from practical point of view as it decreases the driver cost time and balances the routes lengths.

3 Various Approaches to the Problem

Heuristics such as simulated annealing and its improvements [4,35], together with an iterated local search [35] indicate that one can get 'sufficiently good' solutions of the CVRPTW problem within an reasonable amount of time. This approach is based on a limited exploration of the search space quickly, while meta-heuristics abide by the principles of intensification and diversification in their search [27]. Population based meta-heuristic algorithms work well for global searches offering global exploration and local exploitation capability over simple heuristics. In this context, the population-based approach with several evolutionary stages and swarm intelligence-based algorithms like particle swarm optimization, ant colony optimization, artificial bee colony algorithm can outperform simple heuristics when solving VRPTW [8]. Another bio-inspired meta-heuristics applied to the basic CVRP problem is the cuckoo search algorithm which mimics the nesting behavior of the cuckoo bird. The advantage of cuckoo search is that it solves CVRP in an acceptable time [34]. Cuckoo search solves efficiently even more elaborate VRP problems such as the VRP with heterogeneous fleet, mixed backhauls (pickup and delivery) and time windows [23]. Also, the bat algorithm (BA) is a new bio-inspired meta-heuristic based on the echolocation behavior of microbats when probing for their prey in nature. In particular, the BA simulation results show that this approach has a satisfactory performance in solving VRPTW instances [28]. There are also other bio-inspired approaches considered in literature such as artificial immune systems and neural networks. For the VRP problem many contributions have been made using hybrid algorithms: using a greedy method for an initial solution followed by tabu search [31] or based on ant colony optimization algorithm [27] with a large neighborhood search (LNS) algorithm [24]. In general, meta-heuristics have the tendency to yield good results when hybridized with local search methods [8, 27], adding specific problem intelligence in the enhancement of the solutions. A hybrid genetic algorithm was proposed in [32] for multi-depot and periodic VRP. A similar published work to our studied problem can be found in [3] and [14], which implements multi objective vehicle routing problem with time windows and optimizes for two concurrent objectives, namely the total distance and the number of vehicles to obtain the set of Pareto optimal solutions. In [14], enhanced reproduction operators are proposed for the CVRPTW problem. An even closer work to this paper is found in [29], which implements a hybrid multi objective evolutionary algorithm (MOEA) accommodating the sequence-oriented optimization. Also [11] propose an improved MOEA for the VRPTW optimising mainly for two objectives: the number of routes and the total distance traveled. It also introduces as a third objective the completion or delivery time but does not report in its table result the three objectives explicitly but it rather provides a coverage performance metric analysis, which makes it hard to compare it directly to our three objectives results.

The major contribution of this paper is the introduction of strength Pareto evolutionary algorithm SPEA2 to solve for the CVRPTW problem. Also, SPEA2 is improved with a local search namely hill climbing operator to reduce the total distance traveled and therefore ensure a fast convergence. In fact, [36] introduced SPEA2 as an improvement to SPEA. The advantage of the SPEA2 algorithm is that it is an effective evolutionary population search algorithm for obtaining Pareto optimal solutions. SPEA2 is more consistent than NSGA-II in terms of diversity and convergence at the expense of computational complexity [17]. In this work, SPEA2 optimizes for three objectives (total distance, number of routes, and average time per route) concurrently to solve the single depot CVRPTW. This paper also proposes its own evolutionary operators similarly to the another paper [13] where tailored intelligent neighborhood moves to the genetic operators and guides the search toward optimal results irrespective of the quality of the initial solution. Promising results were achieved using the new approach which makes it a good reference for comparison in the presence of three objectives.

4 Our Approach

Variants of local search are the preferred heuristics for otherwise complex problems such as graph partitioning and the vehicle routing [12]. Our proposed methodology tackling the CVPRTW problem is based on the hybrid SPEA2-local search algorithm as detailed in the following subsections:

4.1 Setting the Input Parameters

Based on the extracted information from the `Solomon`'s benchmark 100-customer 56 instances (1987) (see [8] for more details), we populate two data structures namely: Nodes list $Nodes$, and the graph matrix G about the coordinates (i_x, i_y) of the customer i to retrieve the cost $c_{i,j}$. Each node in $Nodes$ contains information related to the customer i: demand m_i, earliest arrival time e_i (opening time), service time f_i and due date of the delivery l_i (closing time). The scheduling has to enforce that the arrival time t_i to a customer i lies within the time window $[e_i, l_i]$. As for the arrival time t_i, the waiting time w_i for the service operation to be opened, and the travel time between two nodes $t_{i,j}$ are computed dynamically during the scheduling process.

A typical solution S to our studied problem is represented as an object with the following fields: an array of routes where each $route_k$ contains the sequence of visited customers, an average time per route, and a total traveled distance. In fact, the minimization of average completion time per route is linked to work-in-progress inventory management [18] which can yield balanced routes lengths.

4.2 Building the Initial Population Using a Greedy Approach

We generate a feasible pool of initial solutions using a *greedy algorithm* (random seeds, i.e. each time we visit the nodes in a different random order). The algorithm loops to visit all nodes - in the random order - that are not previously assigned to any given route until all nodes are assigned to routes. The starting solution has an empty list of routes. In each iteration we create a new route and add to it all customers i, $(i \in \{1\ldots, N\})$ that are not allocated to any route so far keeping all the constraints valid. Our proposed greedy algorithm ensures that the obtained greedy solution meet the constraints: the vehicle capacity constraint and the time windows restrictions. A validate function is implemented to ensure that the solution is feasible i.e. meets the aforementioned constraints. Here, the validate function does not check if the total number of allocated vehicle surpasses the fleet size for a given greedy solution and thus considers the generated greedy solution as valid. The check for the number of used vehicles is postponed till the end of the overall simulation as in the process of improvement of the initial solution, the number of vehicles may decrease. If at the end of the simulation a solution with a fleet size violation is obtained then such solution will be discarded. In our `Solomon's` benchmarks simulations, the fleet size violation never occurred as we see in the presentation of our results in Sect. 5.

We generate a random pool of initial solutions relying on the previously defined greedy algorithm. The challenge is now to advance the search starting from the obtained inception state in a proficient meta-heuristic approach that we base on SPEA2 hybridized with local search. In the following we define a few evolution operators that are building blocks for the overall approach that will be described in details.

4.3 Tweak Operator

The *tweak operator* produces each time a slightly randomly different candidate solution. It is also called a modification procedure that makes search exploitative by exchanging two customers position either inter route insertion swap or intra routes insertion exchange.

More precisely, there are two options for the tweak operators, let S be an initial solution: (a) two randomly selected nodes in S are within the same route – in that case we just swap two nodes inside the route; (b) two randomly selected nodes i and j are in different routes namely the route k_i and route k_j. In this case we firstly move the node i from the route k_i to the route k_j and insert the node i after the node j in the route k_j. This modification S to S^* can decrease the number of routes as after applying it, the route k_i might become empty. In such case we simply remove it from S^*. Using the validate operator, we check a feasibility of the solution: if it is valid we return S^*; otherwise we roll back to the initial solution S. Note that S^* may not be necessarily better than S. We repeat the tweak operation and if no valid S^* is found after a fixed number of trials, we return the initial solution S. In our implementation, we have repeated the trials maximum 9 times.

4.4 Recombination Operator

The *recombination operator* is an "asexual recombination" as illustrated in Fig. 2. One of its advantages is that it explores the search space and avoids the fall into a local minima. The implementation is given as follows:

Given an initial solution S, we select randomly two different routes –or paths– in the solution and perform a fixed-point recombination from the middle of each route, which results in two new offspring routes that can be optionally merged together using a uniform combination. The resulting solution, S^*, is verified for feasibility using the validate operator. If the solution is not feasible then the operator is repeated for a number of times that is equal to the number of routes in S plus a bias value that we tune empirically. At any moment the trial can terminate if S^* is feasible. Otherwise, the solution S is kept.

Recombination Operation

1: Apply fixed-point recombination after determining the middle of each route
  Parent Route 1: {0-6-7-3-2-8-10-0}
  Parent Route 2: {0-4-1-5-9-0}
2: Apply the fixed-point recombination
  Offspring Route1: {0-6-7-3-5-9-0}
  Offspring Route2: {0-4-1-2-8-10-0}
3: We then optionally mesh the offspring randomly to obtain multiple point uniform recombination
  Offspring Route1: {0-4-7-2-5-10-0}
  Offspring Route2: {0-6-1-3-8-9-0}

Fig. 2. Recombination Example: Fixed Point and Uniform [22].

4.5 Fuse Operator: Naïve Merge

The *fuse operator* is also an asexual recombination merging two randomly selected routes of a given solution by naïvely queuing one route after the other to create one path or one offspring route that will be added to the list of routes while removing the two selected parent routes from the solution. However, this move is only efficacious at the beginning of the meta-heuristic deployment. In fact, at early stages it is highly probable that there exists a room for improvement over the initial pool of solutions. However in general, this operator will not improve the existing solution at later stages. This is the reason why we only apply the operator with a low probability in order not to slow down the simulation without improvement. Its implementation details can be described as follows:

Given an initial solution S, we create a new solution S^* in which two random parent routes in S are joined to form a new route: the last non-depot node of the first randomly selected route is connected with the first non-depot vertex of the second randomly selected route and the two parents random routes are moved from S^*. The solution S^* is checked using the validate operator; if it is a valid one, we return S^*. Otherwise we rollback and repeat the trial a number of times equal to the initial total number of routes in S plus a bias or we stop earlier, if generated S^* is valid. If the trial fails, we return the initial solution S.

4.6 Tuning Operator

The *tuning operator* is applied after the genetic operators and helps in performing a local search in order to converge towards an optimal solution or in the worst case return the same initial solution if no improvement is obtained. In fact, the steepest ascent hill climbing with replacement operator [20] is used to reduce the total cost (distance traveled) of a given offspring solution; in other words, to force the evolution of population downhill for this objective. Thus, each time the tuning operator is invoked, it performs either a tweak or a recombination (mutually exclusive) operation with a certain probability to be determined empirically. The rational behind this can be justified by the fact that we need to switch back and forth from one type of operator to another in order to escape local minima, see Fig. 3.

Steepest Ascent Hill Climbing

```
n ← 25                                        ▷ Number of attempts to sample the gradient
TweakRate ← 0.8                               ▷ In this case RecombinationRate is 0.2
CurrentSol ← Some initial candidate solution
repeat
    NewSol ← Tweak(CurrentSol, TweakRate)
    for i = 0 to n -1 do
        Temp ← Tweak(CurrentSol,TweakRate)
        if fitness(Temp) > fitness(NewSol) then     ▷ fitness is the distance
            NewSol ← Temp
    if fitness(NewSol) > fitness(CurrentSol) then
        CurrentSol ← NewSol
until n iterations reached
return CurrentSol
```

Fig. 3. Tuning Operator as a Local Search.

4.7 SPEA2 Fitness Computation and Archive Construction

Our approach to attack the CVRPTW problem is based on the three objectives functions that need to be minimized concurrently, namely the total distance of routes D which impacts the fuel cost, the number of routes K which impact the number of vehicles needed, and the average time per route T_{avg} which impacts the driver's working time.

Formally, and based on Eq. 1 the total distance D is given as in Eq. 12:

$$D = \sum_{i=0}^{N} \sum_{j=0, j \neq i}^{N} \sum_{k=1}^{K} c_{ij} x_{ijk} \tag{12}$$

and based on Eq. 7 the average time per route is equal to:

$$T_{avg} = \sum_{k=1}^{K} \sum_{i=1}^{N} \sum_{j=0, j \neq i}^{N} x_{ijk}(t_{ij} + f_i + w_i)/K. \tag{13}$$

In general, the *fitness function* is very important as it helps guiding the algorithm towards the local optima [36]. In our case, the fitness function is a vector-valued function that rely on the Pareto dominance concept and is defined as:

$$f = (D, K, T_{avg}) \tag{14}$$

Following the definition from [37], an objective vector f_1 dominates another objective vector f_2 ($f_1 \succ f_2$) if no component of f_1 is smaller than the corresponding component of f_2 and at least one component is greater.

To each ach individual solution A_i in the archive A (as introduced in Fig. 4) as well as P_i in population P (refer to Fig. 5) is given a strength value, $Strength(i)$, denoting the number of solutions it dominates as described by Eq. 15.

$$Strength(i) = \sum_{i,j \in P \cup A} [i \succ j], \tag{15}$$

where the bracked notation $[i \succ j]$ is 1 if $i \succ j$ is true and 0 otherwise. Based on this definition, the raw fitness $f(i)$ of an individual solution i is determined as follows [21]:

$$f(i) = \sum_{i,j \in P \cup A, j \succ i} Strength(j). \tag{16}$$

The raw fitness is evaluated by the strengths of its denominators in the archive and in the population, where the strength of a solution is defined as the number of other solutions in the current population that it dominates. Possible ties are resolved using a nearest neighbor density estimation function, $Density(i)$.

$$Density(i) = \frac{1}{\sigma_i^k + 2} \tag{17}$$

where σ_i^k is the Euclidean distance of the objective values between a given solution i and the k_{th} nearest neighbor of the solution, where $k = \sqrt{(M_P + M_A)}$, where M_P is the population size and M_A is the archive size. For every individual in the population, we sort the population by distance to that individual, and take the k_{th} closest individual. The algorithm has a runtime complexity of $O(M_P^2 \log M_P)$ as described in [21].

SPEA2 Archive

```
P ← {P1, . . .,Pm}                                                      ▷ population
O ← {O1, . . .,On}                                          ▷ objective to assess with
a ← desired archive size
A ← Pareto non-dominated front of P                                    ▷ the archive
Q ← P - A
if |A| < a then
    Sort Q by fitness
    A ← A∪ the a − |A| fittest individuals in Q, breaking ties arbitrarily
while |A| > a do
    Closest ← A1
    c ← index of A1 in P
    for all individual Ai ∈ A except A1 do
        l ← index of Ai in P
        for k = 1 to m-1 do
            if DistKthNearest(k, Pl) < DistKthNearest(k, Pc) then
                Closest ← Ai
                c ← l
                break
            else
                if DistKthNearest(k, Pl) > DistKthNearest(k, Pc) then
                    break
        A ← {Closest}
return A
```

Fig. 4. SPEA2 Archive Construction Algorithm [20].

4.8 SPEA2 Algorithm

The *strength pareto evolutionary algorithm* (SPEA2) is a multiple objective evolutionary optimization algorithm that localizes and preserves a front of non-dominated solutions, preferably a set of Pareto optimal solutions. We achieve this goal by constructing an archive as in Fig. 4 and relying on an evolutionary routine as introduced in Sect. 4.9. We propose the deployment of the following SPEA2 algorithm pseudo code in Fig. 5 based on [20].

4.9 Evolve Operator

SPEA2 requires an evolution mechanism, which we base on the following four operators: tweak, recombine, fuse, and tuning. The algorithm applies these operators by taking turn, and using probabilities that are determined empirically (Sect. 5.1). Together,

SPEA2 Algorithm

```
m ← desired population size
a ← desired archive size                                           ▷ typically a = m
P ← {P1, . . .,Pm}            ▷ build initial population using a greedy method Section 4.2
repeat
    AssessFitness(P)
    P ← P ∪ A
    BestFront ← Pareto Front of P
    A ← Construct SPEA2 Archive of size a from P
    P ← Evolve(A), using tournament selection of size 3
until number of generations reached
return BestFront
```

Fig. 5. SPEA2 Algorithm Pseudo Code based on [20].

the diverse and comprehensive set of operators exploit and explore the search space while escaping the local minima.

5 Summary of Our Experimental Results

5.1 Hyperparameters Tuning

Evolutionary algorithms are stochastic in nature and thus very sensitive to hyperparametres. Therefore, one of the main tasks was tuning the problem's parameters. Thus, fixed-point and uniform recombination and we selected the better of the two reported solutions. The experiment was repeated 18 times for each recombination type (fixed point and uniform) and resulted with the identification of two main configurations groups. Table 2 shows the problem's parameters such as the bias is not null which makes the search operators run more iteration and thus improve solutions through intensive neighborhood search (i.e. more exploitative iterative improvement for local searches at the expense of execution time of the whole meta-heuristic solution) which justifies reducing number of generation for our meta-heuristic setting. Where as Table 1 is more exploratory i.e. the number of generation parameter is higher since a bias of zero implies less number of iterations for the deterministic operators, thus pushing our method to aggressively explore the search space from a higher-level framework perspective.

Exceptionally, for the R1 benchmark some obtained results were better using the first configuration while for others the second configuration showed better results, so we ended up choosing the better solution among the two. As for hill climbing the distribution of tweak operator ratios are: 0.8 tweak and 0.2 recombination (to escape local minima). All the configuration parameters were determined empirically relying on a step wise approach, i.e. changing one parameter at a time while fixing others.

Table 1. Parameters settings of SPEA2 for C1, RC1 benchmarks [22].

Parameter	# Gen	Tweak	Recombine	Fuse	popsize	HC	Bias
Values+	260	0.8	0.4	0.1	200	25	0

Table 2. Parameters settings of SPEA2 for C2, RC2 benchmarks [22].

Parameter	# Gen	Tweak	Recombine	Fuse	popsize	HC	Bias
Values*	150	0.8	0.4	0.1	200	25	10

Table 3. Results of Hybrid SPEA2 Applied to Solomon's 100-customer benchmark [22].

Benchmark	Ours			BCO	
	Distance	Routes	Avg Time	Distance	Routes
C101	**828.94**	10	982.89	828.94	10
C102	864.04	10	987.26	**828.94**	10
C103	**828.94**	10	982.89	835.71	10
C104	**824.78**	10	995.40	885.06	10
C105	866.23	11 ▽	916.78	**828.94**	10
C106	833.24	10	983.32	**828.94**	10
C107	836.76	10	983.67	**828.94**	10
C108	831.98	10	983.20	**831.73**	10
C109	**828.94**	10	982.89	840.66	10
C201	**591.56**	3	3197.19	591.56	3
C202	**591.56**	3	3197.19	591.56	3
C203	659.72	4 ▽	2992.22	**593.21**	3
C204	667.84	4 ▽	2561.31	**606.90**	3
C205	**588.88**	3	3196.29	588.88	3
C206	**588.49**	3	3196.16	588.88	3
C207	**588.29**	3	3220.13	590.59	3
C208	**588.49**	3	3196.16	593.15	3
R101+	1669.81	20 ▽	180.85	**1643.18**	20
R102*	1499	18	184.37	**1476.11**	18
R103+	**1240.08**	15 ▽	196.46	1245.86	14
R104+	**1026.47**	12 ▽	201.04	1026.91	11
R105+	1391.59	16 ▽	179.46	**1361.39**	15
R106+	**1261.06**	14 ▽	179.95	1264.50	13
R107+	1127.84	12	195.12	**1108.11**	11
R108+	**990.62**	10	205.77	994.68	10
R109+	1197.33	12	193.51	**1168.91**	13
R110+	1125.04	13 ▽	182.92	**1108.22**	12
R111+	1090.52	12	195.66	**1080.84**	12
R112*	**982.91**	11	193.94	992.22	10
R201	1260.58	6 ▽	766.38	**1197.09**	7
R202	1097.58	6 ▽	837.05	**1092.22**	6
R203	**962.40**	6 ▽	730.76	983.06	5
R204	**807.29**	4 ▽	695.79	845.30	4
R205	1036.02	5	719.31 ▽	**999.54**	5
R206	**941.09**	4 ▽	752.71	955.94	4
R207	**892.61**	4 ▽	715.75	903.59	4
R208	805.50	3	688.20	**769.96**	3
R209	938.48	5 ▽	640.27	**935.57**	5
R210	**981.40**	5 ▽	793.38	988.34	6
R211	**854.82**	5 ▽	630.22	867.95	3
RC101	1682.02	17 ▽	192.15	**1637.40**	15
RC102	1532.44	15 ▽	196.52	**1486.85**	14
RC103	1356.21	12	210.34	**1299.38**	12
RC104	**1190.46**	11	210.5	1200.60	10
RC105	1591.59	16 ▽	198.98	**1535.80**	15
RC106	1441.08	14 ▽	198.36	**1403.07**	13
RC107	1291.70	13 ▽	197.03	**1230.32**	12

continued

Table 3. continued

Benchmark	Ours			BCO	
	Distance	Routes	Avg Time	Distance	Routes
RC108	**1156.92**	11	212.29	1165.17	11
RC201	1345.36	6 ▽	748.61	**1315.57**	7
RC202	1174.23	6 ▽	716.13	**1169.72**	5
RC203	1038.97	5 ▽	777.83	**1010.74**	4
RC204	**883.45**	4 ▽	712.01	890.28	4
RC205	1237.81	7 ▽	721.89	**1221.28**	7
RC206	1147.15	6 ▽	702.32	**1097.65**	6
RC207	1061.85	5 ▽	672.37	**1024.17**	5
RC208	865.07	5 ▽	612.30	**864.56**	4

[a] We list our best solution among the recombination methods (fixed and uniform)
[b] ▽ a solution exists where the number of routes is lower than the reported given a larger total distance

5.2 Results Analysis

We implemented the proposed hybrid SPEA2 algorithm using Python and experimented with various benchmarks on an Intel Xeon CPU E5-2650 with 16 cores. We attempted the `Solomon's benchmark` and received favorable results as shown in Table 3. For each benchmark, we report the solution with the best Pareto optimal answer belonging to one of the 36 Pareto fronts in all 36 runs. Although there are results in literature with lower routes cost, none of the reported solutions optimized for these three objectives: distance, average time per route, and number of vehicles simultaneously.

For each instance, one run took on average around one day, which is not unexpected given the problem's intractability and the complexity of dealing with optimizing three constraints. Thus, it is difficult to compare to reported works. Nonetheless, to have a meaningful reference to compare achieved results, we matched our results with the two objective based CVRPTW problem. In fact, if we compare our solution with the absolute best reported results for two objectives in literature [2] and [14], we obtain either optimal or sub optimal solutions. But for more practical matter we compared our results to the bee colony optimization (BCO) algorithm introduced by [16] where we identified a noticeable improvement in 43 % of the cases as in Table 3.

The overall performance of our solution is good since with few configurations we were able to get good results for all benchmarks. In general, the configuration of parameters in Table 2 (which is more exploitative) gave better solutions than the Table 1 (which is more exploratory) if the best reported number of routes is relatively small (below 9 routes). On the other hand, the first configuration gave better results than the second configuration mainly in case when the number of expected routes is high (based on BCO algorithm findings as a reference). Some exceptions were observed where the second configuration gave better results for given benchmarks of type R1 (refer to Table 3, parameter type */+ for a given R1 benchmark).

Finally, it should be noted that while executing all the simulations, we noticed that the second configuration is faster in converging to a desired solution than the first one. Benchmarks C1, R1 and RC1 have a relatively short scheduling horizon and permit only a few customers per route (around 5 to 10) as in Fig. 6. On the contrary, the benchmarks R2, C2 and RC2 have a long scheduling horizon permitting many customers (sometimes more than 30) to be visited by the same vehicle as in Fig. 7. This is in complete match with the literature [25].

Figure 8, Fig. 9, Fig. 10 and Fig. 11 show the variance analysis of two benchmarks namely `Solomon`_100 R108 and RC201 with fixed and uniform recombination operator. We picked two benchmarks belonging to two different families of benchmarks. The box plot depicts the total distance variation at each run (18 times i.e. 18 Pareto sets) of a given benchmark (given uniform or recombination operator option). Thus, the experiment is repeated 36 times (2×18 times) for each benchmark. The overall distribution of the total distance data and its skewness over the instances are fairly close, which indicate that the experiment is repeatable. Also, we can notice that the R108 benchmark is slightly less repeatable than the R201 benchmark. Moreover, no major difference in the distribution and skewness is noticed between fixed and uniform recombination options.

RC102 Sample Solution
Total Distance : 1532.44
Average Time Per Route : 196.52
Route 1 (Time= 231.32) : 0-92-50-27-29-31-34-93-94-80-0
Route 2 (Time= 219.50) : 0-83-19-23-18-48-21-25-24-0
Route 3 (Time= 237.41) : 0-39-36-40-38-41-43-35-37-72-0
Route 4 (Time= 226.72) : 0-98-73-79-7-46-4-2-100-70-0
Route 5 (Time= 235.49) : 0-33-26-28-30-32-89-0
Route 6 (Time= 228.24) : 0-65-52-99-57-86-74-77-0
Route 7 (Time= 167.06) : 0-64-22-49-20-0
Route 8 (Time= 100.24) : 0-90-0
Route 9 (Time= 133.83) : 0-42-44-61-81-0
Route 10 (Time= 175.13) : 0-85-63-76-51-84-56-66-0
Route 11 (Time= 218.26) : 0-69-88-53-78-17-13-0
Route 12 (Time= 180.00) : 0-1-3-45-5-8-6-55-68-0
Route 13 (Time= 191.93) : 0-12-14-47-16-15-9-10-60-0
Route 14 (Time= 222.19) : 0-82-11-87-59-97-75-58-0
Route 15 (Time= 180.52) : 0-91-95-62-67-71-54-96-0

Fig. 6. RC102 Typical Solution.

RC202 Sample Solution
Total Distance : 1174.23
Average Time Per Route : 716.13
Route 1 (Time= 894.16) : 0-91-95-85-63-33-34-31-29-27-26-28-30-32-89-24-25-77-58-52-0
Route 2 (Time= 929.12) : 0-69-98-88-53-99-57-86-87-9-10-59-97-75-74-13-17-60-100-70-96-93-94-80-0
Route 3 (Time= 775.57) : 0-82-14-47-16-15-11-12-78-73-79-7-6-8-46-4-43-35-37-0
Route 4 (Time= 610.80) : 0-65-83-64-19-23-21-48-18-76-51-84-49-22-20-56-66-0
Route 5 (Time= 670.14) : 0-2-45-5-3-1-42-39-36-44-40-38-41-72-54-68-55-0
Route 6 (Time= 417.00) : 0-92-62-50-67-71-81-61-90-0

Fig. 7. RC202 Typical Solution.

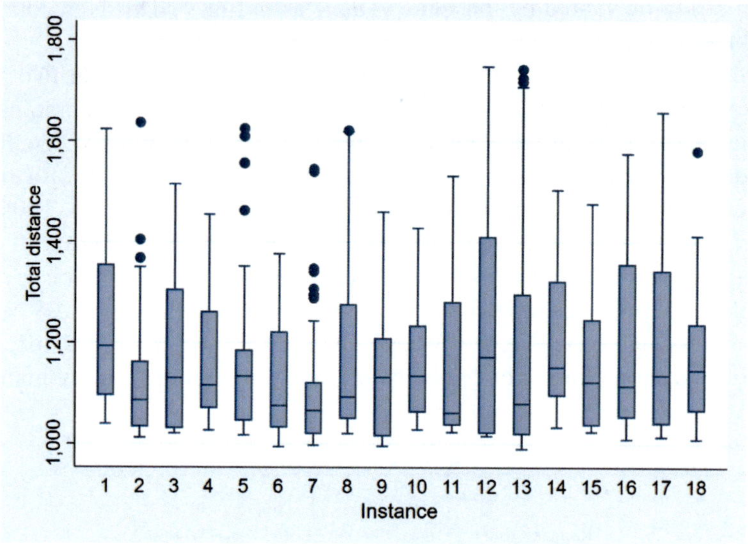

Fig. 8. Variance analysis for Solomon_100 R108 with a uniform recombination operator [22].

For illustration purposes, we select the Pareto front (one out of the 36 instances) containing the lowest total distance point for the benchmark C201 (size 100). A typical Pareto front surface (with 3 objectives) is shown in Fig. 12, where we apply a triangulation to the Pareto points and plot it as a triangular surface. Here in this example, the number of routes varies between 3 and 19, the total distance between 591.56 and 3840, and the average time between 1759.43 and 3197.19. The standard deviation for

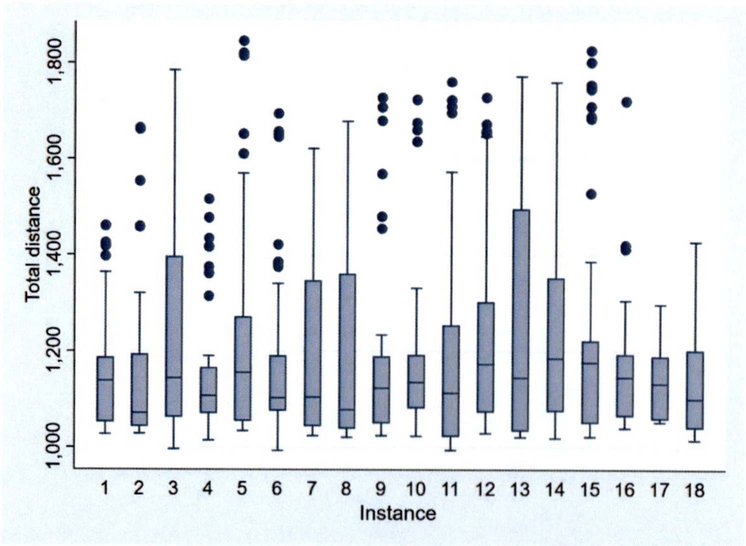

Fig. 9. Variance analysis for Solomon_100 R108 with a fixed recombination operator [22].

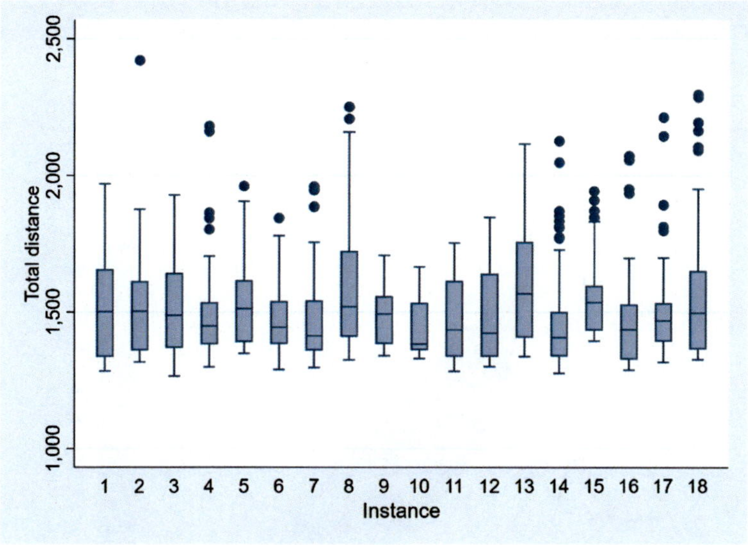

Fig. 10. Variance analysis for Solomon_100 R201 with a uniform recombination operator [22].

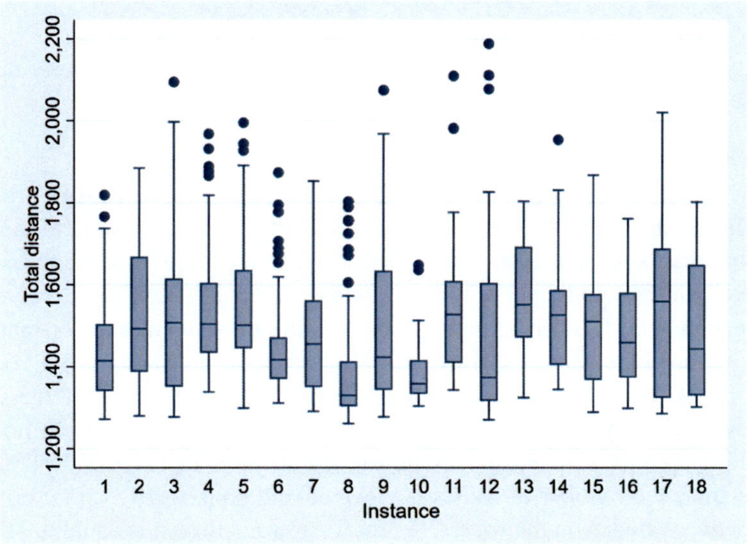

Fig. 11. Variance analysis for Solomon_100 R201 with a fixed recombination operator [22].

total distance is 888.18, for the number of routes it is 4.06 and for the average time per route it is 281.64. In general, we observe a positive correlation between the number of routes and the total distance and an inverse correlation between the number of routes and average time. The hypervolume (HV) indicator is a measure for comparing Pareto sets in term of the closeness and the spread of the solutions with respect to the Pareto front. For Fig. 12, a good value of $HV = 0.85$ is obtained given a reference point $r = 1.3$ (30% larger than the nadir point) [15].

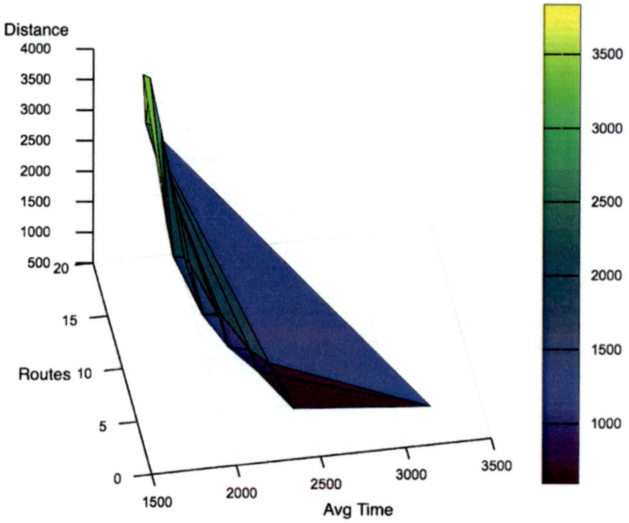

Fig. 12. A visual representation sample of the Pareto front surface obtained for Solomon_100 C201 [22].

Lastly, the visualization of the obtained solutions is very helpful in presenting the final results. In this context, Fig. 13, Fig. 14, and Fig. 15 are topological maps of sample benchmarks (with size 100). More explicitly, C104 has a total distance of 824.78, a number of routes equals to 10, and an average time of 995.40. In the comparison of these 3 benchmarks, C104 is situated in the middle in terms of total distance, number of routes and average time. C201 has the shortest total distance of 591.56, a number of routes equals to 3, and average time of 3197.19. Clearly, in C201 we can see the reduced number of routes and the short distance they cover since the nodes on the routes are juxtaposed. Whereas R103 has the highest total distance of 1240.08 which is justifiable since the number of routes is elevated and equals to 15 and it covers higher distance between nodes on the routes. As for its average time it is equal to 196.46. Concerning the average times, not much can be inferred from the plots since the problem is asymmetric but one can clearly reiterate the inverse correlation between the number of routes and the average time.

6 Conclusions

In the logistic realm, it is incredibly hard task to determine and integrate the most optimized routing schedule given the different imposed constraints or objectives that might even be conflicting each other in some cases. To find a satisfyingly good solution for such an optimization problem, the meta-heuristics seem to fill in this trad-off gap. The

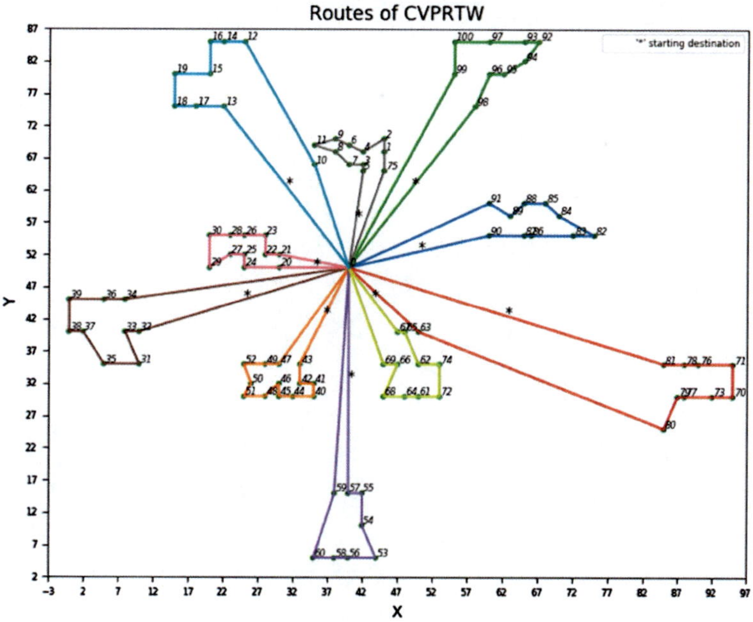

Fig. 13. Network topology for Solomon_100 C104: distance = 824.78, routes = 10, time = 995.40 [22].

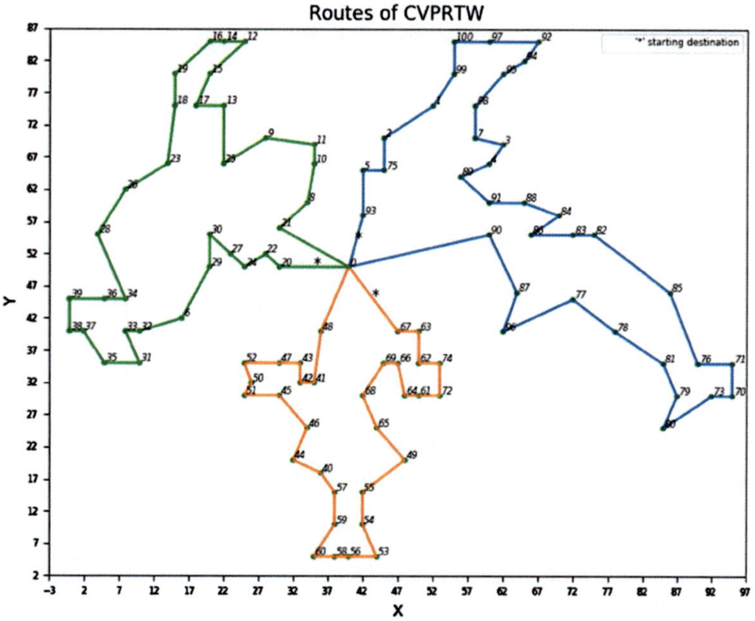

Fig. 14. Network topology for Solomon_100 C201: distance = 591.56, routes = 3, time = 3197.19 [22].

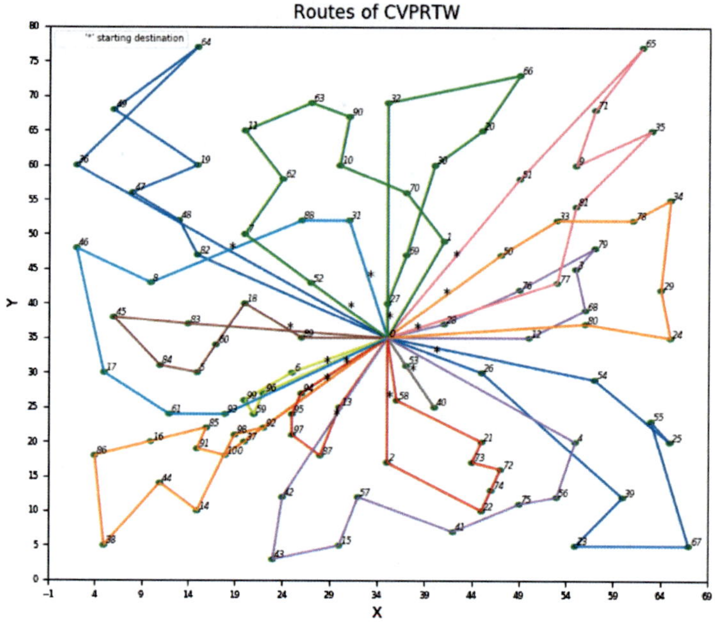

Fig. 15. Network topology for Solomon_100 R103: distance = 1240.08, routes = 15, time = 196.46 [22].

VRP problem became the major problem of E-commerce retailers and its significance even increased during the recent pandemic which amplifies the importance of our contribution to a proposed solution.

In this work, we propose a novel meta-heuristic approach using a hybrid SPEA2 augmented with a hill climbing as a local search tuning operator to solve for the CVRPTW problem by optimizing for three objectives: the total distance traveled, the number of routes and the average time per route. The last objective is well motivating in order to reduce driver cost time and balance the routes.

Satisfactory results are reported when running the widely studied `Solomon's 56 100-customer instances benchmark` (a 43% improvement when compared to a main stream bi-objective BCO algorithm). In fact, considering three objectives improves globally the best two standard objectives since it guides the meta-heuristic search toward favorable search areas. We checked our proposed algorithm for the bi-objectives case and the results do not rebut this claim. On the other hand, the local search is necessary to ensure quality results despite the latency introduced. The complexity added to the solution in order to improve quality of the solution is in accordance with the paradox: quality results versus speed of convergence [26]. As for the simplicity measure, the algorithm is not very hard to configure across different type of classes for this benchmark. Concerning the flexibility of the proposed algorithm, future work should examine scaling this algorithmic solution to handle very large benchmarks or adapt dynamically to changing objectives or deploying it on different variants of VRP such as: the green version of the CVRPTW that accounts for carbon emissions [19], or

VRP with soft timing constraints, which if violated must account for a penalty. Ambitious future work can also account for traffic, i.e. higher cost routes in city centers [27] in what is known as time dependent VRPTW. Another extension could be to target split delivery vehicle routing problem with time windows where the delivery to a customer can be split between any number of vehicles resulting in considerable improvements in terms number of vehicles used and cost [5, 10].

On the other hand, this work can be generalized to address a different class of VRP altogether, namely the VRP with backhauls (pick up and delivery) [30]. Moreover, in some "rich VRP scenarios" it is not necessary to visit all customers, and the distributor chooses to services its customers relying on their attractiveness (either multiple times visit or with multiple planning days) [33]. Also, the proposed meta-heuristic solution can be a building block to a parallel implementation version with different partitioned zones (i.e. multi-depot solution) and thus service a very large number of customers proficiently.

Acknowledgements. We thank dearly Dr. Mohamad Bader-El-Den from the School of Computing at University of Portsmouth for his valuable comments. Numerical computations were done on the Sciama High Performance Compute (HPC) cluster which is supported by the ICG, SEPNet and the University of Portsmouth.

References

1. Berger, J., Barkaoui, M., Bräysy, O.: A route-directed hybrid genetic approach for the vehicle routing problem with time windows. INFOR Inf. Syst. Oper. Res. **41**(2), 179–194 (2003)
2. Bychkov, I., Batsyn, M.: A hybrid approach for the capacitated vehicle routing problem with time windows. In: OPTA-SCL. CEUR Workshop Proceedings (2018)
3. Chiang, T.C., Hsu, W.H.: A knowledge-based evolutionary algorithm for the multiobjective vehicle routing problem with time windows. Comput. Oper. Res. **45**, 25–37 (2014)
4. Czech, Z.J., Czarnas, P.: Parallel simulated annealing for the vehicle routing problem with time windows. In: Proceedings 10th Euromicro Workshop on Parallel, Distributed and Network-Based Processing, pp. 376–383. IEEE (2002)
5. Da Silva, M.D.M., Subramanian, A., Ochi, L.S.: An iterated local search heuristic for the split delivery vehicle routing problem. In: ROADEF-15ème congrès annuel de la Société française de recherche opérationnelle et d'aide à la décision (2014)
6. Dash, M., Anekal, B.I.: Variables reduction in vehicle routing problems. SPC ERA IJBM **2**(6) (2014)
7. De Backer, B., Furnon, V., Prosser, P., Kilby, P., Shaw, P.: Local search in constraint programming: application to the vehicle routing problem. In: Proceedings CP-97 Workshop Industrial Constraint-Directed Scheduling, pp. 1–15. Schloss Hagenberg Austria (1997)
8. Dixit, A., Mishra, A., Shukla, A.: Vehicle routing problem with time windows using metaheuristic algorithms: a survey. In: Yadav, N., Yadav, A., Bansal, J.C., Deep, K., Kim, J.H. (eds.) Harmony Search and Nature Inspired Optimization Algorithms. AISC, vol. 741, pp. 539–546. Springer, Singapore (2019). https://doi.org/10.1007/978-981-13-0761-4_52
9. Dumez, D., Lehuédé, F., Péton, O.: A large neighborhood search approach to the vehicle routing problem with delivery options. Transp. Res. Part B Methodol. **144**, 103–132 (2021)
10. Feillet, D., Dejax, P., Gendreau, M., Gueguen, C.: Vehicle routing with time windows and split deliveries. Technical Paper 851 (2006)

11. Garcia-Najera, A., Bullinaria, J.A.: An improved multi-objective evolutionary algorithm for the vehicle routing problem with time windows. Comput. Oper. Res. **38**(1), 287–300 (2011)
12. Ghosh, D., Chakravarti, N.: A competitive local search heuristic for the subset sum problem. Comput. Oper. Res. **26**(3), 271–279 (1999)
13. Hosny, M.I.: Investigating heuristic and meta-heuristic algorithms for solving pickup and delivery problems. Cardiff University (United Kingdom) (2010)
14. Hsu, W.H., Chiang, T.C.: A multiobjective evolutionary algorithm with enhanced reproduction operators for the vehicle routing problem with time windows. In: 2012 IEEE Congress on Evolutionary Computation, pp. 1–8. IEEE (2012)
15. Ishibuchi, H., Imada, R., Setoguchi, Y., Nojima, Y.: How to specify a reference point in hypervolume calculation for fair performance comparison. Evol. Comput. **26**(3), 411–440 (2018)
16. Jawarneh, S., Abdullah, S.: Sequential insertion heuristic with adaptive bee colony optimisation algorithm for vehicle routing problem with time windows. PLoS ONE **10**(7), e0130224 (2015)
17. King, R.A., Deb, K., Rughooputh, H.: Comparison of NSGA-II and SPEA2 on the multiobjective environmental/economic dispatch problem. Univ. Mauritius Res. J. **16**(1), 485–511 (2010)
18. Li, W., Dai, H., Zhang, D.: The relationship between maximum completion time and total completion time in flowshop production. Procedia Manufact. **1**, 146–156 (2015)
19. Lin, C., Choy, K.L., Ho, G.T., Chung, S.H., Lam, H.: Survey of green vehicle routing problem: past and future trends. Expert Syst. Appl. **41**(4), 1118–1138 (2014)
20. Luke, S.: Essentials of metaheuristics. lulu, 2013. Metaheuristics in large-scale global contінues optimization: a survey (2013). http://cs.gmu.edu/sean/book/metaheuristics
21. Marrouche, W., Farah, R., Harmanani, H.M.: A strength pareto evolutionary algorithm for optimizing system-on-chip test schedules. Int. J. Comput. Intell. Appl. **17**(02), 1850010 (2018)
22. Marrouche, W., Harmanani, H.: A strength pareto evolutionary algorithm for solving the capacitated vehicle routing problem with time windows. In: Proceedings of the 13th International Joint Conference on Computational Intelligence (IJCCI 2021), pp. 96–106 (2021)
23. Meryem, B., Abdelmadjid, B.: Resolving a vehicle routing problem with heterogeneous fleet, mixed backhauls and time windows using cuckoo behaviour approach. Int. J. Oper. Res. **24**(2), 132–144 (2015)
24. Messaoud, E., El Idrissi, A.E.B., Alaoui, A.E.: The green dynamic vehicle routing problem in sustainable transport. In: 2018 4th International Conference on Logistics Operations Management (GOL), pp. 1–6. IEEE (2018)
25. Minocha, B., Tripathi, S., Mohan, C.: Solving vehicle routing and scheduling problems using hybrid genetic algorithm. In: 2011 3rd International Conference on Electronics Computer Technology, vol. 2, pp. 189–193. IEEE (2011)
26. Reimann, M., Doerner, K., Hartl, R.F.: Analyzing a unified ant system for the VRP and some of its variants. In: Cagnoni, S., et al. (eds.) EvoWorkshops 2003. LNCS, vol. 2611, pp. 300–310. Springer, Heidelberg (2003). https://doi.org/10.1007/3-540-36605-9_28
27. Rizzoli, A.E., Oliverio, F., Montemanni, R., Gambardella, L.M.: Ant colony optimisation for vehicle routing problems: from theory to applications. Galleria Rassegna Bimestrale Di Cultura **9**(1), 1–50 (2004)
28. Taha, A., Hachimi, M., Moudden, A.: A discrete bat algorithm for the vehicle routing problem with time windows. In: 2017 International Colloquium on Logistics and Supply Chain Management (LOGISTIQUA), pp. 65–70. IEEE (2017)
29. Tan, K.C., Chew, Y.H., Lee, L.H.: A hybrid multiobjective evolutionary algorithm for solving vehicle routing problem with time windows. Comput. Optim. Appl. **34**(1), 115–151 (2006)

30. Toth, P., Vigo, D.: The vehicle routing problem: society for industrial and applied mathematics. Siam Monographs on Discrete Mathematics and Applications (2001)
31. Udin, K., Gui, R., Rahmawan, A.: Green vehicle routing problem with heterogeneous fleet and time windows. In: Proceedings of the 6th International Conference on Software and Computer Applications, pp. 223–227 (2017)
32. Vidal, T., Crainic, T.G., Gendreau, M., Lahrichi, N., Rei, W.: A hybrid genetic algorithm for multidepot and periodic vehicle routing problems. Oper. Res. **60**(3), 611–624 (2012)
33. Wen, M., Clausen, J., Larsen, J.: Rich vehicle routing problems and applications. DTU Management (2010)
34. Xiao, L., Hajjam-El-Hassani, A., Dridi, M.: An application of extended cuckoo search to vehicle routing problem. In: 2017 International Colloquium on Logistics and Supply Chain Management (LOGISTIQUA), pp. 31–35. IEEE (2017)
35. Yasin, M., Vincent, F.Y.: A simulated annealing heuristic for the green vehicle routing problem. In: Lin, Y.K., Tsao, Y.C., Lin, S.W. (eds.) Proceedings of the Institute of Industrial Engineers Asian Conference 2013, pp. 1261–1269. Springer, Singapore (2013). https://doi.org/10.1007/978-981-4451-98-7_149
36. Zitzler, E.: SPEA2: improving the performance of the strength pareto evolutionary algorithm. Computer Engineering and Communication Networks Lab (TIK) (2001)
37. Zitzler, E., Laumanns, M., Bleuler, S.: A tutorial on evolutionary multiobjective optimization. In: Metaheuristics for Multiobjective Optimisation, pp. 3–37 (2004)

Risk Assessment Modeling Based on a Graded Fuzzy Concept Lattice

Māris Krastiņš[1(✉)], Ingrida Uljane[1,2], and Alexander Šostak[1,2]

[1] Department of Mathematics, University of Latvia, Jelgavas iela 3, Riga 1004, Latvia
mk18032@edu.lu.lv, ingrida.uljane@lu.lv, aleksandrs.sostaks@lumii.lv

[2] Institute of Mathematics and Computer Science University of Latvia, Raiņa bulvāris 29, Riga 1459, Latvia

Abstract. Fuzzy concept analysis is a well developed field of mathematics with many useful applications in various "real-world" problems. On the other hand, while different methods based on fuzzy logic, for example, fuzzy implication systems, are successfully used in the risk assessment based on expert opinion, we are not aware of any work in which the apparatus of fuzzy concept lattices would be used in the study of problems related to risk analysis. The main purpose of this article is to analyze the prospects for using graded fuzzy concept lattices as the basis for a risk assessment model. To make a paper self-contained, we provide all the necessary information concerning fuzzy concept lattices and their graded version. Theoretical models are further considered in the analysis of pandemic scenario which provides an insight into practical modeling of risk management processes and assessment of particular risks.

Keywords: Risk assessment · Risk factors · Pandemic scenario · Fuzzy context · Graded fuzzy concept lattice · Fuzzy preconcept · Gradation of fuzzy preconcept lattices

1 Introduction

The risk assessment process is a crucial part of the risk management. It is carried out in different activities from a simple daily behaviors, like participation in the traffic, up to the different industries, science and managing of a large scale projects. The risk assessment process often relies on statistic data, but in many instances these data should be further combined with expert opinions on specific and previously unknown circumstances and their impact. There are also scenarios of unique and previously unknown conditions when historic statistic data are either not available or could be hardly applicable. Therefore the experts should evaluate any possible developments and propose concrete actions based on their experience and judgments. These provisions have provided a solid basis for many studies on application of fuzzy logic in the risk assessment, see, e.g. [1,2].

For the purposes of constructing the risk assessment model, first we consider the risk hierarchy with more general risks including further specific risks which, in turn,

J. Garibaldi et al. (Eds.): IJCCI 2021, SCI 1119, pp. 144–162, 2023.
https://doi.org/10.1007/978-3-031-46221-4_7

are characterized by risk factors (those can be considered as sub-classes of the specific risks) which are assessed by corresponding risk levels. These risk levels are obtained via evaluations which are usually based on the fuzzy inference systems, with Mamdani and Sugeno systems being the most visible examples, see, e.g. [3]. In our research we propose the risk assessment model based on the fuzzy preconcept lattices which, in our opinion, provides a good illustrative basis for modeling of different scenarios and also allows to study certain mathematical properties embedded in this model.

A wide range of methods are used in the risk assessment processes with fuzzy implication systems being among the most popular. While these standard tools have proven their usefulness, we propose the application of an alternative approach. Namely, in this paper we consider the possibility of using the ideas and technique of fuzzy concept analysis in introducing the risk assessment model. Fuzzy concept analysis is a well developed field of mathematics having many useful applications in various "real-world" problems. Actually there are three versions of what is known now as fuzzy concept analysis. In all these versions the starting point is so called fuzzy context, that is a tuple (X, Y, R, L) where X and Y are sets, interpreted as the set of some objects and the set of some properties, L is a lattice and $R : X \times Y \to L$ is fuzzy relation assigning to an object $x \in X$ and a property $y \in Y$ a valued $R(x, y) \in L$ interpreted as the degree to which the object x has the property y. Further the goal of each version to distinguish pairs (A, B) of fuzzy subsets $A \in L^X$ and $B \in L^Y$ corresponding to each other via the relation R. The semantic difference between the three versions comes down to understanding what the word "corresponding" means. The mathematical difference between them is the way how operators $R^{\uparrow}$ and $R^{\downarrow}$ induced by fuzzy relation $R : X \times Y \to R$ on the powersets L^X and L^Y are interrelated. While the first, actually the original, version of concept analysis initiated by R. Wille [4] is based on the antitone Galois connection between these operator while the other two initiated respectively by Y. Yao [5], see also [6] and Düntch and Gediga [7] are based on the isotone Galois connection between them. In turn the difference between the other two is just "advantage" of objects over attributes or vice versa; so theoretically they are dual and can be obtain by replacing a fuzzy relation $R : X \times Y \to L$ by the fuzzy relation $R^{-1} : Y \times X \to L$.

Anyway, for application of (fuzzy) concept analysis in each specific situation one has to choose which version to use in a model. After reasoning about this question we concluded that for our merits the most appropriate is the original, Wille's approach. Therefore in the sequel when speaking about concept analysis we always assume that it is concept analysis initiated in eighties of the previous century in the paper [4] and whose principles and fundamental results are exposed in the monograph [8]. In the first decade of the 21st century different fuzzy counterparts of the concept analysis were introduced. The most important work in the first decade of the 21st century in fuzzy concept analysis was carried out by R. Bělohlávek, see, e.g. [9–11]. Since its inception, crisp concept analysis has found important applications in the study of different "real-world" problems. Starting with illustrative examples of application of crisp lattices given in [4], there appeared many serious works in which concept lattices were used. Concept analysis has found a very wide range of applications in medical research and teaching of medical students (see, e.g. [12–15]), in particular in the process of basic studies of nurses (see e.g. [16]). Methods of concept analysis are used also in the research of problems related to biology [17, 18], geology [19], social type problems [20], software engineering [21] and in other applied sciences. On the other hand, we found only a few works, where fuzzy concept analysis is used in the research of any practical problems. In our opinion, the problem to use the fuzzy concept lattices that the request in the concept analysis of the *precise correspondence* between a fuzzy set A of objects and a fuzzy set B of attributes in "real-world" situations is (almost) impracticable. Moreover, even if the sets A of objects and B of attributes are crisp, but only the relation R between these elements is fuzzy, then the pair (A, B) cannot satisfy the formal definition of a fuzzy concept, see also our Example 1. In this case one sooner has to deal with the weaker request asking *that the correspondence between A and B must hold up to a certain degree*. In order to provide a theoretical basis for the research in this situation, in [22], we introduced the notion of a graded preconcept lattice, more flexible than the notion of a concept lattice and initiated the study of the properties of graded preconcept lattices. In this paper we provide a further insight in the theoretical considerations and practical application of the theoretical models in the assessment of risks related to spread of pandemic crisis initially outlined in our paper [23].

This paper is organized as follows. In the second section we briefly remind the notions related to lattices, residuated lattices or quantales, fuzzy sets and fuzzy relations. In the third section we define fuzzy preconcepts, introduce partial order relation $\preceq$ on the family of all fuzzy preconcepts of a given fuzzy context (X, Y, R, L) and show that the resulting structure $(\mathbb{P}(X, Y, R, L), \preceq)$ is a lattice. Further in this section we recall the notion of a fuzzy concept, see, e.g. [10] interpret it as a special preconcept and following introduce a lattice structure on the family of all fuzzy concepts of a given fuzzy context. In Sect. 4 we remind the relation of a fuzzy inclusion for fuzzy sets. Basing on this relation we propose a method allowing to determine the grade of the distinction of a fuzzy preconcept from "being a real fuzzy concept" and study the corresponding graded preconcept lattices. In the sixth section we discuss the possible application of fuzzy preconcept lattices in the analysis of pandemic scenario and proceed with practical calculations in modeling the spread of Covid-19 pandemics. In the last, conclusion section, we briefly summarize main results obtained and survey some directions for the future work.

2 Background

2.1 Lattices and Quantales

We use standard terminology accepted in theory of lattices, see, e.g. [24–26]. When speaking about a lattice $(L, \leq, \wedge, \vee)$, we always assume that it is complete with 0_L and 1_L as its bottom and top elements respectively.

The notion of a quantale first appears in Rosenthal's monograph [27]. Here we recall information on quantales that is needed in our work.

Let $(L, \leq, \wedge, \vee)$ be a complete lattice and $* : L \times L \to L$ a binary associative monotone operation. The tuple $(L, \leq, \wedge, \vee, *)$ is called *a quantale* if $*$ distributes over arbitrary joins:

$$a * \left(\bigvee_{i \in I} b_i\right) = \bigvee_{i \in I} (a * b_i) \text{ and } \left(\bigvee_{i \in I} b_i\right) * a = \bigvee_{i \in I} (b_i * a)$$

for all $\{b_i \mid i \in I\} \subseteq L$ and for all $a \in L$.

A quantale $(L, \leq, \wedge, \vee, *)$ is called symmetric if $a * b = b * a$ for every $a, b \in L$.

A quantale $(L, \leq, \wedge, \vee, *)$ is called integral if it has a unit element $1_L \in L$ and it is the top element of the lattice $(L, \leq, \wedge, \vee)$.

When saying *a quantale*, we always assume that it is symmetric and integral.

Operation $* : L \times L \to L$ is called meet-continuous, if it distributes over arbitrary meets, that is

$$a * \left(\bigwedge_{i\in I} b_i\right) = \bigwedge_{i\in I} (a * b_i).$$

for all $\{b_i \mid i \in I\} \subseteq L$ and for all $a \in L$.

In a quantale a further binary operation $\mapsto: L \times L \to L$, the residuum, can be introduced as associated with operation $*$ via the Galois connection:

$$a * b \leq c \Longleftrightarrow a \leq b \mapsto c \text{ for all } a, b, c \in L.$$

A

quantale $(L, \leq, \wedge, \vee, *)$ provided with the derived operation $\mapsto$, that is the tuple $(L, \leq, \wedge, \vee, *, \mapsto)$, is known also as a (complete) residuated lattice [26].

In the following proposition we collect well-known properties of the residuum in a quantale which will be used in the main text:

Proposition 1. see, e.g. [28–30], etc.

(1) $(\bigvee_i a_i) \mapsto b = \bigwedge_i (a_i \mapsto b)$ *for all* $\{a_i \mid i \in I\} \subseteq L$, *for all* $b \in L$;
(2) $a \mapsto (\bigwedge_i b_i) = \bigwedge_i (a \mapsto b_i)$ *for all* $a \in L$, *for all* $\{b_i \mid i \in I\} \subseteq L$;
(3) $1_L \mapsto a = a$ *for all* $a \in L$;
(4) $a \mapsto b = 1_L$ *whenever* $a \leq b$;
(5) $a * (a \mapsto b) \leq b$ *for all* $a, b \in L$;
(6) $(a \mapsto b) * (b \mapsto c) \leq a \mapsto c$ *for all* $a, b, c \in L$;
(7) $a \mapsto b \leq (a * c) \mapsto (b * c)$ *for all* $a, b, c \in L$;
(8) $a * b \leq a \wedge b$ *for any* $a, b \in L$;
(9) $(a * b) \mapsto c = a \mapsto (b \mapsto c)$ *for any* $a, b, c \in L$;
(10) $a_1 \leq a_2 \Longrightarrow a_1 \mapsto b \geq a_2 \mapsto b$ *for any* $a_1, a_2, b \in L$;
(11) $b_1 \leq b_2 \Longrightarrow a \mapsto b_1 \leq a \mapsto b_2$ *for any* $a, b_1, b_2 \in L$.

2.2 Fuzzy Sets

The concept of a fuzzy set, or more precisely a fuzzy subset of a set X was introduced in the celebrated paper by L.A. Zadeh [31]. Namely, according to L.A. Zadeh a fuzzy subset of a set X may be defined as a mapping $A : X \to [0, 1]$. Soon after J.A. Goguen [32], a student of L.A. Zadeh, extended this concept by replacing the unit interval $[0, 1]$ by an arbitrary complete lattice, or more generally, by an arbitrary quantale.

Given a quantale $(L, \leq, \wedge, \vee, *)$, an L-fuzzy subset of a set X is a mapping $A : X \to L$. The lattice and the quantale structures of L are point-wise extended to the set L^X of all L-fuzzy subsets of X. Specifically, the union and intersection of a family of L-fuzzy sets $\{A_i \mid i \in I\} \subseteq L^X$ are defined by their join $\bigvee_{i\in I} A_i$ and meet $\bigwedge_{i\in I} A_i$ respectively. Quantale operation $*$ on L induces operation $* : L^X \times L^X \to L^X$ as follows. Given two L-fuzzy sets $A, B \in L^X$ we define $A * B \in L^X$ by setting

$$(A * B)(x) = A(x) * B(x) \; \forall x \in X.$$

2.3 Fuzzy Relations

The concept of a fuzzy relation was introduced in 1971 by L.A. Zadeh [33] and further studied and applied by different authors, see, e.g. [34].

Let $A \subseteq X$, $B \subseteq Y$. An L-fuzzy relation $R : X \times Y \to L$ is called left connected on (A, B) if $\bigwedge_{y \in B} \bigvee_{x \in A} R(x, y) = 1$. If for every $y \in B$ there exists $x \in A$ such that $R(x, y) = 1_L$, then R is called *strongly left connected on* (A, B). An L-fuzzy relation R is called right connected on (A, B) if $\bigwedge_{x \in A} \bigvee_{y \in B} R(x, y) = 1_L$. If for every $x \in A$ there exists $y \in B$ such that $R(x, y) = 1_L$, then R is called *strongly right connected on* (A, B). An L-fuzzy relation $R : X \times Y \to L$ is called connected (strongly connected) on (A, B) if it is both left and right connected (respectively strongly left and right connected on (A, B)).

In a certain sense opposite to left and right connectedness of a fuzzy relation is the property of left and right disjointeness of a fuzzy relation. An L-fuzzy relation $R : X \times Y \to L$ is called left disjoint on (A, B) if $\bigwedge_{y \in B} \bigvee_{x \in A} R(x, y) = 0$. If for every $y \in B$ there exists $x \in A$ such that $R(x, y) = 0_L$, then R is called *strongly left disjoint on* (A, B). An L-fuzzy relation R is called right disjoint on (A, B) if $\bigwedge_{x \in A} \bigvee_{y \in B} R(x, y) = 0_L$. If for every $x \in A$ there exists $y \in B$ such that $R(x, y) = 0_L$, then R is called *strongly right connected on* (A, B). An L-fuzzy relation $R : X \times Y \to L$ is called disjoint (strongly disjoint) on (A, B) if it is both left and right disjoint (respectively strongly left and right disjoint on (A, B)).

3 Concept Lattices Vs. Preconcept Lattices

3.1 Preconcepts and Preconcept Lattices

Let L be a complete lattice. Further, let X, Y be sets and $R : X \times Y \to L$ be a fuzzy relation. Following terminology accepted in the theory of (fuzzy) concept lattices, see, e.g. [4, 9–11], we refer to the tuple (X, Y, R, L) as a fuzzy context.

Definition 1. *Given a fuzzy context* (X, Y, R, L), *a pair* $P = (A, B) \in L^X \times L^Y$ *is called a fuzzy preconcept.*

On the set $\mathbb{P} = L^X \times L^Y$ of all fuzzy preconcepts we introduce a partial order $\preceq$ as follows. Given $P_1 = (A_1, B_1)$ and $P_2 = (A_2, B_2)$, we set $P_1 \preceq P_2$ if and only if $A_1 \leq A_2$ and $B_1 \geq B_2$. (To substantiate such definition of the order $\preceq$ notice that the increase of the set of objects implies the decrease of the set of attributes that have all these properties.)

Let $(\mathbb{P}, \preceq)$ be the set $L^X \times L^Y$ endowed with this partial order. Further, given a family of fuzzy concepts $\{P_i = (A_i, B_i) : i \in I\} \subseteq L^X \times L^Y$, we define its join (supremum) by $\underline{\vee}_{i \in I} P_i = (\bigvee_{i \in I} A_i, \bigwedge_{i \in I} B_i)$ and its meet (infimum) as $\overline{\wedge}_{i \in I} P_i = (\bigwedge_{i \in I} A_i, \bigvee_{i \in I} B_i)$. The proof of the following theorem is easy and can be found, e.g. in [22].

Theorem 1. $(\mathbb{P}, \preceq, \overline{\wedge}, \underline{\vee})$ *is a complete lattice. Besides, if* L *is an infinitely bi-distributive lattice, then* $(\mathbb{P}, \preceq, \overline{\wedge}, \underline{\vee})$ *is also an infinitely bi-distributive lattice.*

In the sequel we write just $\mathbb{P}$ or $(\mathbb{P}, \preceq)$ instead of $(\mathbb{P}, \preceq, \overline{\wedge}, \underline{\vee})$ if no misunderstanding is possible. We use also notation $(\mathbb{P}, X, Y, L, R)$ in case when it is important to specify the fuzzy context we are working in.

3.2 Operators $R^{\uparrow}$ and $R^{\downarrow}$ on [SPSDOLLAR4DOLLARSPS]-Powersets and Fuzzy Concept Lattices

Let X and Y be sets and let $R : X \times Y \to L$ be a fuzzy relation, where L is a fixed *quantale*. Given a fuzzy context (X, Y, R, L), we define operators $R^{\uparrow} : L^X \to L^Y$ and $R^{\downarrow} : L^Y \to L^X$ as follows:

Definition 2. (see, e.g. [11]) *Given $A \in L^X$ and $B \in L^Y$, we define $A^{\uparrow} \in L^Y$ and $B^{\downarrow} \in L^X$ by setting*
$A^{\uparrow}(y) = \bigwedge_{x \in X}(A(x) \mapsto R(x, y))\ \forall y \in Y,$
$B^{\downarrow}(x) = \bigwedge_{y \in Y}(B(y) \mapsto R(x, y))\ \forall x \in X.$
Considering A as a variable over L^X and B as a variable over L^Y, we get operators $R^{\uparrow} : L^X \to L^Y$ and $R^{\downarrow} : L^Y \to L^X$.

In the crisp case, that is when $A \subseteq X, B \subseteq Y$ and $R : X \times Y \to \{0, 1\}$, this definition is obviously equivalent to the original definition of operators $A \longrightarrow A'$ and $B \longrightarrow B'$ in [4]. Informally these operators can be described as follows. Let a set X of objects with a subset $A \subseteq X$ and a set Y of properties with a subset $B \subseteq Y$ be given. Further, let $R \subseteq X \times Y$ be a relation where $(x, y) \in R$ means "object x has property y". Now $A^{\uparrow}$ is the set of properties $y \in Y$ which have all objects $x \in A$ while $B^{\downarrow}$ is the set of objects $x \in X$ which have all properties $y \in B$. Naturally, the increase of the set of objects implies the decrease of properties satisfied by all these objects and vice versa. This observation is summarized in the following easy provable proposition.

Proposition 2. (cf, e.g. [4] in crisp case, [10])
Operators $R^{\uparrow} : L^X \to L^Y$ and $R^{\downarrow} : L^Y \to L^X$ are decreasing:
$A_1 \le A_2 \Rightarrow A_1^{\uparrow} \ge A_2^{\uparrow}$, $B_1 \le B_2 \Rightarrow B_1^{\downarrow} \ge B_2^{\downarrow}$.

In the sequel we write $A^{\uparrow\downarrow}$ instead of $(A^{\uparrow})^{\downarrow}$ and $B^{\downarrow\uparrow}$ instead of $(B^{\downarrow})^{\uparrow}$. We write also $R^{\uparrow\downarrow}$ for the composition $R^{\downarrow} \circ R^{\uparrow} : L^X \to L^X$ and $R^{\downarrow\uparrow}$ for the composition $R^{\uparrow} \circ R^{\downarrow} : L^Y \to L^Y$.

Proposition 3. (cf, e.g. [4] in crisp case, [10]) $A^{\uparrow\downarrow} \ge A$ *for every* $A \in L^X$ *and* $B^{\downarrow\uparrow} \ge B$ *for every* $B \in L^Y$.

Proposition 4. (cf, e.g. [4] in crisp case, [10])
$A^{\uparrow} = A^{\uparrow\downarrow\uparrow}$ *for every* $A \in L^X$ *and* $B^{\downarrow} = B^{\downarrow\uparrow\downarrow}$ *for every* $B \in L^Y$.

Proposition 5. (cf, e.g. [4] for the crisp case, [10]) *Given a family $\{A_i \mid i \in I\} \subseteq L^X$, we have $\left(\bigvee_{i \in I} A_i\right)^{\uparrow} = \bigwedge_{i \in I} A_i^{\uparrow}$. Given a family $\{B_i \mid i \in I\} \subseteq L^Y$, we have $\left(\bigvee_{i \in I} B_i\right)^{\downarrow} = \bigwedge_{i \in I} B_i^{\downarrow}$.*

3.3 Concepts and Concept Lattices

Referring to the definition of a (fuzzy) concept given in [4, 10] we reformulate it as follows:

Definition 3. *A fuzzy preconcept (A, B) is called a fuzzy concept if $A^{\uparrow} = B$ and $B^{\downarrow} = A$.*

The following simple example indicates the limitations of the direct use of fuzzy concepts.

Example 1. Let a fuzzy context (X, Y, R, L) be given and let $A \subseteq X$.[1] Then for every $y \in Y$ $A^{\uparrow}(y) = \bigwedge_{x \in X} A(x) \mapsto R(x, y) = \bigwedge_{x \in A} R(x, y)$. In the same way we prove that if $B \subseteq Y$, then $B^{\downarrow}(x) = \bigwedge_{y \in B} R(x, y)$. Hence, even in case when $A \subseteq X, B \subseteq Y$ a pair (A, B) can be a concept (either crisp or fuzzy) only in case when R is also crisp on $A \times B$, that is when $R : A \times B \to \{0, 1\}$. This shows the limitation for the use of concept lattices in the case of a fuzzy context and gives an additional evidence in favor of the graded approach to fuzzy concept lattices.

Continuing the previous example we calculate $A^{\uparrow\downarrow}$ and $B^{\downarrow\uparrow}$ in case of crisp sets A and B:

$A^{\uparrow\downarrow}(x) = \bigwedge_{y \in Y} \left(\bigwedge_{x' \in A}(R(x', y) \mapsto R(x, y))\right),$

$B^{\downarrow\uparrow}(y) = \bigwedge_{x \in X} \left(\bigwedge_{y' \in B}(R(x, y') \mapsto R(x, y))\right).$ □

Let $\mathbb{C} = \mathbb{C}(X, Y, L, R)$ be the subset of $\mathbb{P} = \mathbb{P}(X, Y, L, R)$ consisting of fuzzy concepts (A, B) and let $\preceq$ be the partial order on $\mathbb{C}$ induced by the partial order $\preceq$ from the lattice $(\mathbb{P}, \preceq)$. Then $(\mathbb{C}, \preceq)$ is a partially ordered subset of the lattice $(\mathbb{P}, \preceq)$. However, generally $(\mathbb{C}, \preceq)$ is not a sublattice of the lattice $(\mathbb{P}, \preceq, \bar{\wedge}, \underline{\vee})$ and we need to define joins and meets in $(\mathbb{C}, \preceq)$ differently from $\bar{\wedge}$ and $\underline{\vee}$. To do this first we recall the following simple lemma the proof of which can be found in [22].

Lemma 1. *Let (A_1, B_1), (A_2, B_2) be fuzzy concepts. If $A_1 \leq A_2$, then $B_1 \geq B_2$ and if $B_1 \geq B_2$ then $A_1 \leq A_2$.*

Corollary 1. *Let $(A_1, B_1), (A_2, B_2) \in \mathbb{C}$. Then $(A_1, B_1) \preceq (A_2, B_2)$ if and only if $A_1 \leq A_2$ and if and only if $B_1 \geq B_2$.*

Proposition 6. *If $A \in L^X$, then $(A^{\uparrow\downarrow}, A^{\uparrow})$ is the smallest concept containing A as the fuzzy set of objects. If $B \in L^Y$, then $(B^{\downarrow}, B^{\downarrow\uparrow})$ is the smallest concept containing B as the fuzzy set of attributes.*

Proof From Proposition 4 it follows that $(A^{\uparrow\downarrow}, A^{\uparrow})$ is a fuzzy context. Further, from Proposition 3 we know that $A \leq A^{\uparrow\downarrow}$. Assume that there is a fuzzy concept $(A_o, A_o^{\uparrow})$ such that $A \leq A_o \leq A^{\uparrow\downarrow}$. Then $A^{\uparrow} \geq A_o^{\uparrow} \geq A^{\uparrow\downarrow\uparrow} = A^{\uparrow}$ and hence $A_o^{\uparrow} = A^{\uparrow}$. Therefore $A_o = A^{\uparrow\downarrow}$ and $(A_o, A_o^{\downarrow}) = (A^{\uparrow\downarrow}, A^{\uparrow})$. In a similar way we can prove that $(B^{\downarrow}, B^{\downarrow\uparrow})$ is the smallest context containing B as the fuzzy set of attributes. □

Patterned after [10,22], we introduce operations $\curlywedge, \curlyvee : 2^{\mathbb{P}} \to \mathbb{P}$ as follows. Let $\{\mathcal{C} = \{C_i = (A_i, B_i)\} \subseteq \mathbb{C}(X, Y, L, R, \preceq$. We define

$$\curlywedge \mathcal{C} = \curlywedge_{i \in I} C_i = \left(\bigwedge_{i \in I} A_i, \left(\bigvee_{i \in I} B_i\right)^{\downarrow\uparrow}\right)$$

$$\curlyvee \mathcal{C} = \curlyvee_{i \in I} C_i = \left(\left(\bigvee_{i \in I} A_i\right)^{\uparrow\downarrow}, \bigwedge_{i \in I} B_i\right)$$

---

[1] Here and in the sequel we do not distinguish between a crisp set $A \subseteq X$ and its characteristic function $\chi_A : X \to \{0, 1\}$.

The proof of the following theorem can be found e.g. in [22]. For convenience of the reader we sketch the idea of this proof here.

Theorem 2. *Let (X, Y, L, R) be a fuzzy context and let $\preceq$ be the partial order induced from the lattice $\mathbb{P}(X, Y, L, R, \preceq)$. Then $\mathbb{C}(X, Y, R, L \preceq, \curlywedge, \curlyvee)$ is a complete lattice.*

Proof. To prove the theorem we have to show that $\curlywedge : 2^{\mathbb{C}} \to \mathbb{C}$ is operation of meet in $(\mathbb{C}, \preceq)$ and $\curlyvee : 2^{\mathbb{C}} \to \mathbb{C}$ is operation of join in $(\mathbb{C}, \preceq)$. We show it as follows.

1. We have to prove only that $\curlywedge_{i \in I} C_i$ is a fuzzy concept; its minimality will be clear from its definition since $\left(\bigwedge_{i \in I} A_i, \bigvee_{i \in I} B_i\right)$ is the meet of $\mathcal{C}$ as a subset of $\mathbb{P}(X, Y, R, L)$. We show that $\curlywedge_{i \in I} C_i$ is a fuzzy concept by noticing that

$$\left(\bigvee_{i \in I} B_i\right)^{\downarrow\uparrow} = \left(\bigvee_{i \in I} A_i^{\uparrow}\right)^{\downarrow\uparrow} = \left(\bigwedge_{i \in I} A_i\right)^{\uparrow\downarrow\uparrow} = \left(\bigwedge_{i \in I} A_i\right)^{\uparrow};$$
$$\left(\bigwedge_{i \in I} A_i\right)^{\uparrow} = \left(\bigwedge_{i \in I} B^{\downarrow}\right)^{\uparrow} = \left(\bigvee_{i \in I} B_i\right)^{\downarrow\uparrow}.$$

2. We have to prove only that $\curlyvee_{i \in I} C_i$ is a fuzzy concept; its maximality will be clear from its definition since $\left(\bigvee_{i \in I} A_i, \bigwedge_{i \in I} B_i)\right)$ is the join of $\mathcal{C}$ in $\mathbb{P}(X, Y, L, R)$. We show that $\curlyvee_{i \in I} C_i$ is a fuzzy concept by noticing that

$$\left(\bigvee_{i \in I} A_i\right)^{\uparrow\downarrow} = \left(\bigvee_{i \in I} B_i^{\downarrow}\right)^{\uparrow\downarrow} = \left(\bigwedge_{i \in I} B_i)\right)^{\downarrow\uparrow\downarrow} = \left(\bigwedge_{i \in I} B_i\right)^{\downarrow};$$
$$\left(\bigwedge_{i \in I} B_i\right)^{\uparrow} = \left(\bigwedge_{i \in I} A^{\uparrow}\right)^{\uparrow} = \left(\bigvee_{i \in I} A_i\right)^{\uparrow\downarrow}.$$

□

Taking into account that in a fuzzy concept (A_i, B_i) it holds $A_i^{\uparrow} = B_i$ and $B_i^{\downarrow} = A_i$, we get the following corollary from the previous Proposition:

Corollary 2. *Let $\mathcal{C} = \{C_i = (A_i, B_i) \mid i \in I\} \subseteq \mathbb{C}$ be a family of fuzzy concepts. Then*

1. $\curlywedge_{i \in I} C_i = \left(\bigwedge_{i \in I} A_i, \left(\bigwedge_{i \in I} A_i\right)^{\uparrow}\right)$
   is its infimum in the lattice $(\mathbb{C}, \preceq)$.
2. $\curlyvee_{i \in I} C_i = \left(\left(\bigwedge_{i \in I} B_i\right)^{\downarrow}, \bigwedge_{i \in I} B_i\right)$
   is its supremum in the lattice $(\mathbb{C}, \preceq)$.

4 Graded Concept Lattices

4.1 Measure of Inclusion of L-Fuzzy Sets

The gradation of a preconcept lattice presented below is based on the fuzzy inclusion between fuzzy sets. We give here a brief introduction into this field.

In order to fuzzify the inclusion relation $A \subseteq B$ "a fuzzy set A is a subset of a fuzzy set B", we have to interpret it as a certain fuzzy relation $\hookrightarrow$ based on "if - then" rule, that is on some implication $\Rightarrow$ defined on the lattice L. In the result we come to the formula $A \hookrightarrow B = \inf_{x \in X}(A(x) \Rightarrow B(x))$.

In this paper we use implication $\Rightarrow$ defined by means of the residuum: $a \Rightarrow b =_{def} a \mapsto b$.

Definition 4. *By setting* $A \hookrightarrow B = \bigwedge_{x \in X}(A(x) \mapsto B(x))$ *for all* $A, B \in L^X$, *we obtain a mapping* $\hookrightarrow: L^X \times L^X \to L$. *We call* $A \hookrightarrow B$ *by the (L-valued) measure of inclusion of the L-fuzzy set A into the L-fuzzy set B. We denote* $A \cong B =_{def} (A \hookrightarrow B) \wedge (B \hookrightarrow A)$ *and view it as the degree of equality of L-fuzzy sets A and B.*

Proposition 7. (see, e.g. [30, 35]) *Mapping* $\hookrightarrow: L^X \times L^X \to L$ *satisfies the following properties:*

(1) $(\bigvee_i A_i) \hookrightarrow B = \bigwedge_i (A_i \hookrightarrow B)$ *for all* $\{A_i \mid i \in I\} \subseteq L^X$ *and for all* $B \in L^X$;

(2) $A \hookrightarrow (\bigwedge_i B_i) = \bigwedge_i (A \hookrightarrow B_i)$ *for all* $A \in L^X$ *and for all* $\{B_i \mid i \in I\} \subseteq L^X$;

(3) $A \hookrightarrow B = 1_L$ *whenever* $A \leq B$;

(4) $1_X \hookrightarrow A = \bigwedge_x A(x)$ *for all* $A \in L^X$ *where* $1_X : X \to L$ *is a constant function with the value* $1_L \in L$;

(5) $(A \hookrightarrow B) \leq (A * C \hookrightarrow B * C)$ *for all* $A, B, C \in L^X$;

(6) $(A \hookrightarrow B) * (B \hookrightarrow C) \leq (A \hookrightarrow C)$ *for all* $A, B, C \in L^X$;

(7) $(\bigwedge_i A_i) \hookrightarrow (\bigwedge_i B_i) \geq \bigwedge_i (A_i \hookrightarrow B_i)$ *for all* $\{A_i : i \in I\}$, $\{B_i : i \in I\} \subseteq L^X$;

(8) $(\bigvee_i A_i) \hookrightarrow (\bigvee_i B_i) \geq \bigwedge_i (A_i \hookrightarrow B_i)$ *for all* $\{A_i : i \in I\}$, $\{B_i : i \in I\} \subseteq L^X$.

4.2 Conceptuality Degree of a Fuzzy Preconcept

Let (X, Y, R, L) be a fuzzy context and $(A, B) \in \mathbb{P}(X, Y, L, R)$.

Definition 5. *The degree of contentment of the fuzzy set A of objects by the fuzzy set B of attributes or the degree object based contentment of the preconcept* (A, B) *for short is defined by* $\mathcal{D}^{\uparrow}(A, B) =_{def} A^{\uparrow} \cong B$.

Definition 6. *The degree of contentment of the fuzzy set B of of attributes by the fuzzy set A of objects or the attribute based contentment of the preconcept* (A, B) *is defined by* $\mathcal{D}^{\downarrow}(A, B) =_{def} A \cong B^{\downarrow}$.

Definition 7. *The degree of conceptuality of the preconcept* (A, B) *in the preconcept lattice* $\mathbb{P}$ *is defined by* $\mathcal{D}(A, B) = \mathcal{D}^{\uparrow}(A, B) \wedge \mathcal{D}^{\downarrow}(A, B)$.

Considering (A, B) as a variable over $L^X \times L^Y$, we obtain mappings $\mathcal{D}^{\uparrow} : \mathbb{P} \to L$, $\mathcal{D}^{\downarrow} : \mathbb{P} \to L$ and $\mathcal{D} : \mathbb{P} \to L$.

4.3 Examples of Evaluation of Conceptuality Degree for Fuzzy Preconcepts

The aim of this subsection is to illustrate the evaluation of conceptuality degree in the fuzzy context (X, Y, R, L). For simplicity we restrict to the case when A and B are crisp sets.

We illustrate the evaluation of conceptuality degree in the fuzzy context (X, Y, L, R) in some simple situations. To simplify calculations we distinguish the following special conditions. The first one concerns the properties of the product $*$ of the quantale while the next have to do with the fuzzy relation R or with the preconcept (A, B) itself.

$(\dagger_*)$ Operation $*$ has no zero divisors, that is

$$a * b = 0 \Rightarrow a = 0 \text{ or } b = 0 \text{ for any } a, b \in L.$$

The next conditions are applicable only for calculating gradation degrees of crisp pairs of preconcepts $(A, B) \in \mathbb{P}(X, Y, L, R)$. In this case we denote $A^c = X \setminus A$ and $B^c = Y \setminus B$, that is the complements of the sets A and B respectively.

$(\dagger_A^B)$ Relation $R : X \times Y$ is right disjoint on the pair (A, B^c) In particular, this holds if $B = Y$.

$(\dagger_B^A)$ Relation $R : X \times Y$ is left disjoint on the pair (A, B^c). In particular, this holds if $A = X$.

$(\dagger_R)$ Relation $R : X \times Y$ is disjoint on the pair (A, B^c). In particular, this holds if $A = X$ and $B = Y$.

Example 2. Let $A \subseteq X, B \subseteq Y$, let $(L, \leq, \wedge, \vee, *)$ be an arbitrary quantale, $\mapsto: L \times L \to L$ its residuum, and $R : X \times Y \to L$ a fuzzy relation. Then

$$A^\uparrow \hookrightarrow B = \bigwedge_{y\in Y} \left(\bigwedge_{x\in X} (A(x) \mapsto R(x,y)) \mapsto B(y)\right) =$$

$$\bigwedge_{y\in B^c} \left(\bigwedge_{x\in X} (A(x) \mapsto R(x,y)) \mapsto 0\right) = \bigwedge_{y\in B^c} \left(\bigwedge_{x\in A} (R(x,y) \mapsto 0)\right);$$

$$B \hookrightarrow A^\uparrow = \bigwedge_{y\in Y} \left(B(y) \mapsto \left(\bigwedge_{x\in X} (A(x) \mapsto R(x,y))\right)\right) =$$

$$\bigwedge_{y\in Y} \left(B(y) \mapsto \bigwedge_{x\in A} R(x,y)\right) = \bigwedge_{y\in B} \bigwedge_{x\in A} R(x,y);$$

$$\mathcal{D}^\uparrow(A,B) = \left(\bigwedge_{y\in B^c} \left(\bigwedge_{x\in A} (R(x,y) \mapsto 0)\right)\right) \wedge \left(\bigwedge_{x\in A, y\in B} R(x,y)\right).$$

From the above it easily follows that if either condition $(\dagger_*)$ or condition $(\dagger_{R_A^B})$ holds, then $B \hookrightarrow A^\uparrow = 1$ and hence $\mathcal{D}^\uparrow(A,B) = \bigwedge_{x\in A, y\in B} R(x,y)$.

In a similar way we prove that

$$A \hookrightarrow B^\downarrow = \bigwedge_{x\in A, y\in B} R(x,y),\ B^\downarrow \hookrightarrow A = \bigwedge_{x\in A^c} \left(\bigwedge_{y\in B} (R(x,y) \mapsto 0)\right),$$

$$\mathcal{D}^\downarrow(A,B) = \bigwedge_{x\in A^c} \left(\bigwedge_{y\in B} (R(x,y) \mapsto 0)\right) \wedge \left(\bigwedge_{x\in A, y\in B} R(x,y)\right)$$

and

$\mathcal{D}^{\downarrow}(A,B) = \bigwedge_{x\in A, y\in B} R(x,y)$ in case when either $(\dagger_*)$ or $(\dagger_{R_{AB}})$ holds.

Thus in case of crisp object and attribute sets A, B and under assumption that either $(\dagger_*)$ or $(\dagger_R)$ holds, the degree of the conceptuality of the pair (A,B) is $\mathcal{D}(A,B) = \bigwedge_{x\in A, y\in B} R(x,y)$. □

Example 3. Let now (X,Y,R,L) be a fuzzy context where $L = [0,1]$, $a \in (0,1)$, $X_a \subseteq X$, $B \subseteq Y$ and a fuzzy set $A : X \to [0,1]$ be defined by

$$A(x) = \begin{cases} a & \text{if } x \in X_a \\ 0, & \text{if } x \notin X_a \end{cases}.$$

Then

$$B \hookrightarrow A^{\uparrow} = \bigwedge_{y\in Y} (B(y) \mapsto A^{\uparrow}(y)) = \bigwedge_{y\in B} A^{\uparrow}(y) = \bigwedge_{y\in B, x\in X_a} (a \mapsto R(x,y));$$

$$A^{\uparrow} \hookrightarrow B = \bigwedge_{y\in B^c} \left(\bigwedge_{x\in X_a} (a \mapsto R(x,y))\right) \mapsto 0$$

$$\mathcal{D}^{\uparrow}(A,B) = \left(\bigwedge_{y\in B, x\in X_a} (a \mapsto R(x,y))\right) \wedge \left(\bigwedge_{y\in B^c} \left(\bigwedge_{x\in X} (a \mapsto R(x,y))\right) \mapsto 0\right),$$

and hence

$$\mathcal{D}^{\uparrow}(A,B) = \bigwedge_{y\in B, x\in X_a} (a \mapsto R(x,y))$$

if condition $(\dagger_*)$ or condition $(\dagger_{R_A^B})$ is satisfied.

In a similar way, in order to calculate $\mathcal{D}^{\downarrow}(A,B)$, we have

$$A \hookrightarrow B^{\downarrow} = \bigwedge_{x\in X} (A(x) \mapsto B^{\downarrow}(x)) = \bigwedge_{x\in X_a} \left(a \mapsto \bigwedge_{y\in B} R(x,y)\right);$$

$$B^{\downarrow} \hookrightarrow A = \left(\bigwedge_{x\in X_a} \left(\bigwedge_{y\in B} (R(x,y) \mapsto a)\right)\right) \wedge \left(\bigwedge_{x\in X_a^c} \left(\bigwedge_{y\in B} (R(x,y) \mapsto 0)\right)\right);$$

$$\mathcal{D}^{\downarrow}(A,B) = \left(\bigwedge_{x\in X_a^c} \left(\bigwedge_{y\in B} (R(x,y) \mapsto 0)\right)\right) \wedge$$
$$\left(\bigwedge_{x\in X_a} \left(\bigwedge_{y\in B} (R(x,y) \mapsto a) \wedge (a \mapsto \bigwedge_{y\in B} R(x,y))\right)\right).$$

Under assumption of $(\dagger_*)$ or $(\dagger_R)$ this formula can be simplified and we get

$$\mathcal{D}(A,B) = \bigwedge_{x\in X_a} \left(\bigwedge_{y\in B} (R(x,y) \mapsto a) \wedge (a \mapsto \bigwedge_{y\in B} R(x,y))\right).$$

Example 4. Let now $A \subseteq X$, $Y_b \subseteq Y$, $L = [0,1]$, $b \in (0,1)$ and $B : Y \to L = [0,1]$ be defined by

$$B(y) = \begin{cases} b & \text{if } y \in Y_b \\ 0, & \text{if } y \notin Y_b \end{cases}.$$

Then, under assumption of $(\dagger_*)$ or $(\dagger_R)$, calculating similar as in the previous example, we get:

$$\mathcal{D}(A,B) = \left(\bigwedge_{y\in Y_b} (b \mapsto \bigwedge_{x\in A} R(x,y))\right) \wedge \left(\bigwedge_{y\in Y_b} \left(\bigwedge_{x\in A} (R(x,y) \mapsto b)\right)\right).$$

Example 5. We demonstrate the previously obtained formulas for calculating $\mathcal{D}(A, B)$ in case of the Example 3 for the three basic t-norms $*$ on $[0, 1]$: $*_\wedge = \wedge$ - the minimum t-norm, $*_L$ - the Łukasiewicz t-norm and $*_P$ - the product t-norm, see, e.g. [36].

(1) *Łukasiewicz t-norm* has zero divisors. Therefore, to simplify situation we will consider the case when $X_a = X$, $B = Y$, that is in case when assumption $(\dagger_{AB})$ is satisfied. Then from the above formulas we have

$$\mathcal{D}^{\uparrow}(A, B) = \left(\bigwedge\nolimits_{x \in X, y \in Y} (1 - a + R(x, y))\right) \wedge 1;$$

$$\mathcal{D}^{\downarrow}(A, B) = \left(\bigwedge\nolimits_{x \in X} (1 - \bigwedge\nolimits_{y \in Y} R(x, y) + a)\right) \wedge 1;$$

$$\mathcal{D}(A, B) = \bigwedge\nolimits_{x \in X, y \in Y} (1 - |a - R(x, y)|).$$

(2) *The product t-norm* has no zero-divisors, that is, it satisfies assumption $(\dagger_*)$. Hence, under this assumption we can apply formulas obtained in Example 3 and have

$$\mathcal{D}(A, B) = \left(\bigwedge\nolimits_{x \in X_a} (a \mapsto \bigwedge\nolimits_{y \in B} R(x, y))\right) \wedge \left(\bigwedge\nolimits_{x \in X_a} \left(\bigwedge\nolimits_{y \in B} (R(x, y) \mapsto a)\right)\right).$$

To describe $\mathcal{D}(A, B)$ for the product t-norm in this situation we denote
$X_1 = \{x \in X \mid a < \bigwedge_{y \in B} R(x, y)\}$,
$X_2 = \{x \in X \mid a \geq \bigwedge_{y \in B} R(x, y)\}$. Then

$$\mathcal{D}(A, B) = \left(\bigwedge_{x \in X_1} \frac{a}{\bigwedge_{y \in B} R(x, y)}\right) \wedge \left(\bigwedge_{x \in X_2} \frac{\bigwedge_{y \in B} R(x, y)}{a}\right).$$

(3) *The minimum t-norm* has no zero-divisors, that is, it satisfies assumption $(\dagger_*)$. Therefore, using formulas obtained in Example 3 and notations from the previous paragraph, we have

$$\mathcal{D}(A, B) = \begin{cases} \bigwedge_{x \in X_2, y \in Y} R(x, y) & \text{if } X_2 \neq \emptyset; \\ a & \text{otherwise} \end{cases}.$$

□

Example 6. Here we will sketch calculation of $\mathcal{D}^{\uparrow}(A, B)$ and $\mathcal{D}^{\downarrow}(A, B)$ in case when A and B are 3-valued fuzzy sets. For simplicity of exposition we assume that condition $(\dagger_*)$ is satisfied. Besides we restrict to calculation only the basic expressions, that is $A^{\uparrow} \hookrightarrow B$, $B \hookrightarrow A^{\uparrow}$, $A \hookrightarrow B^{\downarrow}$ and $B^{\downarrow} \hookrightarrow A$.

Let X, Y be sets and $R : X \times Y \to [0, 1]$ be a fuzzy relation. Let $0 < a < 1$ and let $X = X_0 \cup X_a \cup X_1$ where the sets X_0, X_a, X_1 are disjoint. Further, let $0 < b < 1$ and let $Y = Y_0 \cup Y_b \cup Y_1$ where the sets Y_0, Y_b, Y_1 are disjoint. We define fuzzy sets $A : X \to [0, 1]$ and $B : Y \to [0, 1]$ by

$$A(x) = \begin{cases} 0 & \text{if } x \in X_0 \\ a & \text{if } x \in X_a \\ 1 & \text{if } x \in X_1 \end{cases}, \quad B(y) = \begin{cases} 0 & \text{if } y \in Y_0 \\ b & \text{if } y \in Y_b \\ 1 & \text{if } y \in Y_1 \end{cases}.$$

Then, reasoning in the same way as in the previous example, we have

$$A^{\uparrow}(y) = (\bigwedge_{x\in X_1} R(x,y)) \wedge (\bigwedge_{x\in X_a}(a \mapsto R(x,y)))\,;$$
$$B \hookrightarrow A^{\uparrow} = \left(\bigwedge_{y\in Y_1, x\in X_1} R(x,y)\right) \wedge \left(\bigwedge_{y\in Y_1, x\in X_a}(a \mapsto R(x,y))\right) \wedge$$
$$\left(\bigwedge_{y\in Y_b, x\in X_1}(b \mapsto R(x,y))\right) \wedge \left(\bigwedge_{y\in Y_b, x\in X_a}(b \mapsto (a \mapsto R(x,y)))\right);$$
$$A^{\uparrow} \hookrightarrow B = \left(\bigwedge_{y\in Y_1, x\in X_1} R(x,y)\right) \wedge \left(\bigwedge_{y\in Y_1, x\in X_a}(a \mapsto R(x,y))\right) \wedge$$
$$\left(\bigwedge_{y\in Y_b}(\bigwedge_{x\in X_1}(R(x,y) \mapsto b))\right) \wedge \left(\bigwedge_{y\in Y_b}(a \mapsto (\bigwedge_{x\in X_a}(R(x,y) \mapsto b)))\right);$$

$$B^{\downarrow}(x) = \left(\bigwedge_{y\in Y_1} R(x,y)\right) \wedge \left(\bigwedge_{y\in Y_b}(b \mapsto R(x,y))\right);$$
$$B^{\downarrow} \hookrightarrow A = \left(\bigwedge_{x\in X_1, y\in Y_1} R(x,y)\right) \wedge \left(\bigwedge_{x\in X_1, y\in Y_b}(b \mapsto R(x,y))\right) \wedge$$
$$\left(\bigwedge_{x\in X_a}(\bigwedge_{y\in Y_1} R(x,y) \mapsto a)\right) \wedge \left(\bigwedge_{x\in X_a}\left(b \mapsto (\bigwedge_{y\in Y_b} R(x,y) \mapsto a)\right)\right);$$

$$A \hookrightarrow B^{\downarrow} = \left(\bigwedge_{y\in Y_1, x\in X_1} R(x,y)\right) \wedge \left(\bigwedge_{x\in X_1, y\in Y_b}(b \mapsto R(x,y))\right) \wedge$$
$$\left(\bigwedge_{x\in X_a, y\in Y_1}(a \mapsto R(x,y))\right) \wedge \left(\bigwedge_{x\in X_a}\left(a \mapsto (b \mapsto \bigwedge_{y\in Y_b} R(x,y))\right)\right).$$

4.4 Graded Preconcept Lattices

Given a fuzzy context (X, Y, R, L), let $(\mathbb{P}(X, Y, R, L), \preceq)$ be the corresponding fuzzy preconcept lattice, and let $\mathcal{D}^{\uparrow}$, $\mathcal{D}^{\downarrow}$ be the operators defined in the previous subsection.

Definition 8. *The tuple* $(\mathbb{P}, \preceq, \mathcal{D}^{\uparrow}, \mathcal{D}^{\downarrow})$ *is called graded fuzzy concept lattice of the fuzzy context* (X, Y, R, L).

In the next theorems we provide the most important properties of the operators of $\mathcal{D}$-gradation and $\mathcal{D}$-graded fuzzy preconcept lattices.

Theorem 3. *Let* $\mathbb{P} = (\mathbb{P}, \preceq \curlyvee, \bar{\curlywedge})$ *be a fuzzy preconcept lattice. Given a family of fuzzy preconcepts* $\mathcal{P} = \{P_i = (A_i, B_i) \mid i \in I\} \subseteq \mathbb{P}$ *it holds*

$$\mathcal{D}^{\uparrow}(\underline{\curlyvee}_{i\in I} P_i) \geq \bigwedge_{i\in I} \mathcal{D}^{\uparrow}(P_i).$$

Thus the degree of object based contentment $\mathcal{D}^{\uparrow}$ *of the union of fuzzy preconcepts is not less then the infimum (in L) of the degrees of object based contentments* $\mathcal{D}^{\uparrow}$ *of the separate fuzzy preconcepts.*

Proof. The proof follows from the following series of (in)equalities that are justified by applying Proposition 5 and Proposition 7:

$$\mathcal{D}^{\uparrow}(\underline{\curlyvee}_{i\in I}(A_i, B_i)) = (\bigvee_{i\in I} A_i)^{\uparrow} \cong (\bigwedge_{i\in I} B_i) = \left(\bigwedge_{i\in I} A_i^{\uparrow}\right) \cong (\bigwedge_{i\in I} B_i) =$$
$$\left(\left(\bigwedge_{i\in I} A_i^{\uparrow}\right) \hookrightarrow (\bigwedge_{i\in I} B_i)\right) \wedge \left(\left(\bigwedge_{i\in I} A_i^{\uparrow}\right) \hookleftarrow (\bigwedge_{i\in I} B_i)\right) \geq$$
$$\left(\bigwedge_{i\in I} A_i^{\uparrow} \hookrightarrow B_i\right) \wedge \left(\bigwedge_{i\in I}(A_i^{\uparrow} \hookleftarrow B_i)\right) = \bigwedge_{i\in I} \mathcal{D}^{\uparrow}(A_i, B_i).$$

□

Theorem 4. *Given a family of fuzzy preconcepts* $\mathcal{P} = \{P_i = (A_i, B_i) \mid i \in I\} \subseteq \mathbb{P}$ *it holds*

$$\mathcal{D}^{\downarrow}(\bar{\wedge}_{i\in I} P_i) \geq \bigwedge_{i\in I} \mathcal{D}^{\downarrow}(P_i).$$

Thus the degree of attribute based contentment $\mathcal{D}^{\downarrow}$ *of the meet of fuzzy preconcepts in the lattice* $(\mathbb{P}, \preceq, \bar{\wedge}, \underline{\vee})$ *is not less than the infimum of degrees of attribute based contentment* $\mathcal{D}^{\downarrow}$ *of the separate fuzzy preconcepts.*

The proof follows from the next series of (in)equalities that are justified by applying Proposition 5 and Proposition 7:

$$\mathcal{D}^{\downarrow}(\bar{\wedge}_{i\in I}(A_i, B_i)) = (\bigwedge_{i\in I} A_i) \cong (\bigvee_{i\in I} B_i)^{\downarrow} = (\bigvee_{i\in I} A_i) \cong \left(\bigwedge_{i\in I} B_i^{\downarrow}\right) =$$
$$\left((\bigwedge_{i\in I} A_i) \hookrightarrow \left(\bigwedge_{i\in I} B_i^{\downarrow}\right)\right) \wedge \left((\bigwedge_{i\in I} A_i) \hookleftarrow \left(\bigwedge_{i\in I} B_i^{\downarrow}\right)\right) \geq$$
$$\left(\bigwedge_{i\in I}\left(A_i \hookrightarrow B_i^{\downarrow}\right)\right) \wedge \left(\bigwedge_{i\in I}\left(A_i \hookleftarrow B_i^{\downarrow}\right)\right) = \bigwedge_{i\in I} \mathcal{D}^{\downarrow}(A_i, B_i).$$

□

5 Risk Assessment and Fuzzy Preconcept Lattices

5.1 Risk Assessment Model

The risk assessment process often allows a combination of statistical data and qualitative evaluations based on expert opinions. When dealing with any risks of rare occurrence the whole society usually faces unique circumstances with no historic patterns. Such circumstances are often called the tail events in the terminology of the risk management. These events are characterized by low possibility, but huge impact. Pandemic scenario is among the most obvious examples. Therefore the spread of Covid-19 pandemic provided a unique opportunity in studying possible pandemic developments from different angles. While a lot of statistical data have been collected, we should still combine the statistics with expert based opinions on possible further developments related to sudden virus outbreaks and mutations in order to propose the actions for, e.g. national governments or other decision making institutions. In its essence, the whole Covid-19 crisis management was dependent on fuzzy logic based decisions and provided favorable preconditions for enhancing the fuzzy logic methods. Several medical studies based on fuzzy logic applications have been carried out, see, e.g. [37], [38]. From a different perspective, it is always important to introduce the models for taking different decisions under these circumstances, such as modeling of availability of medical services, limiting or restricting public activities and services etc. We propose the analysis of data containing estimated total maximum number of infected inhabitants vs estimated total maximum number of hospitalized inhabitants at any fixed time moment as explained in the Table 1. It should be admitted that the estimated total number of infected inhabitants can be replaced with any other parameter characterizing the spread of pandemics, e.g. estimated cumulative cases, while the estimated total maximum number of hospitalized inhabitants at any fixed time moment is the most important parameter in the model characterizing the crisis severity. In many instances such model could be expanded further to the analysis of mortality rate, but this stage illustrates the efficiency of the healthcare system and its ability to save the lives of inhabitants. In the example we also consider the values of total number of inhabitants of particular country and the maximum admissible number of hospitalized inhabitants, but these values are used just for the illustrative reference purposes as the benchmark values for assessment of the severity of pandemics. We introduce the model containing the following objects and attributes: set X contains estimated quantities of infected inhabitants and set Y contains estimated quantities of hospitalized inhabitants. In this context we treat X as a set of objects expressed in a vector format $x = (x_1, x_2, \ldots, x_m)$ and Y as a set of attributes expressed in a vector format $y = (y_1, y_2, \ldots, y_n)$. Furthermore, we introduce a fuzzy relation $R(x_i, y_j) = \alpha_{ij}$ which is a matrix containing corresponding expert opinion:

$$R(x_i, y_j) = \begin{pmatrix} \alpha_{11} & \alpha_{12} & \cdots & \alpha_{1n} \\ \alpha_{21} & \alpha_{22} & \cdots & \alpha_{2n} \\ \vdots & \vdots & \ddots & \vdots \\ \alpha_{m1} & \alpha_{m2} & \cdots & \alpha_{mn} \end{pmatrix}$$

This model can be applied for various purposes, including illustration of expert assessments, comparing of these assessments against real statistics, adjusting these assessments depending on new circumstances and taking decisions on implementation of public restrictions. For the sake of clarity, it should be noted that the proposed setup can be used for developing the model for any risk assessment, but we have chosen the tail events and Covid-19 crisis in particular for highlighting the importance of expert opinion. In the following subsection we show how the fuzzy preconcept lattice can be applied for estimation of the possible impact of Covid-19 crisis on the healthcare system in Latvia.

Table 1. The data for modeling of pandemic scenario. The following data are provided in the respective columns: 1 - Total Number of Country Inhabitants w, 2 - Estimated Total Maximum Number of Infected Inhabitants x_i, 3 - Estimated Total Maximum Number of Hospitalized Inhabitants y_j, 4 - Maximal Admissible Number of Hospitalized Inhabitants z.

1	2	3	4
	x_1	y_1	
	x_2	y_2	
w	$\vdots$	$\vdots$	z
	x_m	y_n	

5.2 Assessment of Possible Covid-19 Impact on the Healthcare System in Latvia

The management of Covid-19 crisis created almost similar challenges in all countries around the world. Different public restrictions had been implemented and lifted based

Table 2. The data for modeling of pandemic scenario in Latvia. The following data are provided in the respective columns: 1 - Total Number of Inhabitants, 2 - Estimated Total Maximum Number of Infected Inhabitants, 3 - Estimated Total Maximum Number of Hospitalized Inhabitants, 4 - Maximal Admissible Number of HospitalizedInhabitants.

1	2	3	4
	50,000		
	80,000	500	
1,883,000	110,000	1,000	3,000
	140,000	1,500	
	170,000		

on actual rates of infected inhabitants, risks of new Covid-19 mutations and vaccination rates. In our opinion, the management of the public restrictions can be based on the proposed risk assessment model. Therefore the corresponding rates for Latvia provided in the Table 2 can be replaced with corresponding numbers for any country.

In the example we link the estimated total maximum number of infected inhabitants x with estimated total maximum number of hospitalized inhabitants y by applying the following fuzzy relation reflecting hypothetical expert opinion[2]:

$$R(x, y) = \begin{pmatrix} 0.9 \; 0.9 \; 0.4 \\ 0.9 \; 0.8 \; 0.4 \\ 0.6 \; 0.5 \; 0.3 \\ 0.5 \; 0.4 \; 0.3 \\ 0.4 \; 0.3 \; 0.2 \end{pmatrix}$$

This matrix is further analyzed in such way that we consider the values of $R(x, y)$ for each particular row. First of all we analyze scenario when the values of x are considered as precise and calculate the values of $\mathcal{D}(A, B)$ as provided in the Example 5. In such case $a = 1$ and we obtain the following values for corresponding rows for the Łukasiewicz t-norm, the product t-norm and the minimum t-norm (all results are equal for the corresponding t-norms in this case, indices of $\mathcal{D}(A, B)$ denote the corresponding rows for which the corresponding values have been calculated): $\mathcal{D}_1(A, B) = 0.4$, $\mathcal{D}_2(A, B) = 0.4$, $\mathcal{D}_3(A, B) = 0.3$, $\mathcal{D}_4(A, B) = 0.3$, $\mathcal{D}_5(A, B) = 0.2$.

Furthermore we can assume that the values of x are not precise. In such case we fuzzify these values by applying different values of a. The following example shows the values of $\mathcal{D}(A, B)$ calculated for $a = 0.8$. For the Łukasiewicz t-norm we obtain the following values: $\mathcal{D}_1(A, B) = 0.6$, $\mathcal{D}_2(A, B) = 0.6$, $\mathcal{D}_3(A, B) = 0.5$, $\mathcal{D}_4(A, B) = 0.5$, $\mathcal{D}_5(A, B) = 0.4$. For the product t-norm we obtain the following values: $\mathcal{D}_1(A, B) = 0.5$, $\mathcal{D}_2(A, B) = 0.5$, $\mathcal{D}_3(A, B) = 0.375$, $\mathcal{D}_4(A, B) = 0.375$, $\mathcal{D}_5(A, B) = 0.25$. For the minimum t-norm we obtain the following values:

[2] The values closer to 0 indicate lower possibility while the values closer to 1 indicate higher possibility.

$\mathcal{D}_1(A, B) = 0.4$, $\mathcal{D}_2(A, B) = 0.4$, $\mathcal{D}_3(A, B) = 0.3$, $\mathcal{D}_4(A, B) = 0.3$, $\mathcal{D}_5(A, B) = 0.2$. Another example shows the values of $\mathcal{D}(A, B)$ calculated for $a = 0.4$. Consequently, for the Łukasiewicz t-norm we obtain the following values: $\mathcal{D}_1(A, B) = 0.5$, $\mathcal{D}_2(A, B) = 0.5$, $\mathcal{D}_3(A, B) = 0.8$, $\mathcal{D}_4(A, B) = 0.9$, $\mathcal{D}_5(A, B) = 0.8$. For the product t-norm we obtain the following values: $\mathcal{D}_1(A, B) = 0.44$, $\mathcal{D}_2(A, B) = 0.44$, $\mathcal{D}_3(A, B) = 0.67$, $\mathcal{D}_4(A, B) = 0.75$, $\mathcal{D}_5(A, B) = 0.5$. For the minimum t-norm we obtain the following values: $\mathcal{D}_1(A, B) = 0.4$, $\mathcal{D}_2(A, B) = 0.4$, $\mathcal{D}_3(A, B) = 0.3$, $\mathcal{D}_4(A, B) = 0.3$, $\mathcal{D}_5(A, B) = 0.2$.
These results reflect differences in the impact of the t-norms used in the above calculations. On the practical side we can foresee that application of Łukasiewicz t-norm returns higher values indicating possibility of more severe pandemic impact. At the same time we should admit that these results should not be interpreted as correct or incorrect, but just as illustration of different expert opinions and application of different t-norms in their analysis.

6 Conclusions

The research has been focused on the application of an alternative method in risk assessment based on application of graded fuzzy concept lattices. Graded fuzzy concept lattices were introduced and studied in the paper [22], their basic properties are expounded also in our earlier papers. The use of graded concept lattices could be more appropriate for practical tasks, in particular, those ones that are related to risk assessment, than fuzzy concept lattices due to the possibility to establish a more flexible relationship between the fuzzy set of objects and corresponding fuzzy set of attributes. In this paper, fundamentals of the theory of graded fuzzy concept lattices have been provided. We analyzed the application of graded fuzzy concept lattices in the analysis of pandemic spread by introducing the model of estimated total number of infected inhabitants vs. estimated levels of hospitalized inhabitants which have been assessed based on expert opinion. The application of different t-norms in the calculation of the degree of conceptuality revealed differences in the impact of the selected t-norms on the assessment of the crisis severity. The obtained results can be further applied in the implementation of public restrictions aimed at containment of Covid-19 type pandemics. The application of the proposed model in the analysis of other risks, e.g., implementation of international sanctions (embargoes), including assessment of their efficiency, and the aggregation of multiple opinions analyzed by means of fuzzy preconcept lattices are subject to our further research.

References

1. Chan, H.K., Wang, X.: Fuzzy Hierarchical Model for Risk Assessment: Principles, Concepts, and Practical Applications. Springer Verlag, London (2013). https://doi.org/10.1007/978-1-4471-5043-5
2. Jones, W.: Project Risk Assessment: Fuzzy Logic Approach to Project Risk Assessment. VDM Verlag, Düsseldorf (2009)
3. Lilly, J.H.: Fuzzy Control and Identification, 1st edn. John Willey & Sons Inc, Hoboken (2010)

4. Wille, R.: Concept lattices and conceptual knowledge systems. Comput. Math. Appl. **23**(6–9), 493–515 (1992)
5. Yao, Y.: Concept lattices in rough set theory. In: Dick, S., Kurgan, L., Pedrycz, W., Reformat, M. (eds.) Proceedings of 2004 Annual Meeting of the North American Fuzzy Information Processing Society (NAFIPS 2004), pp. 796–801. IEEE, Piscataway, N.J. (2004)
6. Yao, Y., Chen, Y.: Rough set approximations in formal concept analysis. In: Peters, J.F., Skowron, A. (eds.) Transactions on Rough Sets V. LNCS, vol. 4100, pp. 285–305. Springer, Heidelberg (2006). https://doi.org/10.1007/11847465_14
7. Düntsch, I., Gediga, G.: Approximation operators in qualitative data analysis. In: de Swart, H., Orlowska, E., Schmidt, G., Roubens, M. (eds.) Theory and Application of Relational Structures as Knowledge Instruments, pp. 216–233. Springer, Heidelberg (2003). https://doi.org/10.1007/11964810
8. Ganter, B., Wille, R.: Formal Concept Analysis: Mathematical Foundations. Springer Verlag, Berlin (1999). https://doi.org/10.1007/978-3-642-59830-2
9. Bělohlávek, R.: Fuzzy galois connections. Math. Log. Q. **45**, 497–504 (1999)
10. Bělohlávek, R.: Concept lattices and order in fuzzy logic. Ann. Pure Appl. Logic **128**, 277–298 (2004)
11. Bělohlávek, R., Vychodil, V.: What is the fuzzy concept lattice? In: Proceedings of the CLA 2005 International Workshop on Concept Lattices and their Applications, pp. 34–45. Olomouc, Czech Republic (2005)
12. Hu, J., Chen, D., Liang, P.: A novel interval three-way concept lattice model with its application in medical diagnosis. Mathematics **7**(1), 103 (2019)
13. Keller, B.J., Eichinger, F., Kretzler, M.: Formal concept analysis of disease similarity. In: AMIA Joint Summits on Translational Science Proceedings, pp. 42–51. American Medical Informatics Association (2012)
14. Liu, X., Hong, W., Song, J., Zhang, T.: Using formal concept analysis to visualize relationships of syndromes in traditional Chinese medicine. In: Zhang, D., Sonka, M. (eds.) ICMB 2010. LNCS, vol. 6165, pp. 315–324. Springer, Heidelberg (2010). https://doi.org/10.1007/978-3-642-13923-9_34
15. Yazdani, S., Hoseini, M.: Clinical reasoning in medicine: a concept analysis. J. Med. Educ. **16**(3), 154–162 (2018)
16. Rodhers, B.L., Jacelon, C.S., Kraft, K.A.: Concept analysis and the advance of nursing knowledge: state of the science. J. Nurs. Scholarsh. **50**(4), 451–459 (2018)
17. Raza, K.: Formal concept analysis for knowledge discovery from biological. Int. J. Data Min. Bioinform. **18**(4), 281–300 (2017)
18. Hashikami, H., Tanabata, T., Hirose, F., Hasanah, N.: An algorithm for recomputing concepts in microarray data analysis by biological lattice. J. Adv. Comput. Intell. Intell. Inform. **17**(5), 761–771 (2013)
19. Bělohlávek, R.: Fuzzy concept analysis in Geology. In: Demico, R.V., Klir, G.J. (eds.) Fuzzy Logic in Geology, pp. 191–237. Academic Press, San Diego, California, USA (2004)
20. Missaoui, R., Kuznetsov, S.O., Obiedkov, S. (eds.): Formal Concept Analysis of Social Networks. LNSN, Springer, Cham (2017). https://doi.org/10.1007/978-3-319-64167-6
21. Tonella, P.: Formal concept analysis and software engineering. In: ICSE 2004: Proceedings of the 26th International Conference on Software Engineering (Abstracts), pp. 743–744. IEEE Computer Society, Washington, DC, USA (2004)
22. Šostak, A., Uljane, I., Krastiņš, M.: Gradations of fuzzy preconcept lattices. Axioms **10**(1), 41 (2021). https://doi.org/10.3390/axioms10010041
23. Krastiņš, M., Uljane, I., Šostak, A.: Application of graded fuzzy preconcept lattices in risk analysis. In: Proceedings of the 13th International Joint Conference on Computational Intelligence, pp. 177–184 (2021). https://doi.org/10.5220/0010656500003063

24. Birkhoff, G.: Lattice Theory, 3rd edn. AMS, Providence, RI (1995)
25. Gierz, G., Hofmann, K.H., Keimel, K., Lawson, J.D., Mislove, M., Scott, D.S.: Continuous Lattices and Domains. Cambridge University Press, Cambridge (2003)
26. Morgan, W., Dilworth, R.P.: Residuated lattices. Trans. Am. Math. Soc. **45**, 335–354 (1939)
27. Rosenthal, K.I.: Quantales and Their Applications (Pitman Research Notes in Mathematics Series). Longman Scientific and Technical (1990)
28. Höhle, U.: M-valued sets and sheaves over integral commutative CL-monoids. In: Rodabaugh, S.E., Höhle, U., Klement, E.P. (eds.) Applications of Category Theory to Fuzzy Subsets, pp. 33–72. Springer, Netherlands, Dordrecht (1992). https://doi.org/10.1007/978-94-011-2616-8_3
29. Höhle, U.: Commutative residuated l-monoids. In: Höhle, U., Klement, E.P. (eds.) Non-Classical Logics and their Applications to Fuzzy Subsets: A Handbook of the Mathematical Foundations of Fuzzy Set Theory, pp. 53–106. Springer, Netherlands, Dordrecht (1995). https://doi.org/10.1007/978-94-011-0215-5_5
30. Han, S.-E., Šostak, A.: On the measure of M-rough approximation of L-fuzzy sets. Soft. Comput. **22**(12), 3843–3855 (2018)
31. Zadeh, L.A.: Fuzzy sets. Inf. Control **8**(3), 338–353 (1965)
32. Goguen, J.A.: L-Fuzzy sets. J. Math. Anal. Appl. **18**, 145–174 (1967)
33. Zadeh, L.A.: Similarity relations and fuzzy orderings. Inf. Sci. **3**, 177–200 (1971)
34. Valverde, L.: On the structure of F-indistinguishability operators. Fuzzy Sets Syst. **17**(3), 313–328 (1985)
35. Han, S.-E., Šostak, A.: M-valued measure of roughness for approximation of L-fuzzy sets and its topological interpretation. In: Merelo, J.J., Rosa, A., Cadenas, J.M., Dourado, A., Madani, K., Filipe, J. (eds.) Computational Intelligence. SCI, vol. 620, pp. 251–266. Springer, Cham (2016). https://doi.org/10.1007/978-3-319-26393-9_15
36. Klement, E.P., Mesiar, R., Pap, E.: Triangular Norm. Springer, Dordrecht (2000). https://doi.org/10.1007/978-94-015-9540-7
37. Orozco-del-Castillo, M.G., et al.: Fuzzy logic-based COVID-19 and other respiratory conditions pre-clinical diagnosis system. In: Mata-Rivera, M.F., Zagal-Flores, R., Barria-Huidobro, C. (eds.) WITCOM 2020. CCIS, vol. 1280, pp. 402–419. Springer, Cham (2020). https://doi.org/10.1007/978-3-030-62554-2_29
38. Shaban, W.M., Rabie, A.H., Saleh, A.I., Abo-Elsoud, M.A.: Detecting COVID-19 patients based on fuzzy inference engine and deep neural network. Appl. Soft Comput. J. **99**, 106906 (2021)

Check for updates

Improving Simulation Realism in Developing a Fuzzy Modular Autonomous Driving System for Electric Boats

Emanuele Ferrandino, Antonino Capillo(✉), Enrico De Santis, Fabio M. F. Mascioli, and Antonello Rizzi

Department of Information Engineering, Electronics and Telecommunications (DIET), University of Rome "La Sapienza", Rome, Italy
{emanuele.ferrandino,antonino.capillo,enrico.desantis, fabiomassimo.frattalemascioli,antonello.rizzi}@uniroma1.it

Abstract. In this work an Autonomous Driving System is presented with the attempt to define a qualified standard for a specific class of electric boats. The design involved both Classical Computation and Computational Intelligence techniques (mainly Fuzzy Logic and Artificial Neural Networks) and several version of the same ADS modular architecture have obtained progressively replacing the classical modules with the intelligent ones. The modular architecture is organized in "vertical" control levels which establish a control hierarchy and "horizontal" layers that follow the direction of data processing (from real-time perception of the environment to propulsion control). Compared to what was done in our previous works, the architecture proposed for the Autonomous Driving System for a specific type of boat has been updated and it is more detailed here. Furthermore, the ad-hoc Simulator developed to test and validate the ADS was improved in realism degree and other simulations have been performed. Outcoming data and post-analyses are reported here together with our conclusions.

Keywords: e-Boats · Autonomous driving system · Modular architecture · Target oriented navigation · Dynamic obstacle avoidance · Dockside avoidance · Motion control · Virtual anchor · Fuzzy controller · Fish schooling behavior · Hydrodynamic modeling · Simulations

1 Introduction

The self-driving problem has become an important topic in research and industry in recent times. Its main application is on land vehicles, especially private cars. In other sectors, such as the naval one, systems that automate only certain phases of navigation have long been present. This is due to an environment which, although equally rich in elements, still remains complex to manage, as deduced from [2] and highlighted in [3]. A vessel does not stand on a surface and it is not completely submerged in fluid but it is semi-submerged. Furthermore, in the sea there are no vertical or horizontal signs but there are luminous and sound ones or even communications through flags and other symbols. Again, at sea there are no roads and parking lots but routes and ports (structures that can consist of a simple pier or many docks). Despite these differences,

J. Garibaldi et al. (Eds.): IJCCI 2021, SCI 1119, pp. 163–184, 2023.
https://doi.org/10.1007/978-3-031-46221-4_8

there are rules of precedence at sea when two boats meet as well as between two cars at a crossroads. To manage all this, a pilot is not enough, hence a crew is needed. Perhaps for this very reason, applying the problem of self-driving in a marine context, rather than in a road one, is certainly complicated and challenging from many points of view.

The subject of this study is the Valentino III, a solar-hybrid electric boat [9], which is the core of a project that involves the development of a marine sustainable mobility system – managed by the *Pole for Sustainable Mobility* (Po.Mo.S.) in collaboration with the Department of Information Engineering, Electronics and Telecommunications (DIET) of the University of Rome *La Sapienza* – as a part of an intermodal sustainable mobility system that, in turn, is part of a major European project, named *LIFE for Silver Coast* (LIFE16 ENV/IT/000337), affecting the town of Orbetello and the territories of Monte Argentario and Isola del Giglio (in Italy) which The e-boat is intended for public transport, the touristic service and environmental monitoring in the lagoon of Orbetello. As a lagoon, the environment of the boat is a very particular marine environment. The lagoon is a closed body of water near the sea. The lagoon of Orbetello is impracticable for many boats (the average depth is 1 (m) with a maximum excursion of 0.4 (m)). The smaller boats are the only ones able to navigate in this environment. The Valentino III is shown in Figs. 1 a three-hull boat with a draft of a few centimeters and a paddle propulsion (as in a ferry boat) instead of a propeller one. In this way, it will be possible to transport many people together from one side of the lagoon to the other while also respecting the seabed. The original propulsion system, consisting of four paddle wheels with controllable paddle pitches and rotational speeds, gives to the boat 5 degrees of freedom (the greatest maneuverability for a boat). Such feature is inspired by the nature and in particular by aquatic animals with four limbs such as turtles, which can additionally submerge and emerge.

Fig. 1. Valentino III Concept: side view.

In our previous work, [3], we established that the best approach in designing a self-driving system for boats is the modular one and we designed an ad-hoc architecture to manage all navigating conditions (navigating in water open, avoid collisions with other boats, navigate near and inside ports). Compared to the previous version of the Simulator, the hydrodynamics of the boat [1] and the water currents are now modeled. For this reason all simulations performed in [3] are repeated giving new and more realistic results. In addition, test results are also reported on the lower and higher control level (in the past only the middle control level was considered) as we designed the motion control and compared the ADS with FSM-based high control level with one based on Fuzzy Controller [24].

In Sect. 2 each function of the ADS architecture is described and it is shown the design of each controller. In Sect. 3 are formulated the evaluation metrics of the ADS performances which provide a tool to compare the results. Simulation results are shown and discussed in Sect. 4 and in Sect. 5 the conclusions are reported.

2 Proposed Autonomous Driving System Architecture

In this Section the ADS architecture proposed for Valentino III class of boats is presented. It is a modular architecture that interfaces with various smart sensors (such as a camera, sonars, a GPS and a IMU), with a HMI and with the Propulsion System. The architecture is shown in Fig. 2.

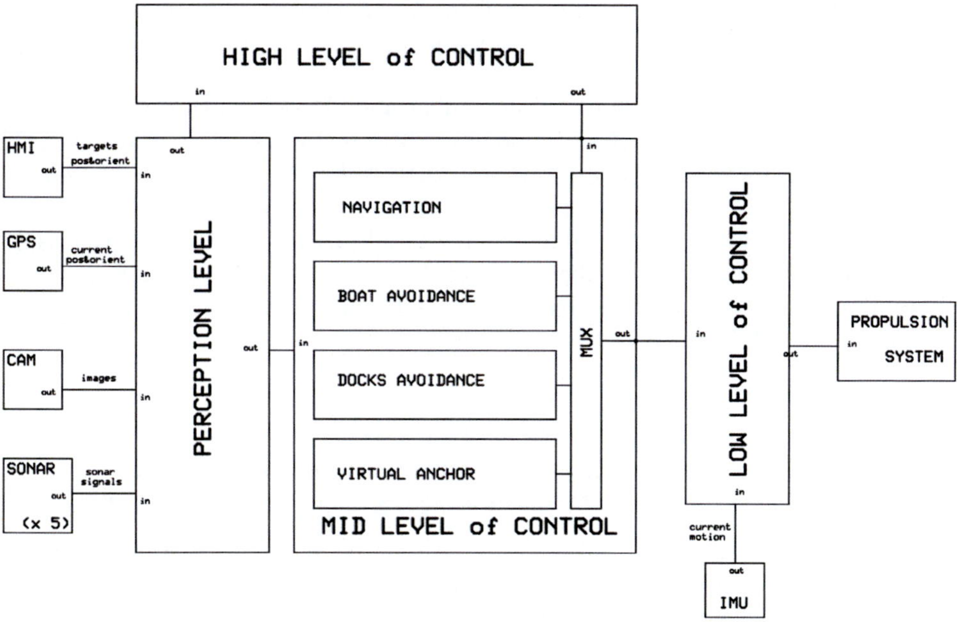

Fig. 2. ADS architecture.

The ADS architecture is organized both with a vertical hierarchy and an horizontal data processing direction. Furthermore a part of the architecture consists of four parallel pipelines.

Vertical Hierarchy. The vertical hierarchy consists of three level of control:

- Low-Level of Control: the lowest level of control of the architecture is responsible to apply and correct the motion of the boat (it is equivalent to the last module of a classical single pipeline ADS architecture).
- Mid-Level of Control: the middle level of control is composed by several parallel pipelines. Each pipeline takes perception signals (from sensors and/or other modules) in input and provides the set point (a maneuver) to the lower level of control.
- High-Level of Control: the highest level of control acts on the selection of the pipeline that must provides the set point to the lowest level of control. To do this, both perception signals and internal state (from the middle level of control) must be considered.

It should be noted that the communication between the control levels is not unidirectional and is not as ordered as the levels themselves. In other words, from a simple three-level hierarchy the highest level is expected to communicate only towards the middle level and that the middle level is expected to communicate only towards the lower level. In the three-level hierarchy presented, indeed, the highest level is placed between the two lowest levels (the middle and the high) and receives communications from the middle level. The latter, on the other hand, communicates only with the low level. Furthermore, it is not excluded that, for future developments, a communication will also be implemented from the lowest to the highest and/or the middle level.

Horizontal Organization. The horizontal organization regards pipelines composition. Pipelines are organized in series of modules, where the first module is always a Perception Level Module and the last module is always the Motion Control Module. So, data are constrained to being processed beginning from Perception-Level, passing for the Mid-Level of Control and ending at Low-Level of Control. Another data path is present: the one which involves the High-Level of Control. Such data-path starts from Perception-Level, passes for the High-Level of Control and ends in the multiplaxer placed between Mid and Low-Level of Control.

Pipelines. A further division is in Mid-level Controller, which distinguishes the Pipelines Each pipeline is devoted to implement a specific driving function. Therefore, only one pipeline at a time can provide input at the low-level of control, and the selection of that pipeline depends on the situation recognized by the high-level of control.

- Navigation Pipeline: this pipeline operates when the boat is sailing in open waters and no obstacles or particular events are detected.
- Boat Avoidance Pipeline: this pipeline operates when the boat is sailing in open waters and another boat is detected and treated as an obstacles to be avoided.
- Docks Avoidance Pipeline: this pipeline operates when the boat is sailing inside a port and has to enter or leave it.
- Virtual Anchor Pipeline: this behavior operates when the boat does not have to sail but must maintain the current position or trim because the objective has been reached or because the captain has requested it through a specific command or because the boat has to wait for the passage of other boats.

In the following sub-sections are detailed the three level of control and, for the mid-level of control, are detailed the four pipelines.

2.1 LLC Design: Motion Control

The motion control represent the Low-Level of Control - LLC and consists of a single module, the Motion Controller. The Motion Controller has the double task of maintain the current motion (acquired by the IMU) close to the set point (given by the MLC) and map it into the control signals of the propulsion system. These tasks is carried out with a period of 1 (s).

Mapping. Paddle wheels with controlled pitch angles define the following mapping between the input space and the output space of the Motion Controller:

$$v_{surge}, v_{sway}, \omega_{yaw}, \omega_{pitch}, \omega_{roll} \quad \mapsto \quad \omega_1, \omega_2, \omega_3, \omega_4, \psi_1, \psi_2, \psi_3, \psi_4 \tag{1}$$

where v_{surge} and v_{sway} are the surge and sway speeds, respectively; ω_{yaw}, ω_{pitch} and ω_{roll} are the boat's yaw, pitch and roll speeds, respectively; ω_i and ψ_i, for $i = 1, ..., 4$, are the paddle wheels' rotation speeds and pitch angles, respectively. Note that the problem is under-defined and than it is not possible to solve it in closed form. For this reason, we propose a feed-forward NN-based Direct Controller (DC) architecture trained by a back-propagation (BP) algorithm for dynamically mapping the motion control signals to the paddle wheels' rotation speeds and pitch angles. Back-propagation signals are the effective *surge*, *sway*, *yaw*, *pitch* and *roll* of the boat, provided by the Inertial Measurement Unit (IMU). In this way, small perturbations (such as surface currents in inland waters) will also be compensated.

From literature ([19,20] and [21]), NN-based Direct Control shown to be particularly effective in solving unknown nonlinear and non-affine problems. On the other hand, the DC architecture causes an instability of the plant response at the beginning of the training. The controller is a three-layered network in which sigmoidal activation functions are used. The plant represents the complex environment. The controller parameters are updated on the basis of the BP algorithm, but in this particular architecture the error is calculated on the output of the plant and not on the output of the network itself (as is done in the classic BP algorithm). The error is back-propagated through the plant at each time step in order to update the controller parameters. To do this it is necessary to know the Jacobian of the plant ([22] and [23]) or its approximation with the signs of the elements of the Jacobian. Fortunately, in our case the Jacobian of the plant is known. In fact, the hydrodynamic model used to describe the boat motion in the simulator also outcomes the Jacobian coefficients, $\frac{\delta y}{\delta u}$, where y (which represents the measurements made by the IMU) is the output of plant and the input of the ANN and u (which represents the signal of control of the Propulsion System that is a part of the Plant) is input of the Plant and the output of the ANN (since ANN and plant constitute a loop). The goal of NN training is to generate the signals u such that the error, $e = r - y$, is close to 0. This architecture offers the advantage of being able to carry out a continuous online training. In this way, the controller can adapt to the constant changes of the environment.

By the way, since a well trained ANN ideally causes a null error between reference signal and measurement, and the training itself involves a non-negligible computational Reward, currently this part is not implemented in the simulator. Indeed, the focus of the present work is on the controllers of the ADS architecture.

P Control. Given the presence of noise in the environment in which the boat moves, it is necessary to approach motion control. This control must reduce the error between the desired movement and the actual movement. In this way, the boat will not be completely at the mercy of the surface water currents and will keep a course as close as possible to that calculated by the HLC.

Beyond the above mapping solving, the motion controller must implement an optimal control technique. The latter takes input signals from an inertial measurement system consisting of three accelerometers and three gyroscopes and the MLC. In a simulation environment it is not crucial to implement the mapping but it is necessary to implement a control that operates as described above. For this purpose, a P control can be implemented.

The application of the P controller at each instant converges the error vector to a minimum value. To obtain an optimal limit and convergence speed, the parameter of the P controller must be adjusted following an optimization criterion. In fact, there is no particular parameter value that satisfies every application. There are many methods of adjusting the parameters, among which the manual one and the Ziegler-Nichols method are the most frequent. Here an ad hoc method was approached which contemplates increasing Kp from 0 to 1 and evaluating the performance obtained by the boat in a set of simulations for each value of Kp. The set of simulations that have achieved the highest performance contains the Kp value that will be passed on to the next stages of the implementation.

2.2 Navigation Pipeline

The Navigation pipeline is responsible for generating a (default) route and piloting the boat to its destination. It is specialized in processing GPS and HMI data. The pipeline consists of the Perception, an Autopilot and the Motion Controller Perception and Autopilot run with a period of 10 s. Within and synchronously with previous modules (in the MLC), the Motion Controller runs ten times. The Perception in turn is composed by the Route planning – that is not automatic yet: the way points (that make up the route) and the target can be inserted by the HMI and registered into a table. –, and a further pre-processing part. The last is intended for computing the current target distance and relative position (from the current position and orientation and the current target position and orientation, which are provided by GPS and HMI, respectively). The Autopilot, on the other hand, is a crisp controller that takes the current target distance and relative position and returns a maneuver as output. The output of the Autopilot consists of a surge and a yaw speed. The sway speed is set to zero because side motion is not usual during navigation, while roll and pitch speeds are not considered since the kind of hull. The pre-processing part represents an interface of this pipeline (inside the MLC) with the HLC. In particular, it provides to the HLC a representation of the current navigation conditions. The pro-processing is implemented as a pre-combinatorial logic or as a digital filtering depending on the implementation of the HLC. In first case, the HLC takes as inputs some boolean state – `INPUT1` (final target reaching) and `INPUT2` (current target reaching) – whereas, in second case, the HLC takes as inputs some "analog" (continuous time-discrete) signals – the final target distance.

2.3 Boat Avoidance Pipeline

The Boat Avoidance pipeline is specialized in processing data from the vision system and it can re-plan a route fraction in order to avoiding collision with other boats. After the action of the Boat Avoidance pipeline, the actual route will change from the default one. The time constant of each pipeline is of 10 s. The present pipeline is composed by Perception, Boat Avoidance Controller - BAC and Motion Controller Perception is in turn composed by the Boat Detection System and Boat Tracker. The Perception provides its boolean state, `INPUT5` – which switch with the boat detection condition activation –, or the currently detected boat distance depending on the HLC implementation. It is assumed that CNN-based Boat Detection System, filled with images (that are sampled from a real-time video stream from the video camera), is pre-trained to recognize boats and return only the bounding box of the nearest detected boat (as seen in [6]). In fact, the BAC can manage only one boat at a time. The bounding boxes (of the same detected boat) are processed by a Boat Tracker which returns the relative position and orientation of the detected boat, and eventually the distance too. Such orientation is relative to the autonomous boat orientation. The Boat Avoidance Controller must calculate adequate maneuvers to avoid collision with the currently detected boat. This operation must be subject to the precedence at sea rules which are referenced by [5] and [8]. It should be noted that the BAC doesn't need for detected boat distance since such rules don't talk about. Similarly to the Autopilot (and within the same criteria), the Boat Avoidance Controller returns only surge and yaw speeds.

Crisp BAC. The crisp version of the controller is characterized by the discretization of inputs and outputs. The first input, the relative position of the detected boat, can belong to only one of three sets, which are labeled as "left", "front" and "right". The second input, the relative orientation of the detected boat, can belong to only one of four sets, which are labeled as "same", "opposite", "to the right" and "to the left". These distinctions are essential to be able to apply the known rule set that from linguistic rule set has been translated into an `if-else` statement (i.e., crisp rules). In this way, only one crisp rule can be activated for each situation. In total, we have 12 distinct situations and therefore we also have 12 rules for this controller. The first output, the surge speed, is discretized in three values: "null" – 0 (m/s) –; "mid" (1/2 of the maximum cruiser speed); "high" (maximum cruising speed). The second output, the yaw speed is discretized in three values: "weak left" (1/8 of a turn), "null" and "weak right" (1/8 of a turn).

Fuzzy BAC. The fuzzy version of the controller is characterized by the fuzzyfication of inputs and rules and defuzzyfication of outputs [7]. The relative position of the detected boat is characterized by a term set of three Membership Functions (MFs) labeled as "left-side", "frontal" and "right-side" (similarly to the crisp controller). The relative orientation of the detected boat is characterized by a term set of five MFs, which are labeled as "same" "opposite 1", "opposite 2", "on right", "on left" and "same". The MFs "opposite 1" and "opposite 2" represent similar situations and are located at the extremes of the same term set. Note that, closing this term set on itself (as if to form a circle), the two MFs make a unique MF equal to the others. Therefore, this term set is analogous to the discretization made for the same input of the crisp controller. The construction of these two term sets implies that the fuzzy controller has 15 rules. The fuzzy controller has been designed as a mamdani-type fuzzy controller to better manage the linguistic rules. The surge speed is characterized by a term set of three MFs: "null", "low" and "high". The yaw speed is characterized by a term set of three MFs: "left", "null" and "right". Each output is obtained by applying the centroid defuzzyfication method. In this way, more rules can be activated at the same time and collaborate in generating the final maneuver. In conclusion, this fuzzy controller consists in the fuzzyfication of its crisp version. This design choice is dictated by the fact that there are linguistic rules known *a priori* and by the fact that we want to generate maneuvers as smooth as possible. In Fig. 3 the Boat Avoidance Fuzzy Controller architecture is shown.

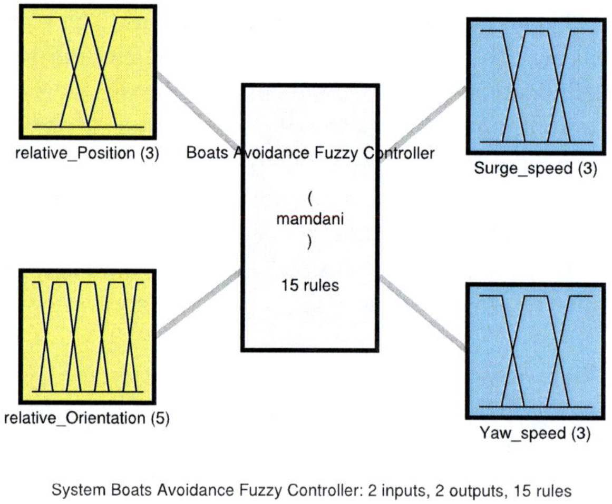

Fig. 3. Fuzzy Boat Avoidance Controllers.

2.4 Docks Avoidance Pipeline

The Docks Avoidance pipeline processes data from sonars and from MLC itself. In particular the Docks Avoidance Controller receives the "route misalignment" from the Perception Module A. The pipeline is composed by the Perception, the Docks Avoidance Controller - DAC and the Motion Controller Perception consists of the Docks Detector and a pre-processing part (which interfaces with the HLC). The latter can be implemented as a pre-combinatorial logic – in order to return two boolean state, `Input3` and `Input4` – or as an algebric function – in order to return the minimum dockside distance between the distances processed by the Docks Detector. Indeed, the Docks Detector converts the signals from five sonars into three distances: frontal distance, starboard (left side) distance and port (right side) distance. These three measures together with the route misalignment are the inputs of the DAC. The approach taken here is to consider the environment the boat navigates as unknown as done for many AMR (see [11]). The Docks Avoidance Controller processes the necessary maneuvers to avoid collision with the docks while performing target-oriented navigation in order to entry or exit ports.

Crisp DAC. The first input, the relative position of the target, can belong to only one of three sets: "left", "front" and "right". Each of the three sonar distance, can belong to only one of three sets: "near", "far" and "nothing". In total, there are 81 distinct situations and then 81 rules for the crisp controller. Here too, the activation of one of the 81 rules is managed by an `if-else` statement. To define the crisp rule set all situations were carefully considered and the most reasonable maneuver assigned to each rule. The first output, the surge speed, is discretized into three values: "null" – 0 (m/s) –, "low" (1/4 of the maximum cruising speed) and "mid" (1/2 of the maximum speed cruise). The second output, the yaw speed, is discretized into five values: "strong left" (1/4 of a turn), "weak left" (1/8 of a turn), "null", "weak right" (1/8 of a turn) and "strong right" (1/4 of a turn).

Fuzzy DAC. The fuzzy version of the controller is characterized by the fuzzyfication of inputs and rules and defuzzyfication of outputs. The relative position of the target is characterized by a term set of three MFs: "left", "front" and "right". The three sonar distances are characterized by term sets with three MFs: "near", "far" and "nothing". The speed surge is characterized by a term set of three MFs: "null", "low" and "high". The yaw speed is characterized by a term set of five MFs: "very left", "left", "null", "right" and "very right". Also in this case, the fuzzy controller is of the mamdani-type and the centroid defuzzyfication method is applied to each output. Finally, the same rule set as the crisp version of the controller is adopted here to define the fuzzy rule base. In Fig. 4 the architecture of the Docks Avoidance Fuzzy Controller is shown.

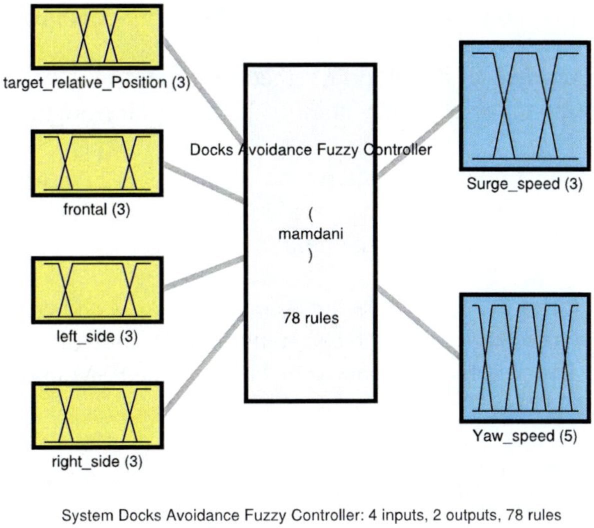

Fig. 4. Fuzzy Docks Avoidance Controllers.

2.5 High Level of Control: Pipeline Selection

HLC receives inputs from the perception modules whereas the output is used as pipeline selector; the multiplexer returns the maneuver for controlling the motion of the boat that is supplied by the pipeline associated with the output of the HLC. Since the number of pipeline are finite and the multiplexer accepts only discrete values for selecting, the HLC output shall be discrete. The HLC has been implemented in two ways: as a Finite State Machine - FSM and a Fuzzy Controller - FC.

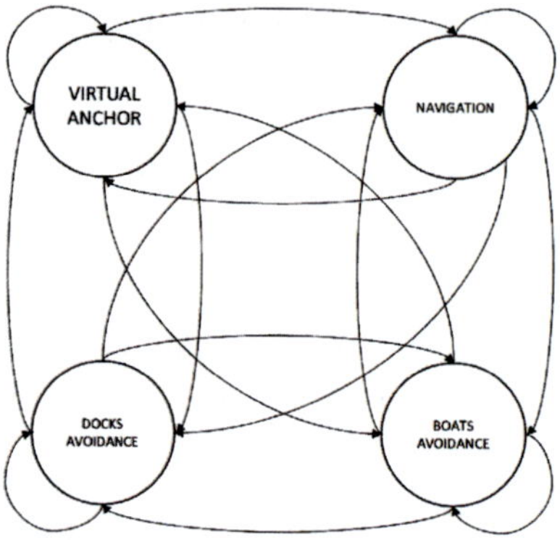

Fig. 5. FSM Transition Diagram.

FSM-Based HLC. The FSM has four state and one output (the state itself). The inputs of the FSM are five binary variables synthesized starting from the inputs of the ADS (measurements carried out by the sensors or results of pre-combinatorial logic):

- Input1: target reached (if it is 0) or assigned (if it is 1).
- Input2: proximity to target or way point (if it is 1) or far from the target or way point (if it is 0).
- Input3: one or more docks detected by the sonar (if it is 1) or not (if it is 0).
- Input4: collision with dock or other boat detected (if it is 1) or not (if it is 0).
- Input5: another boat detected in the field of vision (if it is 1) or not (if it is 0).

Transition rules are omitted whereas transition diagram is shown in Fig. 5. Note that the FSM is full-connected (each state can transition into one of the remaining state or remain in itself). This involved the easy translation of the FSM into a Fuzzy Controller since the future state only depends on the inputs.

FC-Based HLC: The HLC architecture changes as the alternative implementation method The Fuzzy Controller - FC is a type I Mamdani FIS. Figure 6 illustrates its architecture. The inputs are "Distance from the target", "Distance from another boat" and "Distance from a dockside".

The FIS output represents the state of the ADS and takes the labels "Virtual Anchor", "Navigation", "Boat Avoidance" or "Docks Avoidance". The abscissa of the output term set is also normalized but has no physical meaning, but decision making. To obtain this the FIS output shall be discretized with a threshold decision maker.
A FIS rule looks like the following:

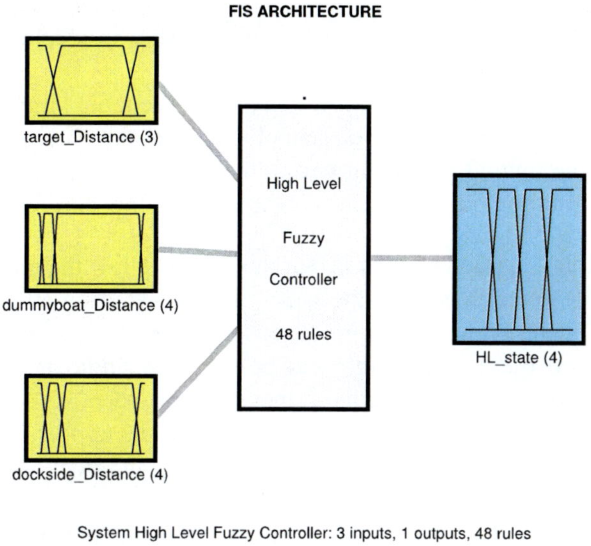

Fig. 6. FC based HLC.

```
IF target_distance IS "reached" AND
    dummy_boat_distance IS "not detected" AND
        dockside_distance IS "close"
THEN state IS "Virtual Anchor"
```

Fuzzy rules are handwritten by translating the transition rules of the FSM– so the Fuzzy-HLC is a pre-trained Fuzzy Controller.

3 Evaluation Metrics

The definition of evaluating metrics is essential to understand if the ADS is working as desired. In particular, the autonomous boat must approach (and not move away from) the target until it reaches it and avoid any type of obstacles (other boats and docks). A performance indicator and several statistics will serve to evaluate the performances obtained in terms navigation and to compare Classical and Fuzzy approaches.

3.1 Fish Schooling Behavior Inspired Reward Function

Fish Schooling Behavior describes the social behavior of fish moving in schools. It has inspired several engineers to develop useful techniques to study the fish themselves (as reported in [12] and [13]), to build a more realistic virtual reality application (as in [14]), to create realistic robot fish such as described in [15], but also to create intelligent transport systems (as in [16]) and search algorithms (as in [17] and [18]). However, the behavior of the school of fish consists of three principles, which are described in [10], namely:

- **Attraction:** the mutual attraction between the fishes in a school and between the school and with a common goal.
- **Repulsion:** the mutual repulsion between fish in a school, which allows each fish to have enough room to move and the school's repulsion to common dangers.
- **Alignment:** the mutual alignment between fish in a school and the school's alignment with a common direction, allowing them to follow current, for example.

These principles, applied to each fish of the school, allow the group to remain cohesive during the movements, achieve the same goal and avoid situations of danger for the individual and for the group. Since the three principles correspond to the appropriate behaviors of each fish in a school, they can be used in a single agent system as well as in a multi-agent system. Here the three principles have been applied to a single agent, the boat, and have been rewritten, according to the present application, as the attraction for the target attitude, repulsion from obstacles (boats and docks) and alignment with the route.

Furthermore, this behavior is suitable for describing the performance of navigation. Since boat and docks avoidance are performed by two distinguished sub-system, the repulsion term is divided in two sub-terms. So, the Reward Function is formulated as the convex combination, with constant convex coefficients, of four terms:

- Attraction to the target (AT);
- Alignment to the route (AL);
- Repulsion from other boats (RB);
- Repulsion from the docks (RD).

Each term is formulated as a sigmode function of the independent variable. In particular, the four terms are well described by the Equations in 5

$$f^{AT}(t) = \frac{1}{1 + \exp -k^{at} \cdot (x^{at}(t) - x_0^{at})} \tag{2}$$

$$f^{RB}(t) = \frac{1}{1 + \exp -k^{rb} \cdot (x^{rb}(t) - x_0^{rb})} \tag{3}$$

$$f^{RD}(t) = \frac{1}{1 + \exp -k^{rd} \cdot (x^{rd}(t) - x_0^{rd})} \tag{4}$$

$$f^{AL}(t) = 1 - \frac{1}{1 + \exp -k^{al} \cdot |x^{al}(t) - x_0^{al}|} \tag{5}$$

where: j is the iteration index of the simulation; $x^{at}(t)$ denotes the difference between the previous distance from the target and the current one and $x_0^{at} = 20\,(\text{m})$ denote the distance for which the function f^{AT} return a value equal to $0.5\,(\text{p.u.})$ (if the boat go back, $x^{at}(t)$ is negative, but it must be less than x_0^{at} to obtain a Reward less than 0.5); $x^{rb/d}(t) - x_0^{rb/d}$ denotes the difference between the actual distance from an obstacle (which can be a dummy boat or a dockside) and the minimum distance (which is set to $80\,(\text{m})$ for a dummy boat and to $60\,(\text{m})$ for a dockside); $x^{al}(t) - x_0^{al}$ denotes the difference between the current route misalignment and the maximum misalignment (which set to 90 (deg)). Note that the three terms are sigmoid functions. The coefficients $k^{at} = 0.25$, $k^{rb} = 0.08$, $k^{rd} = 0.1$ and $k^{al} = 4$ are empirically set scale factors.

So, the formula for the Reward, computed at the current iteration is the following:

$$r(t) = 0.33 \times f^{AT}(t) + 0.16 \times f^{RB}(t) + 0.16 \times f^{RD}(t) + 0.33 \times f^{AL}(t) \quad (6)$$

where the constant coefficients are so set in order to divide the weight equally into three weights of 0.33 (p.u.) (attraction, repulsion and alignment, according to our interpretation of fish schooling behavior) and divide one of the three weights into two equal parts of 0.16 (p.u.). In this way, it's given to Attraction and Alignment Terms a grater weight then the ones assigned to Boat and Docks Repulsion Terms as taken singularly.

The overall Reward for a given simulation can be used to resume the overall obtained performance. It is calculated as in the Eq. 7:

$$R = \frac{\sum_{t=1}^{t_{stop}} r(t)}{t_{stop}} \quad (7)$$

where t_{stop} is the iteration at which the simulation stops because the autonomous boat has reached the target. This formulation constitutes the first metric, which makes it possible to evaluate the ADS on the assumption that it successfully complete a simulation.

3.2 Stall, Collision and Success Probabilities

Beyond the cases in which a simulation ends successfully, collision events (with another boats or with docks) or stall events (i.e., events in which the autonomous boat does not reach the target by the end of the simulation without implying collisions) can occur. If the autonomous boat collides with something the simulation stops and the overall Reward is reset to -1 whereas, if the autonomous boat does go into a stall, the overall Reward is reset to 0 (as shown in Eq. 8). By counting the events of collision, stall and success among all the simulations performed (as illustrated in Eq. 9), it is possible to calculate the respective probabilities. This second metrics provides more qualitative evaluation of the ADS.

$$R = \begin{cases} 0, & \texttt{If Input1= 0 AND j = P} \quad (Stall) \\ -1, & \texttt{If Input4 = 1} \quad (Collision) \\ Eq.7, & \texttt{otherwise} \quad (Success) \end{cases} \quad (8)$$

where P is the maximum number of iteration for all simulations. It is equal to 540 iteration with a sample period of 10 (s) (dt).

$$SuccessProbability = \frac{SuccessCounter}{O} \quad (9)$$

where `Success Counter` contains the number of successfully completed simulations and O is the total number of performed simulations.

4 Simulation Results

In this Section, results are related to 3 different implementation phases. In the first phase, an implementation for the LLC (which only involves the Motion Controller) is found by searching for a Kp (Proportional Coefficient) that optimizes the Motion Controller performance. In the second phase, an implementation for the MLC (mainly Boat and Docks Avoidance Controller) is found by choosing the controllers Logic (Crisp or Fuzzy) that is better in terms of performances. In the third phase, an implementation for the HLC is found by choosing Finite State Machine or Fuzzy Controller approach to generate ADS states.

4.1 Phase 1

16 different simulation settings are identified in this phase: 2 different water current intensity reference levels $\times$ 8 different sources. The effective water current intensity level for a certain simulation is obtained by adding a white noise to the reference level. For all simulation start and target position are fixed. For each setting 100 simulation are performed in order to have a good amount of data to produce statistics (Table 1).

Table 1. Performance of the ADS with Proportional Motion Control.

Kp	Stall P. (%)	Success P. (%)	Mean Reward $(p.u.)$	Reward Variance $(p.u.)^2$
0.00	0	100	0.966854	0.000047
0.01	0	100	0.963096	0.000117
0.02	0	100	0.967583	0.000054
0.03	0	100	0.957037	0.000292
...	...	...	...	...

Note that the P Control achieves a better overall performance than without any P Control (first raw). In particular, for $Kp = 0.02$ we obtain 100% of the successes on 100 simulations with an average Reward (calculated according to the Eq. 7) equal to 0.967583 and a very low variance.

4.2 Phase 2

In this phase the results are reported in Tables and Figures. The latter are very interesting since performance is also evaluated here in terms of smoothness of maneuvers, which are visible from the figures. Tables reports statistics on events probability and on Reward value assigned at each simulation. Furthermore, in this phase, pseudo-random generated Dummy Boats are introduced (with linear routes and no control a part of motion control). Dummy boats can be 0, 2, 4, 8 or 16. The generation is so that only 1 dummy boat can be generated at a certain instant and they shall be uniformly distributed in the simulation time-span. So, the number of settings increases to 65 (1 simulation was performed without water currents and without dummy boats + 64 simulations which differ for water current intensity, source and for maximum number of dummy boats dynamically generated). For each setting from #2 to #65, 100 simulations are performed in order to have a good amount of data to produce statistics. It is not useful repeat 100 times the first setting, since the scenario is static and without perturbations. So, for a map 641 simulations are performed. Since the maps are 6, the total number of simulations performed in this phase is 3846.

The Figures also show the autonomous boat change internal state autonomously cause its interactions with the environment. While dummy boats trace a light blue route, the autonomous boat traces a route that changes color according to its current state: red if it is in Virtual Anchor state; blue if it is in Navigation state; light green if it is in Boat Avoidance state; orange if it is in Docks Avoidance state.

Table 2. Performance Statistics of the ADS with Crisp-MLCs.

Map	Collision P. (%)	Stall P. (%)	Success P. (%)	Mean Reward $(p.u.)$	Reward Variance $(p.u.)^2$
0	28.56	0.00	71.44	0.932464	0.001204
1	43.06	0.00	56.94	0.905778	0.001060
2	61.62	7.80	30.58	0.846370	0.001036
3	59.13	5.15	35.72	0.859683	0.001712
4	55.85	0.94	43.21	0.855143	0.002189
5	57.25	3.43	39.32	0.886785	0.000658
General	50.91	2.89	46.20	0.881037	0.001310

Crisp-Mid Level Controllers based ADS. Figures 7 reports navigation of the same autonomous boat in six scenarios. In those scenarios are involved all maps and no dummy boats or water currents. Such simulations (Setting #1, Simulation number #1) are performed so make sure that ADS can drive the boat in all the ports and in the absence of other obstacles or disturbances from the environment. It works also as an indication of the feasibility of driving in such scenarios.

Tables 2 show collision, stall and success probabilities and means and variances of the performance indicator – calculated only on successfully completed simulations – for the Crisp-MLC based ADS. The calculations of probabilities, mean and variance were performed separately on each map and a general statistic was calculated by averaging the values of each map.

Performances of the this first complete version of the ADS (with FSM based HLC and Crisp Controllers based MLC) shown a not used potential, since the difference in Success probability terms for different maps (and different settings) going from 71 % to 30.58 %. Generally and not mean Reward is almost as high as that of the simulations performed without obstacles. This confirm us the design is well thought out.

Fuzzy-Mid Level Controllers Based ADS. In this paragraph we reports some scenario (Fig. 9) that are more interesting than to review the one-shot simulation performed by this ADS version on all maps with the first setting. In those scenario the differences between Crisp and Fuzzy BAC and DAC based ADSs from the point of view of the maneuvers are highlighted.

For example in Map 0 Fuzzy was noted that Fuzzy Boat Avoidance Controller computes smarter maneuvers than Crisp one: the steering maneuvers are executed in a more closed way, i.e. the fuzzy controller action has a lower impact on the route length. In Map 5 (that is characterized with a complicated docks structure) and in a scenario with the maximum number of dummy boats it is more probable for a boat to collide with a dockside or another boat. Crisp Controllers doesn't manage this scenario very well

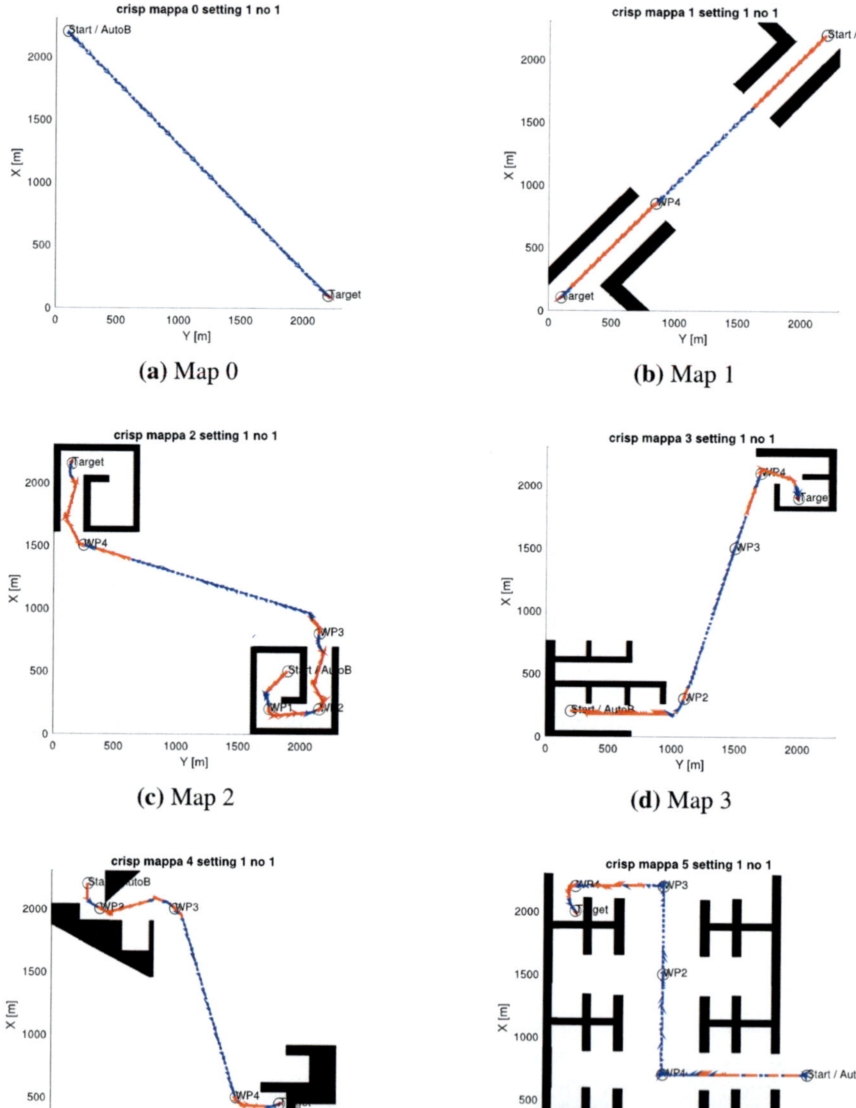

(a) Map 0 (b) Map 1

(c) Map 2 (d) Map 3

(e) Map 4 (f) Map 5

Fig. 7. FSM HLC + Crisp MLC based ADS in one-shot simulations on all Maps with setting 1.

cause its sharp maneuvers. On the other hand, it is shown the same scenario in which the Fuzzy Controllers allow the boat to reach the target without colliding.

Again, in a detail of Map 3, the Crisp and Fuzzy Docks Avoidance Controllers are compared. It is interesting note as the Crisp DAC fall into stalls very frequently. However Fuzzy DAC seems to work better. Finally, looking at Map 1 we want to reaffirm the importance of smoother maneuvers even for DAC: many maneuver changes (as for

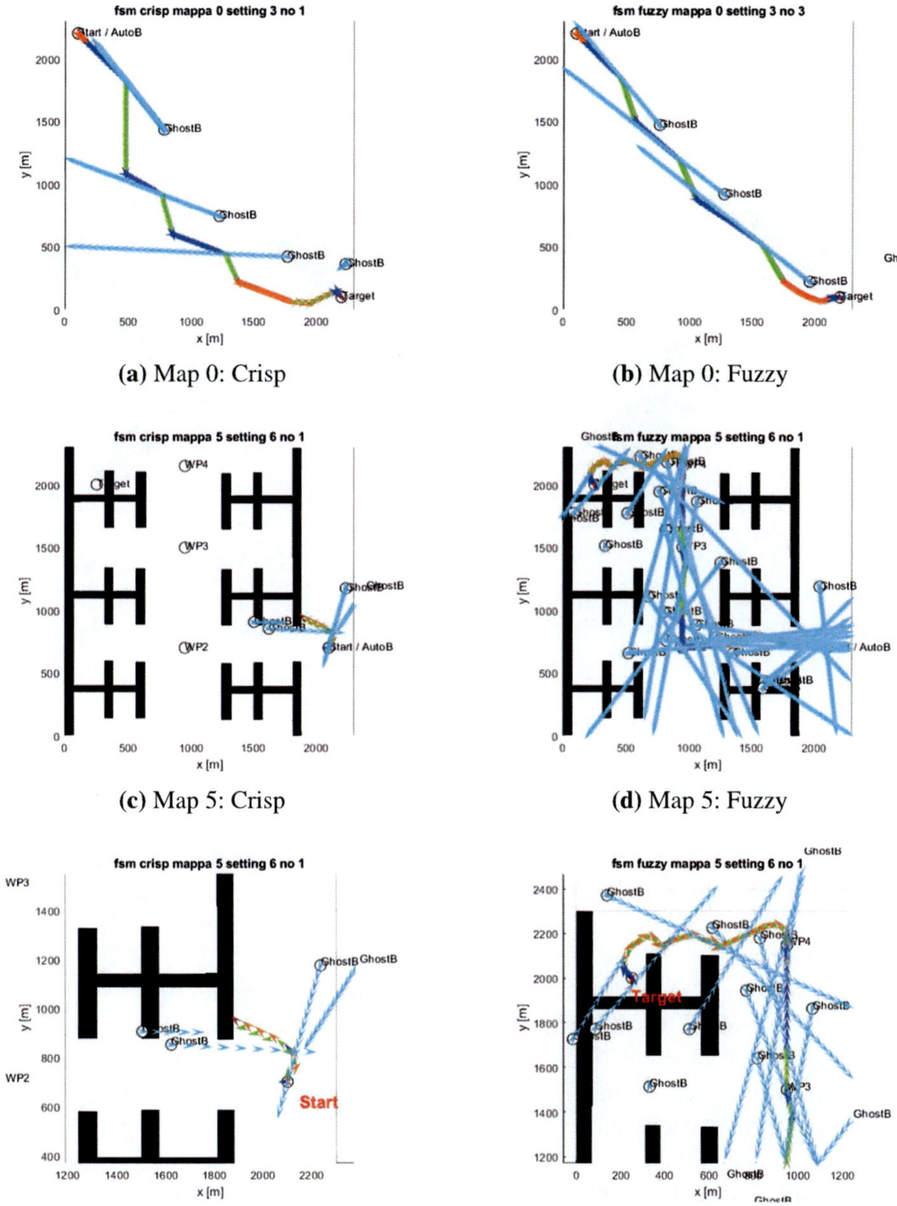

(a) Map 0: Crisp

(b) Map 0: Fuzzy

(c) Map 5: Crisp

(d) Map 5: Fuzzy

(e) Map 5 (Detail): Crisp

(f) Map 5 (Detail): Fuzzy

Fig. 8. FSM HLC + Crisp vs Fuzzy BAC based ADSs in some interesting simulations.

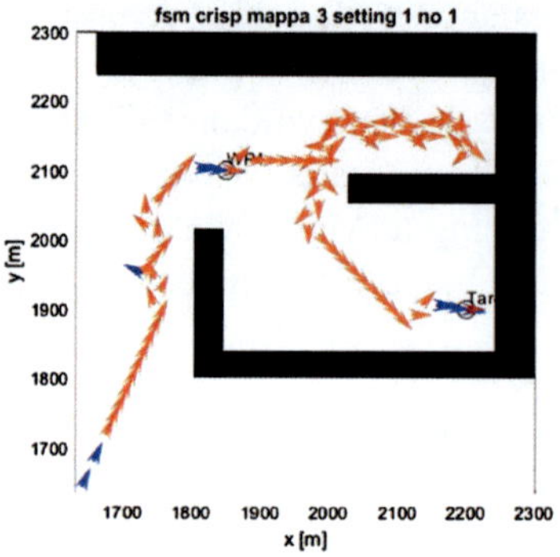

(a) Map 1: Crisp

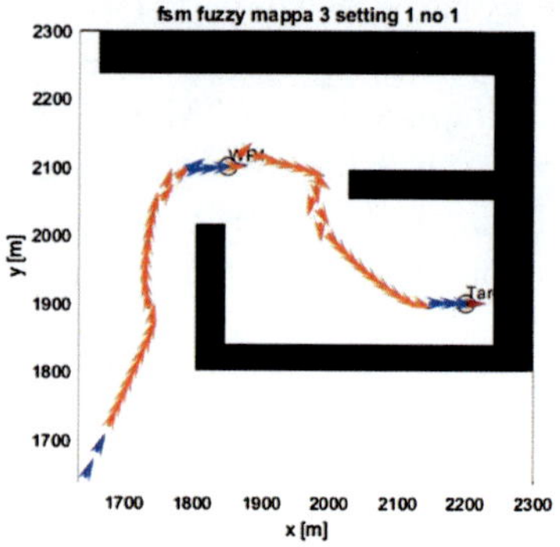

(b) Map 1: Fuzzy

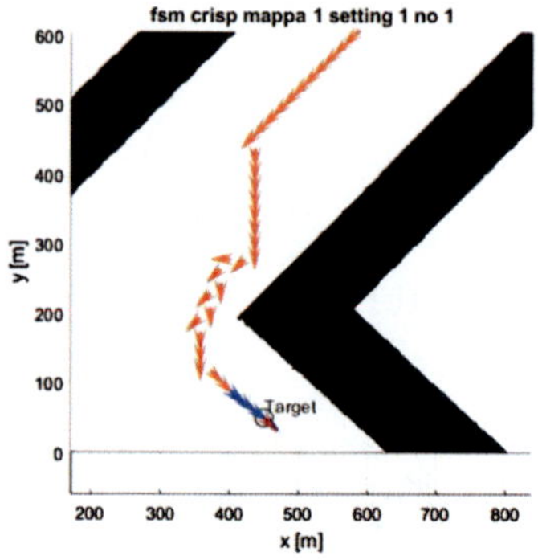

(c) Map 3: Crisp

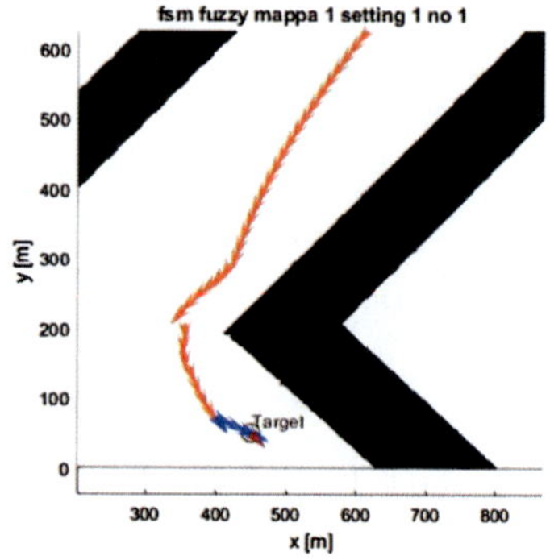

(d) Map 3: Fuzzy

Fig. 9. FSM HLC + Crisp vs Fuzzy DAC based ADSs in some interesting simulations.

Crisp DAC) implies a grater energy consumption from Propulsion System in addition to cause the route length increasing. Fuzzy DAC produces more gentle and pleasant maneuvers for passengers (Fig. 8).

Tables 3 show collision, stall and success probabilities and means and variances of the performance indicator – calculated only on successfully completed simulations – for the Fuzzy-MLC based ADS. The calculations of probabilities, mean and variance were performed separately on each map and a general statistic was calculated by averaging the values of each map.

Here, Collision Probabilities increase and Stall Probabilities decrease wrt that obtained in Crisp-MLC based ADS. Generally and not, performances have gotten worse but, it was noted, from simulations, the maneuvers are more smooth (which contributes to a pleasant driving experience). From Table 4 is also clear that the main reason of increased collision events are the docks. It is our belief that the cause is the state transition from dock avoidance to boat avoidance and vice versa. A further step regards the implementation of the HLC which define what pipeline must drive the autonomous boat. It is our duty to investigate in the search for an implementation of the latest and highest level of control that completes what has been achieved in this phase.

Phase 2 Conclusions. From this phase it appears that the Crisp version of the ADS is better (in statistical terms) than the Fuzzy one. Despite this evidence, it is the Author's idea that even better performance can be achieved by fuzzyfing even the high level of control.

Table 3. Performance Statistics of the ADS with Fuzzy-MLCs.

Map	Collision P. (%)	Stall P. (%)	Success P. (%)	Mean Reward $(p.u.)$	Reward Variance $(p.u.)^2$
0	23.87	0.00	76.13	0.888253	0.001817
1	62.87	0.00	37.13	0.725741	0.000450
2	93.76	0.00	6.24	0.693188	0.000268
3	84.09	0.15	15.76	0.707900	0.001232
4	84.24	0.16	15.60	0.748674	0.001078
5	79.56	0.00	20.44	0.745327	0.000275
General	71.40	0.05	28.55	0.751514	0.000853

4.3 Phase 3

In this phase the ADSs with FSM-based HLC and ADS with Fuzzy-HLC are compared. Results are shown for each solutions in terms of figures and statistics. At the end of this phase, the best ADS complete version is chosen by comparison of the results.

Finite State Machine-High Level Controller Based ADS. The results about this paragraph are the same the ones reported in paragraph 4.2 since, the ADS with Fuzzy Mid Level Controllers already adopts the FSM. This Sub-Section is present in order to highlight that the ADS resulting from Phase 2 is tested also in Phase 3 (by comparison with the Fuzzy-HLC and the FQL-MLC based ADSs).

Table 4. Performance Statistics of the ADS with Fuzzy-HLC.

Map	Collision P. (%)	Stall P. (%)	Success P. (%)	Mean Reward $(p.u.)$	Reward Variance $(p.u.)^2$
0	22.31	0.00	77.69	0.893777	0.001411
1	27.77	0.16	72.07	0.710519	0.000357
2	19.81	23.25	56.94	0.695100	0.000321
3	20.28	19.19	60.53	0.710150	0.001086
4	31.98	7.64	60.38	0.725681	0.001028
5	34.17	11.54	54.29	0.740661	0.000495
General	26.05	10.30	63.65	0.745981	0.000783

Fuzzy-High Level Controller Based ADS. It must be noted that the performances in terms of Success probability are improved wrt the Phase 2. The General Success P. amount to about 64% and Mean Reward is about the same of the FSM-HLC and Fuzzy-MLC based ADS. This means that the fuzzyfication of the HLC has produced a better state transition (because it is able to manage the uncertainty on the scenario presented) without compromising the smoothness of the maneuvers, as the two systems work at different levels.

Phase 3 Conclusions. It is concluded that the fuzzyfication of controllers at both High Level of Control and Mid Level of Control is a good way to improve the performances of the presented ADS.

5 Conclusions

The paper proposed to design and apply an Autonomous Driving System to a class of e-boats. Before a real implementation of such complicate and complex kind of system architecture, it is mandatory to test it in a virtual environment. So, further to an ADS, an ad-hoc Simulator was developed in MATLAB environment (using several Toolboxes). A particular attention was given to the original Propulsion System of the boat under study. A personal contribute was done to the concept of the Propulsion System mechanics and a control technique is also proposed. Simulations outcome is an ADS ready for being ported on a dedicated HW on board of the boat and being validated with real-world test phase. The work done so far is aimed to a standardization (for naval industry) of the ADS architecture and techniques. Furthermore, a *Natural-Inspired* approach is adopted to both the performance evaluation method and on Propulsion System design.

A modular and multi-pipeline ADS is presented, which is both vertically and horizontally organized in Levels of Control and data-path, respectively. Furthermore, the ADS designed with the pipeline approach instead of the End2End one –since it allows to *design a priori* many modules and is less computationally expensive, as declared in [4]- presents several modular pipelines. Each one is responsible for a precise navigation function/state. The whole ADS is governed by an High-Level Controller designed as a Finite State Machine (as it is a recurrent approach) or as a Fuzzy Controller. A form of learning is presented in order to perform (as a black box) the mapping of motion control signals in Propulsion System control signals. The reason is to solve a non-linear unknown mapping. It is obtained with an ANN in a Direct-Control architecture with the plant. By means of a BP algorithm variant and the Jacobian Matrix of the plant, a on-line and continous *Specialized Leaning* can be performed.

Fuzzy Logic is the main framework of the final ADS implementation. In fact, it demonstrates to give the ADS better performances both in terms of *Success Probability* and *Reward Function*, which is inspired by the *Fish Schooling Behavior* since its likeness with the boat self-driving problem. Firstly, Classical Controllers in Mid-Level of Control are substituted with Fuzzy ones. Such operation has highlighted (by simulation results) that Fuzzy Controllers for boat and docks avoidance tasks makes maneuvers more smooth and closed to the obstacles (which decreases the length of the route) at the cost of a worst General Success Probability, 28.55%, compared with the Classical ones,

46.20%, (mainly caused by Collision events). Secondly, Finite State Machine –from whose design it is clear that there is no dependence on the current state and so it can be translated into a controller without internal feedback– is substituted with a Fuzzy Controller, for which FSM inputs decrease from 5 to 3 (simplifying the HLC design) and are fuzzyfied. Also output is fuzzy and so a threshold decision maker is used to derive the current ADS navigation state (Virtual Anchor Docks Avoidance, Navigation, Boat Avoidance). Those further implementation phase revels the best results in terms of Success Probability, 63.65%, even better of the totally Crisp ADS (46.20%) version and at the cost of only 0.13 (p.u.) in General Mean Reward. The rule base of each Fuzzy Controller is pre-trained by translating crisp rules in fuzzy ones. Rules are so generally valid for all simulated scenarios.

Acknowledgements. The Department of Information Engineering, Electronics and Telecommunications (DIET) at University of Rome *La Sapienza* and the *Po.Mo.S* Laboratories would like to thank the EU for financial support to environmental and climate action projects like LIFE for Silver Coast (LIFE16 ENV/IT/000337). Such a help is crucial to achieve natural and historical preservation of Italy, especially of touristic areas.

References

1. Fossen, T.I.: Handbook of Marine Craft Hydrodynamics and Motion Control. Wiley (2021)
2. Elkins, L., et al.: The Autonomous Maritime Navigation (AMN) project: field tests, autonomous and cooperative behaviors, data fusion, sensors, and vehicles. J. Field Robot. **27**, 790–818 (2010)
3. Ferrandino, E., Capillo, A., Santis, E. D., Mascioli, F., Rizzi, A.: A modular autonomous driving system for electric boats based on Fuzzy Controllers and Q-Learning. In: Proceedings of the 13th International Joint Conference on Computational Intelligence (2021). https://doi.org/10.5220/0010678100003063
4. Grigorescu, S., Trasnea, B., Cocias, T., Macesanu, G.: A survey of deep learning techniques for autonomous driving. J. Field Robot. **37**(3), 362–386 (2020)
5. Nauticando. http://www.nauticando.net/lezioni-di-nautica/precedenze-in-mare/. Accessed 15 July 2022
6. Akiyama, T., Kobayashi, Y., Kishigami, J., Muto, K.: CNN-based boat detection model for alert system using surveillance video camera. In: 2018 IEEE 7th Global Conference on Consumer Electronics (GCCE), pp. 669–670 (2018). https://doi.org/10.1109/GCCE.2018.8574704
7. Siddique, N., Adeli, H.: Computational Intelligence: Synergies of Fuzzy Logic, Neural Networks and Evolutionary Computing, pp. 166–167. Wiley (2013)
8. US Coast Guard: Navigation Rules: International-Inland. U.S. Coast Guard Marine Transportation Systems Directorate and Navigation Center of Excellence (2017)
9. Ferrandino, E., Capillo, A., Mascioli, F., Rizzi, A.: Nanogrids: a smart way to integrate public transportation electric vehicles into smart grids. In: Proceedings of the 12th International Joint Conference on Computational Intelligence - CI4EMS, INSTICC, pp. 512–520. SciTePress (2020). https://doi.org/10.5220/0010110005120520
10. Siddique, N., Adeli, H.: Computational Intelligence: Synergies of Fuzzy Logic, Neural Networks and Evolutionary Computing, pp. 7–10. Wiley (2013)
11. Hong, J., Tang, K., Chen, C.: Obstacle avoidance of hexapod robots using fuzzy Q-learning. In: 2017 IEEE Symposium Series on Computational Intelligence (SSCI), pp. 1–6 (2017). https://doi.org/10.1109/SSCI.2017.8280907

12. Brehmer, P., Sarré, A., Gonzalez, L., Cotel, P., Hermand, J.-P., François, G.: Aggregative and schooling behaviour of small pelagic fish schools through echo type characteristics. In: 2013 IEEE/OES Acoustics in Underwater Geosciences Symposium, pp. 1–6 (2013). https://doi.org/10.1109/RIOAcoustics.2013.6683992.
13. Labuguen, R.T.: Automated fish fry counting and schooling behavior analysis using computer vision. In: 2012 IEEE 8th International Colloquium on Signal Processing and its Applications, pp. 255–260 (2012). https://doi.org/10.1109/CSPA.2012.6194729
14. Fujiwara, C., Onishi, N., Tateyama, T., Hiratsuka, S., Sakakibara, K., Chen, Y.-W.: Behavior generation of fish school by the potential field method in Virtual Aqua Museum. In: 2012 6th International Conference on New Trends in Information Science, Service Science and Data Mining, ISSDM 2012, pp, 495–498 (2012)
15. Swain, D.T., Couzin, I.D., Leonard, N.E.: Real-time feedback-controlled robotic fish for behavioral experiments with fish schools. Proc. IEEE **100**(1), 150–163 (2012)
16. Lai, L., Qu, S.: Research on application of fish swarm behaviour in intelligent. transportation system. In: 7th Advanced Forum on Transportation of China, AFTC 2011, pp. 195–200 (2011). https://doi.org/10.1049/cp.2011.1403
17. Aguercif, T., Tighzert, L., Mendil, B., Fonlupt, C.: Rate learning-based fish school search algorithm for global optimization. In: 2017 6th International Conference on Systems and Control (ICSC), pp. 520–525 (2017). https://doi.org/10.1109/ICoSC.2017.7958733
18. Cai, L., Sun, Q.: A regular expression grouping algorithm based on artificial fish school algorithm. In: 2017 7th IEEE International Conference on Electronics Information and Emergency Communication (ICEIEC), pp. 556–559 (2017). https://doi.org/10.1109/ICEIEC.2017.8076627
19. Noriega, J.R., Wang, H.: A direct adaptive neural-network control for unknown nonlinear systems and its application. IEEE Trans. Neural Netw. **9**(1), 27–34 (1998). https://doi.org/10.1109/72.655026
20. Khan, E.R., Ogunfunmi, T.: A multilayered neural net controller for servo systems. In: 1991 IEEE International Joint Conference on Neural Networks, vol. 2, pp. 1577–1581 (1991). https://doi.org/10.5120/12428-9042
21. Park, J.-H., Huh, S.-H., Kim, S.-H., Seo, S.-J., Park, G.-T.: Direct adaptive controller for non-affine nonlinear systems using self-structuring neural networks. IEEE Trans. Neural Netw. **16**(2), 414–422 (2005). https://doi.org/10.1109/TNN.2004.841786
22. Wilamowski, B.M., Chen, Y., Malinowski, A.: Efficient algorithm for training neural networks with one hidden layer. In: Proceedings of the International Joint Conference on Neural Networks, IJCNN 1999 (Cat. No. 99CH36339), vol. 3, pp. 1725–1728 (1999). https://doi.org/10.1109/IJCNN.1999.832636
23. Guo, Cherkassky:A solution to the inverse kinematic problem in robotics using neural network processing. In: International 1989 Joint Conference on Neural Networks, vol. 2, pp. 299–304 (1989). https://doi.org/10.1109/IJCNN.1989.118714
24. Ferrandino, E., Capillo, A., De Santis, E., Massimo, F.M.F., Rizzi, A.: Modular autonomous driving system for electric boats based on Fuzzy Controllers and Q-learning. In: Bäck, T., et al. (eds.) Proceedings of the 13th International Joint Conference on Computational Intelligence, IJCCI 2021, Online Streaming, 25–27 October 2021, pp. 185–195. SCITEPRESS (2021). https://doi.org/10.5220/0010678100003063

Facing Graph Classification Problems by a Multi-agent Information Granulation Approach

Enrico De Santis, Giuseppe Granato, and Antonello Rizzi(✉)

Department of Information Engineering, Electronics and Telecommunications, University of Rome "La Sapienza", Via Eudossiana 18, 00184 Rome, Italy
{enrico.desantis,giuseppe.granato,antonello.rizzi}@uniroma1.it

Abstract. In this paper, we investigate an agent-based prototypical framework for graph classification by offering a comparison with a standard graph neural network on several heterogeneous datasets. The design of the algorithm is based on a swarm of agents orchestrated via evolutionary optimization in charge of finding meaningful substructures from unstructured training data. The algorithm is tailored to face pattern recognition problems by a suitable embedding from the graph domain to Euclidean space, also exploiting local metric learning. Results on five open-access datasets of fully labeled graphs show interesting performances in terms of accuracy, counterbalanced by relatively high computational complexity.

Keywords: Multi-agent systems · Graph embedding · Supervised learning · Structural pattern recognition · Graph neural networks

1 Introduction

It is well-known that designing an intelligent systems able to reach the high performances of the biological brain in recognizing patterns remains a difficult goal. Pattern recognition is a multidisciplinary field that involves cognitive science, psychology, artificial intelligence, and other disciplines. It helps to understand human perception and the ability to gain new knowledge. It also allows for the creation of automated systems (automated pattern recognition systems) in a plethora of applications, such as medical diagnosis, bioinformatics, industrial inspection, smart grids, personal identification and natural language processing tasks. An open issue in cognitive science (and, by extension, in automated pattern recognition) is understanding the relationship between low-level sensory and neural phenomena and higher-level understanding and conceptual thinking [37,39]. In addition, since the world around (both virtual and real) is a complex environment and thanks to the capabilities of artificial sensor systems together with the computational power of modern computers, it is now possible to collect a huge amount of data which, in its most genuine form, has a structured nature. In other words, in modern pattern recognition systems, we are dealing with data in the form of videos, images, sequences, graphs, time series and not just n-tuples of real numbers or matrices. Each of these data has its own internal structure, which captures a specific type of information. Furthermore, the surrounding environment, conceived as a complex system, is

J. Garibaldi et al. (Eds.): IJCCI 2021, SCI 1119, pp. 185–204, 2023.
https://doi.org/10.1007/978-3-031-46221-4_9

hierarchically structured, stochastic, dynamic and composed by numerous interacting elements, in which relationships are often as important as the elements themselves that constitute the system [50]. In this framework, two main orders of scientific problems arise: i) the problem of representing information that has an heterogeneous and hierarchical nature and ii) the ability of an automatic system to follow a stochastic environment that changes over time with a given dynamic. In fact, a major challenge in both automated pattern recognition and cognitive science is the "representation" one [38, 74], i.e., how information or data is modeled so that the cognitive system can perform tasks such as semantic processing. Self-organizing phenomena in the physical world may also be relevant for understanding cognitive processes, and some properties of cognitive systems resemble those of complex systems. From a computational point of view Granular Computing (GrC) is a term that covers theories, methodologies, techniques and tools that make use of "information granules" in complex problem solving [82]. Information granules are atomic units that naturally give rise to hierarchical structures, allowing the same problem or system to be perceived at different levels of specificity depending on the complexity of the problem, available resources and particular needs [3]. Some authors consider GrC as a platform for the analysis and design of human-centric intelligent systems [15]. If GrC is a general approach to computation and information representation, there are two main paradigms for synthesizing automatic recognition systems with learning capabilities (both supervised and unsupervised): the classical paradigm, based on the construction of a suitable feature space and the "connectionist" one, based on Artificial Neural Networks (ANNs) and modern deep learning, known today also as Geometric Deep Learning [23]. The classical paradigm tends to work with a static features space and, at most, can be assisted by an automatic feature selection system. In addition, it is characterized by a dissimilarity metric which may as well be local (i.e., tailored to different classes or clusters) [32]. The connectionist paradigm, based on an associative representation inherent in neural connections and on vector representation from the early stage of the processing chain, is capable of generating a hierarchy of features in each layer of the neural network. This feature is significant in new generation networks based on convolutional kernels such as Convolutional Neural Networks (CNNs) and Graph Neural Networks (GNNs) [77]. Remaining on the question of "representation" and considering an instance of a classification problem defined on an input domain of objects described by heterogeneous characteristics, it is interesting to have a system capable of explicating, given a set of heterogeneous features, which ones are specific for a given class. That is to say, for a given class of objects the color and weight may be important, while the temperature and the ability to reflect light may be relevant for another class. The two paradigms described above try to solve this problem in different ways, given the different nature of the processing chain. While an ANN tends to learn these differences in a diluted way within the synaptic weights, a classical system based on a measure of dissimilarity requires an approach known as "*local* metric learning" [32]. That is, learning a set of class-tailored manifolds, which together build up a composite manifold. It can be shown, as will be explained in this work, that a classical approach, within a suitable composite processing pipeline based on GrC, can show interesting properties regarding the interpretability of the learned models (and output-produced symbols) in line with the Explainable AI paradigm [24, 43, 55]. Conversely, explainability issues related to artificial neural architectures are known [80]. The cognitive scientist Peter Gärdenfors in his book "Conceptual Spaces, the geometry of thought" [39], in dealing with the problem of representation, identifies three level of information processing. Besides the *symbolic* level (where information is represented classically with discrete symbols) and the *associations* level (where information is represented by connections, i.e. as in ANNs), he assumes a *conceptual* level that is based on a geometrization of the features space (Gärdenfors calls it "perceptual space"), that can be obtained by suitable partitions computed though clustering algorithms. Moreover, he states that *the symbolic level depends on the conceptual one* [39]. As will be further explained, in this work we frame the "representation" problem in a similar framework, adopting the GrC approach [88]. Specifically, in solving a classification problem in a starting non-metric domain (that is, the graph domain), and given a suitable dissimilarity measure, through the symbolic histogram [9] technique – built up on a partition (conceptual level) generated by a clustering algorithm on the given unconventional domain –, rough graph data is embedded in a (local) vectorial space, where suitable standard classification algorithms can safely work. In other words, the embedding is computed through the learning of a (local) alphabet of symbols – information granules – represented by cluster representatives. In this sense, in our approach to graph classification *the symbolic level depends on the conceptual one*. It is worth noting that, as it will be extensively explained, for "local embedding" we mean that the "information granule" is constructed in a class-aware fashion, that is several embeddings are available for the same piece of information. Beside the representation problem, in modern information processing systems (of which data come from a complex time-varying environment), in dealing with huge datasets there is the need of acting in a parallel way and using a divide-and-conquer paradigm. Hence, in this study it is proposed an approach to the graph classification problem based on a multi-agent paradigm, where the changing dynamics in data is faced by an evolutionary approach, which drives a swarm of agents. Specifically, in this paper we further deepen our analysis on Evolutive Agent Based Clustering/Classification (E-ABC, for short) for solving a classification problem in the graph domain. In E-ABC, each agent sees only a small portion of the dataset, to be randomly drawn at training time, and different swarms of agents work on class-stratified data (i.e., each swarm processes only patterns belonging to a given class) [10,11]. Further, two evolutionary optimization routines take care of driving the evolution of each agent and the cooperation between agents, respectively. E-ABC has been originally proposed in [57] as a clustering algorithm based on the multi-agent paradigm conceived to work in Euclidean spaces. In [41], it has been extended towards supervised problems by letting the resulting clusters to compose a decision cluster classifier [34] and applied to the classification of faulty states in a real-world smart grid [31]. In [40], E-ABC has been upgraded with a multimodal optimization approach [70,78] in order to foster the exploration of different subspaces, hence allowing its local metric learning capabilities [32]. In [13], E-ABC has been extended towards the graph domain by letting agents to find meaningful subgraphs in a Granular Computing setting [14,69,83,84] to later build a (Euclidean) embedding space where classification is performed [6–12,54,56,58,59]. Whilst in our previous work [13] we compared Graph E-ABC against other Granular

Computing-based approaches (namely [10, 59]), in this work we extend our analysis on Graph E-ABC by comparing its performances against a GNN.

Before concluding, we provide the reader with the distinctive characteristics behind the proposed approach:

1. E-ABC (and its graph-based counterpart: Graph E-ABC) can work on data of any nature, therefore also on structured data such as graphs, thanks to the definition of a suitable – and sufficiently abstract – parametric dissimilarity measure and an ad-hoc data loader;
2. The multi-agent architecture – fulfilling the divide-and-conquer approach – is designed for distributed environments, such as clusters of workstations, cloud environments or multicore/manycore computing devides;
3. Thanks to the specific design related to the sub-structures search together with the adoption of a local metric learning approach, E-ABC fulfills the Explainable AI paradigm, that is, providing the user with a good degree of model interpretability;
4. The evolutionary paradigm underlying the swarms of agents allows pursuing time-varying, therefore dynamic, environments. Specifically, E-ABC can pursue time-varying objective functions with a good degree of adaptability.

It is worth noting that, at this developing stage, E-ABC is not a single monolithic algorithm but it is conceived as a template with which complex classification problems can be solved thought several toolboxes. If, at general level, the rationale behind E-ABC is fixed, several building blocks can be changed, such as the nature of the evolutive metaheuristics, the specific clustering algorithm, the reward function, the objective function, the set of classifiers and, of course, the data loader.

This paper is structured as follows: in Sect. 2 we review some current research papers focused on multi-agent approaches for pattern recognition and machine learning; in Sect. 3 we sketch a point-by-point comparison between multi-agent systems and complex systems, arguing how the former can be a suitable paradigm to model the latter; in Sect. 4 we briefly describe the E-ABC algorithm (referring the interested reader to our previous work [13] for a more exhaustive description) and in Sect. 5 we describe the GNN that serves as baseline competitor in the experimental results, which will be reported in Sect. 6. Finally, Sect. 7 concludes the paper.

2 Related Works

The flexibility of multi-agent systems leaded to different research works where such paradigm has been employed for building supervised and unsupervised learning systems. For example, in [2] a multi-agent approach has been used for local graph clustering in which each agent performs a random walk on a graph with the main constraint that such agents are "tied" together by a rope, forcing them to be close to each other. In [25] a set of self-organising agents by means of ant colony optimisation has been applied to anomaly detection and network control. In [26] each agent runs a different clustering algorithm in order to return the best one for the dataset at hand. In [27] agents negotiate one another rather than being governed by a master/wrapper process (e.g. evolutive algorithm) in order to improve the clustering solution. In [47] ant colony optimisation has been used in order to organise agents, where each ant "walks" on the dataset, building connections amongst points. In [65] each agent represents a set of data points and agents link to each other, thus leading to clustering. In [66] a genetic algorithm has been used where the agents' genetic code is connection-based: each agent is a clustering result whose genetic code builds a (sub)graph which can be interpreted as a cluster. In [68] the multi-agent approach collapses into two agents: a first agent runs a cascade of principal component analysis, self organizing maps and k-means in order to cluster data and a second agent validates such results: the two agents interactively communicate with each other.

As supervised problems are concerned, in [18] a multi-agent algorithm has been proposed in which agents perform a Markovian random walk on a weighted graph representation of the input dataset. Each agent builds its own graph connection matrix amongst data points, weighting the edges according to the selected distance measure parameters, and performs a random walk on such graph in order to discover clusters. This algorithm has been employed in [19] for an unsupervised identification of frequent behaviours of mobile network subscribers starting from a set of call data records. In [67], the authors propose the Granular Agent Evolutionary Classification algorithm, where agents are in charge of clustering patterns with similar attributes. Further, different agents lie in different class-aware subsystems and each subsystem is characterized by a set of classification rules extracted by the information collected by the granular agents. Due to the advent of large corpora, multi-agent systems have been applied for text and document classification [1,63].

The MAS paradigm has also been successfully applied in the field of Computer Vision. In [28], the authors designed an agent-based system for Scene Graph generation for visual genome understanding where objects are agents that have to maximize the quality of the generated scene graph. In [42], the authors proposed an image recognition system for real-time applications where different agents collaborate for extracting relevant features from the acquired images in order to determine the age and gender of individuals in an office building. The MAS paradigm has found application also in the field of bioinformatics [30] for protein structure prediction by distributing each relevant task among different agents in a decentralized way. The distributed aspect that characterizes multi-agent systems has led various authors to investigate their application in the field of Big Data. In [51], the authors proposed a general-purpose actor-based system for distributed data mining addressing the problem of social data analysis such as topic detection, troll detection and hoax detection. Conversely in [35], the authors introduce a multi-agent co-evolutionary adaptive strategy for attribute reduction for big medical datasets concerning infant brain Magnetic Resonance Imaging. Finally, in [62], the authors propose a distributed multi-agent framework for classification, where the authors assume that each agent can see only a subset of features as the data is conceived to be decentralized.

3 Complex and Multi-agent Systems

Complex systems and multi-agent systems are related since complex systems often consist of many interacting elements. In turn, a multi-agent system is a system composed of multiple interacting intelligent agents. These agents can be autonomous and have their own goals and behaviors. Moreover, their interactions can result in the emergence of complex system-level behavior. In other words, multi-agent systems can be seen as a paradigm to model and study complex systems. Referring to Fig. 1, complex systems have several characteristics that distinguish them from simpler systems [73]:

- *Large Number of Interacting Components*: Complex systems are composed by a large number of interacting components, such as agents or sub-systems, that can influence each other's behavior;
- *Nonlinearity*: The behavior of complex systems is often nonlinear, meaning that small changes in the system can lead to large and unexpected outcomes;
- *Emergence*: Complex systems exhibit emergent behavior, meaning that global patterns or properties arise from the interactions of individual components, rather than being explicitly programmed;
- *Adaptivity*: Complex systems are able to adapt and change in response to their environment;
- *Hierarchy*: Complex systems often have a hierarchical structure, with different levels of organization and control;
- *Self-organization*: Complex systems have the ability to self-organize, meaning that the system can spontaneously organize itself without the need for external control;
- *Robustness and Fragility*: Complex systems can be robust, meaning they can maintain their structure and function despite perturbations or disturbances, but also fragile, meaning they can collapse or be severely affected by small perturbations;
- *Uncertainty*: Complex systems often operate in uncertain or dynamic environments, making predictions about their behavior difficult;
- *Diversity*: Complex systems are composed by a diverse set of components, each with their own characteristics and behaviors, that interact in a non-trivial way.

As modeling paradigm of complex systems, several characteristics of a multi-agent system distinguish them from other types of systems [45]. While there is no clear agreement on what multi-agent systems are, scholars have established some defining characteristics [46], namely – see also Fig. 2:

- *Autonomy*: Each agent in a multi-agent system is autonomous, meaning that it can make its own decisions and act independently of other agents;
- *Interaction*: Agents in a multi-agent system interact with each other, either directly or indirectly, in order to achieve a common goal or to complete a task;
- *Decentralization*: The decision-making process in a multi-agent system is decentralized, meaning that no single agent has complete control over the system. Instead, each agent makes decisions based on its own goals and beliefs;
- *Heterogeneity*: Multi-agent systems often consist of agents that are different in terms of their abilities, goals, and beliefs. This heterogeneity can lead to complex and dynamic interactions between agents;
- *Adaptability*: Multi-agent systems are often designed to be adaptable to changing conditions, so that the agents can respond to new situations and challenges;
- *Scalability*: Multi-agent systems can be scaled to handle a large number of agents, which makes them useful for analyzing and designing complex systems;

– *Robustness*: Multi-agent systems are often designed to be robust, meaning that they can continue to function even if some agents fail or are removed from the system.

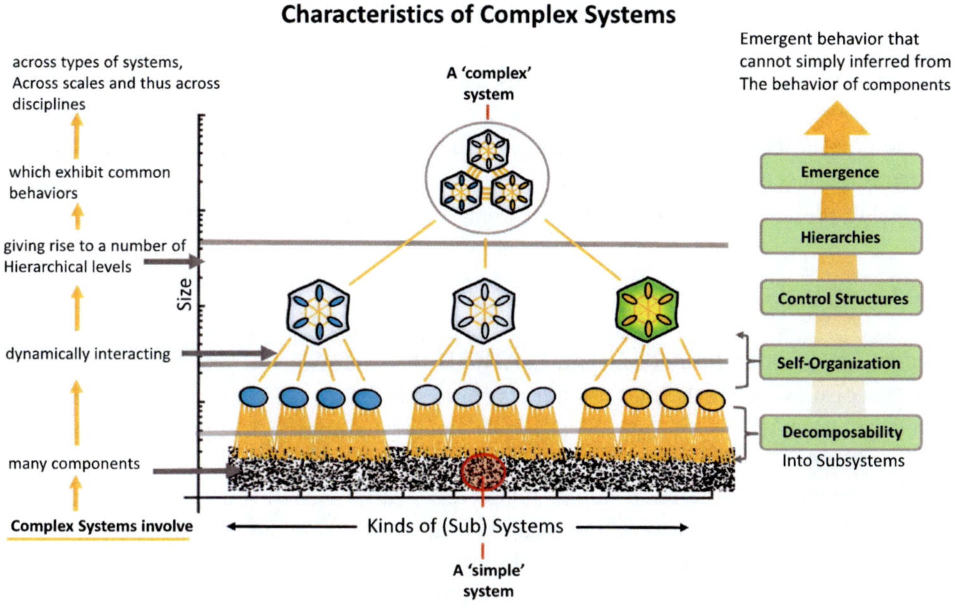

Fig. 1. Generic scheme of a complex system.

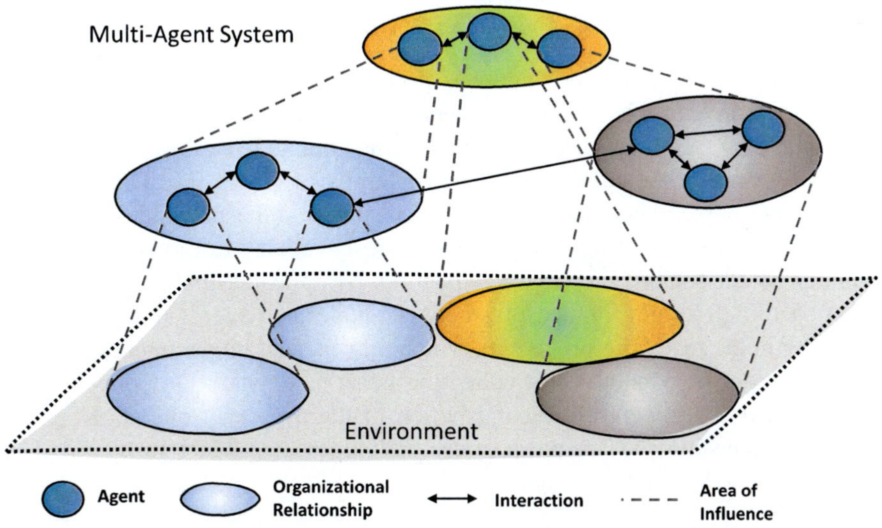

Fig. 2. Generic scheme of a multi-agent system [48].

Graph E-ABC, the version of E-ABC that can operate with unconventional mathematical spaces – such as the graph space – has, to some extent, all of the above characteristics. In particular, i) robustness, as multiple agents can insist on the same portion of data, each capable of extracting specific features, ii) scalability, as the proposed approach is designed to be performed through a parallel computation paradigm; iii) Graph E-ABC has a certain degree of heterogeneity, as different agents work on different tasks; iv) there is a good degree of autonomy and decentralization, as multiple swarms of agents coordinate evolutively to achieve a common goal, working on different portions of the dataset and focusing on heterogeneous characteristics. This is achieved through a specific consensus mechanism. The next section will illustrate the main components of the Graph E-ABC approach and the basic logic underlying its main operation.

4 Graph E-ABC

One of the main features of Graph E-ABC is its swarm-based structure. Let S be the number of classes for the classification problem at hand, Graph E-ABC spawns S different swarms $\Sigma = \{\mathcal{S}_1, \ldots, \mathcal{S}_S\}$, where agents belonging to the ith swarm $\mathcal{S}_i$ will process patterns belonging to the ith class only.

Specifically, each agent a belonging to the swarm $\mathcal{S}_i$ will start its data mining process by selecting uniformly at random W graphs amongst the ones in the training set whose class label is i. On each selected graph, the agent performs a random walk starting from a seed node randomly selected, yielding W subgraphs. After this data collection phase, the agent runs a clustering procedure driven by the BSAS algorithm [75] which depends on the following parameters:

- the maximum number of allowed clusters Q;
- a dissimilarity threshold θ in order to decide whether to include a given pattern into one of the already-existing clusters or to create a new cluster;
- a dissimilarity measure $d(\cdot, \cdot)$ between any two patterns.

Recalling that the clustering procedure is performed on a set of W graphs, the dissimilarity measure must be capable of encoding the (dis)similarity between any two graphs. Due to its speed and flexibility, we leverage the node Best Match First (nBMF) dissimilarity measure, belonging to the wider family of Graph Edit Distances (GED). nBMF exploits a greedy heuristic to match most similar nodes first and then proceeds in matching induced edges between any two previously-found pair of matched nodes (full algorithmic details can be found in [59, Appendix A]). The nBMF dissimilarity measure employs custom dissimilarity measures between nodes and edges[1] to find most similar nodes/edges in the matching procedure, which may as well be parametric: for the sake of generalization, let $\boldsymbol{\gamma}$ be a vector collecting those parameters, if applicable. Finally, nBMF can differently weight the contribution of the three edit operations (insertion, deletions and substitutions) between nodes and edges: hence, let $\mathbf{w}$ be a real-valued vector containing those six weights.

[1] The two dissimilarity measures between nodes and edges, say $d_v : \mathcal{L}^v \times \mathcal{L}^v$ and $d_e : \mathcal{L}^e \times \mathcal{L}^e$ with $\mathcal{L}^v$ and $\mathcal{L}^e$ being the set of nodes and edges label, have to be defined by the end-user according to the problem to face, and thus on the input data space.

Given these ingredients, the agent runs BSAS on the collected subgraphs, yielding a partition $\mathcal{P} = \{\mathcal{C}_1, \ldots, \mathcal{C}_k\}$ of k clusters. Each cluster $\mathcal{C}$ in the partition $\mathcal{P}$ is evaluated by means of a suitable quality index $r(\mathcal{C})$ that jointly consider its cardinality f_{ca} and compactness f_{co}, formally defined as:

$$f_{\mathrm{co}}(\mathcal{C}) = 1 - \frac{1}{|\mathcal{C}| - 1} \sum_{g \in \mathcal{C}} d(g, g^*) \tag{1}$$

$$f_{\mathrm{ca}}(\mathcal{C}) = \frac{|\mathcal{C}|}{W} \tag{2}$$

where g^* is the cluster representative of $\mathcal{C}$, i.e. the medoid of the cluster [60,61]. Starting from Eqs. (1)–(2), the reliability $r(\mathcal{C})$ reads as:

$$r(\mathcal{C}) = \eta \cdot f_{\mathrm{co}}(\mathcal{C}) + (1 - \eta) \cdot f_{\mathrm{ca}}(\mathcal{C}) \tag{3}$$

where $\eta \in [0, 1]$ is a trade-off parameter that weights the importance of compactness against cardinality. The medoid of the clusters whose reliability is greater than a threshold τ will compose a set of candidate symbols, namely the main output from the agent a data mining stage.

Once each agent $a \in \mathcal{S}_i$ has finished its data mining task, we have a class-specific bucket of symbols $\mathcal{H}_i$ obtained by merging the contribution of all agents belonging to $\mathcal{S}_i$.

Starting from $\mathcal{H}_i|_{i=1}^{S}$, a dedicated swarm $\mathcal{Z} = \{z_1, \ldots, z_K\}$ of K individuals is in charge of extracting suitable granules of information to build a vectorial embedding space where classification will be performed. Specifically, for each agent $z \in \mathcal{Z}$:

1. The agent evaluates the maximum number of symbols per class $t = T/S$ that can be extracted according to a user-defined bound T;
2. The agent explores the buckets $\mathcal{H}_i$ and extracts uniformly at random at most t symbols;
3. When all classes are explored, the selected symbols are collected into the multi-class alphabet of symbols $\mathcal{A}_z$.

Hence, the output of the swarm $\mathcal{Z}$ is a set of K alphabets $\mathcal{A}_1, \ldots, \mathcal{A}_K$ that serve as the pivotal entities to move the classification problem from the graph domain towards the Euclidean domain.

According to the symbolic histogram strategy [9], K different embedding spaces are evaluated starting from the graphs in the training set and the validation set and K different classifiers are trained on the respective training set and validated on the respective validation set. This ensemble of K classifiers, alongside their respective performances on the K embedding spaces, are the main output of this stage, which ends a single full iteration of Graph E-ABC.

Nonetheless, Graph E-ABC is an iterative algorithm that exploits two different evolutionary metaheuristics to drive the evolution of the two swarms Σ and $\mathcal{Z}$. The former is driven by a $(\mu + \lambda)$ selection scheme [16], whereas the latter relies on a custom-made evolutionary-like optimization metaheuristic.

Specifically, the evolution of Σ sees each agent as identified by a genetic code that contains:

1. the maximum number of allowed clusters Q for the BSAS clustering algorithm;
2. the threshold τ for discarding unpromising clusters;
3. the six weights for the nBMF dissimilarity measure $\mathbf{w}$;
4. the set of parameters characterizing the dissimilarity measures between nodes and edges γ, if applicable.

and aims at optimizing an objective function that jointly considers the reliability of its clusters and the quality of its symbols (see later Eq. (5)). The latter, in turn, reads as a function of the accuracy (on the validation set) of the classifier that exploited those symbols for building the embedding space. Formally, the symbol quality Q_s starts by the definition of a lookup table where each accuracy value (which is, by definition, bounded in $[0, 1]$) is mapped with a reward value $t \in [t_{\text{min}}, t_{\text{max}}]$, where the two bounds t_{min} and t_{max} are user-defined. Then:

1. for the kth alphabet $\mathcal{A}_k$
2. for each symbol $s \in \mathcal{A}_k$
3. find the accuracy ω of the classifier that exploited s for carrying out the classification task and, thanks to the lookup table, find the corresponding reward t
4. update the symbol quality according to the following rule

$$Q_s^{(\text{new})} = Q_s^{(\text{old})} + t \tag{4}$$

5. repeat from step 2, i.e., for all symbols in $\mathcal{A}_k$
6. repeat from step 1 for all alphabets, i.e., for $k = 1, \dots, K$.

The reasoning behind the symbol quality Q_s is that symbols that belong to fruitful embedding spaces, which in turn yield high performance in terms of classification accuracy, should be properly rewarded (i.e., the higher ω, the higher the corresponding t). However, it is also important to give an appropriate reward to the agent that discovered those symbols. To this end, let us further define the agent quality Q_a. Recall that each agent a runs a BSAS clustering procedure yielding a partition $\mathcal{P}$; however, only clusters having a suitable reliability score (cfr. Eq. (3)) are entitled to be part of the alphabets. Then:

1. for the ith swarm $\mathcal{S}_i \in \Sigma$
2. for each agent $a \in \mathcal{S}_i$
3. evaluate the mean reliability score $\overline{r}$ by considering only clusters discovered by a whose reliability score is greater than τ
4. evaluate the mean symbol quality $\overline{Q}$ for the same clusters
5. evaluate the agent quality as

$$Q_a = \rho \cdot \overline{Q} + (1 - \rho) \cdot \overline{r} \tag{5}$$

where $\rho \in [0, 1]$ weights the importance between the symbols qualities and their reliabilities[2]

[2] In particular, in the early generations of the evolution, we give more importance to the right-hand term in order to initially synthesize well-formed clusters. Next, the quality of an agent shall be better described as its ability in finding informative symbols according to Q_s, since the final scope of the whole procedure is devoted to synthesize meaningful alphabets employed for the classification problem. Indeed, this information is contained in symbols qualities Q_s whose values are actually backtracked in the agent quality Q_a.

6. repeat from step 2, i.e., for all agents in $\mathcal{S}_i$
7. repeat from step 1 for all swarms, i.e., for $i = 1, \ldots, S$.

Conversely, as anticipated, the evolution of the swarm $\mathcal{Z}$ follows a custom-made evolutionary-like optimization. Indeed, if the optimization of Σ can be performed by a standard metaheuristic since the candidate solution is encoded by a plain real-valued vector, the same does not hold for $\mathcal{Z}$. In this case, each candidate solution is an alphabet, hence a set of candidate symbols, where (in turn) each symbol is a subgraph. Therefore, it is necessary to define custom genetic-like operators (mutations, crossover, union) to work on such non-conventional candidate solutions. Full details on such genetic operators can be found in [13].

At the end of both optimizations, the test phase can take place. The test set will be embedded against the best K alphabets and fed to K classifiers. The ensemble is equipped with a winner-takes-all policy to emit the class label for a given test pattern.

5 Graph Neural Network

Graph Neural Network (GNN) architectures [21,72,79,81,85,86] typically implement convolution layers resembling the well-known convolutional neural networks for image recognition applications [5]. In this case, convolutional neural networks leverage on trainable localized filters which scan the images in order to extract high-level features by analyzing pixel relationships in the grid-like structure. On the other hand, graphs are characterized by a generally irregular structure that makes the convolution operation a challenging aspect [85]. One approach consists in defining the convolution operation in the spectral domain according to the eigenvalues and eigenvectors of the graph (normalized) Laplacian matrix $\overline{\mathbf{L}}$ [17,20,22,33], formally defined as:

$$\overline{\mathbf{L}} = \mathbf{D}^{-\frac{1}{2}}\mathbf{L}\mathbf{D}^{-\frac{1}{2}} = \mathbf{I}_n - \mathbf{D}^{-\frac{1}{2}}\mathbf{A}\mathbf{D}^{-\frac{1}{2}} \tag{6}$$

where $\mathbf{I}_n$ is the $n \times n$ identity matrix, $\mathbf{D} \in \mathbb{R}^{n \times n}$ and $\mathbf{A} \in \mathbb{R}^{n \times n}$ are the degree and the adjacency matrix of a graph G with $|\mathcal{V}| = n$ nodes and $\mathbf{L} = \mathbf{D} - \mathbf{A}$ is the (unnormalized) Laplacian matrix. Let also be $\mathcal{L}^v = \mathbb{R}^d$ and $\mathcal{L}^e = \emptyset$, then let $\mathbf{x} \in \mathbb{R}^d$ be a *node features* vector defined for a vertex $v \in \mathcal{V}$ also known as *graph signal*[3]. The convolution operation $\mathbf{f} * \mathbf{x}$ between a filter $\mathbf{f}$ and the graph signal $\mathbf{x}$ in the spectral domain is then defined as follow:

$$\mathbf{f} * \mathbf{x} = \mathbf{U}\mathbf{f}\mathbf{U}^\top\mathbf{x} \tag{7}$$

where $\mathbf{U}$ is the matrix whose columns are the eigenvectors of $\mathbf{L}$. All Graph Convolutional Neural Networks spectral methods share the definition of convolution operation provided in Eq. (7) differentiating each other on how the filter $\mathbf{f}$ (whose coefficient are learned by the network) is implemented [29,87].

However, it is still nontrivial to transfer the spectral-based graph convolutional network models learned on one graph to another graph whose eigenfunctions are different.

[3] For the sake of simplicity, the discussion is limited to graph with only node labels, albeit in principle it can be extended to a more general formulation addressing also edge labels [4].

To address this issue, filtering and convolution operations are generalized in spatial-based approaches. Typically, these methods perform recursive steps of node features propagation towards the neighbor vertices according to the topological structure. Then, each node aggregates all the information gathered from its neighbors in order to evaluate its own new feature vector [49,64,81]. Formally, the node feature representation $\mathbf{x}_i^{(l+1)}$ of ith vertex in the $l+1$ layer is evaluated as follows [4]:

$$\mathbf{x}_i^{(l+1)} = \Gamma^{(l+1)}\left(\mathbf{x}_i^{(l)}, \Psi\left(\{\gamma^{(l+1)}\left(\mathbf{x}_j^{(l)}\right) | j \in \Omega(i)\}\right)\right) \tag{8}$$

where $\Gamma^{(l+1)}$ and $\gamma^{(l+1)}$ are arbitrary non linear transformation of the input data e.g. multi-layer perceptron, Ψ is a permutation invariant function e.g. sum, mean, or max, and $\Omega(i)$ is the neighbor indices of node i. In Eq. (8) it is possible to spot two relevant operations: $\Psi\left(\{\gamma^{(l+1)}\left(\mathbf{x}_j^{(l)}\right) | j \in \Omega(i)\}\right)$ resembles an aggregation phase where the information about the neighbor node features are combined together according to Ψ in a message $\mathbf{m}_{\Omega(i)}$. Later on, the update phase is carried on by combining the current node feature vector $\mathbf{x}_i^{(l)}$ with the message $\mathbf{m}_{\Omega(i)}$ according to $\Gamma^{(l+1)}$. The whole graph representation in the geometric space can then be obtained through pooling operations by combining in a suitable way the representation vectors of all nodes in the graph [53].

In order to be able to perform a benchmark comparison with a state-of-the-art GNN, we employed an architecture built on top of $n^{(\text{conv})} = 2$ graph convolution (GCN) layers, $n^{(\text{dense})} = 2$ dense layers sized respectively 32 and 16 units based on the Rectified Linear Unit (ReLU) activation function, and a single dense sigmoid layer to yield the final prediction. Specifically, we have chosen $n^{(\text{conv})} = 2$ graph convolution layers to take into account second order neighborhood of each node during the processing phase. The whole GNN architecture adopted in this work is depicted in Fig. 3. The learning phase has been carried by means of the Adam optimizer, an early stop criterion based on a patience value, and a specific dropout value to reduce the possibility of overfitting [76].

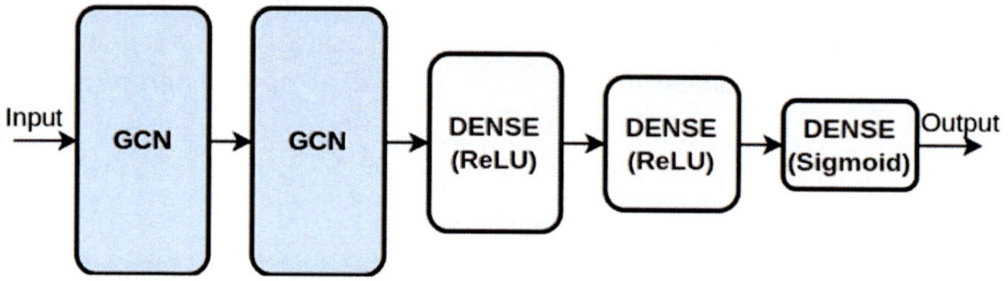

Fig. 3. Reference Graph Convolutional Network high level architecture used in this work.

6 Experimental Results

For the experimental stage, five datasets taken from the IAM Dataset Repository [71] are considered. Brief statistics about the datasets can be found in Table 1, along with the reference paper in which each dataset has been originally presented, to which we refer the interested reader for more information. Specifically, for each dataset, we show the number of graphs in the whole dataset, the average number of nodes and edges amongst graphs in the dataset, the number of classes for the classification problem, whether the classes are balanced or not and the dataset application domain. These datasets are composed by attributed graphs on both nodes and edges, whose detailed description can be found in [59, Appendix B].

Some pre-processing is required to deploy Graph Convolutional Networks on the considered datasets due to the tight constraint on building feature vectors for encoding nodes and edges attributes (see Sect. 5). The three Letters datasets did not require any further pre-processing step since their attributes are already in vectorial form and can conveniently be stacked in a matrix. Conversely, AIDS and GREC do have some categorical attributes (notably, the chemical symbol for AIDS and the type of nodes and edges for GREC): for those features in both datasets we performed a numerical mapping from their respective categorical domain towards integer values.

Table 1. Dataset statistics.

Dataset Name	# Graphs	Avg. # Nodes	Avg. # Edges	# Classes	Balanced	Domain	Reference
AIDS	2000	15.69	16.20	2	No	Chemoinformatics	[71]
Letter-L	2250	4.7	3.1	15	Yes	Computer Vision	[71]
Letter-M	2250	4.7	3.2	15	Yes	Computer Vision	[71]
Letter-H	2250	4.7	4.5	15	Yes	Computer Vision	[71]
GREC	1100	11.5	12.2	22	Yes	Electronics	[36,71]

Parameter settings for Graph E-ABC include:

- agents extract candidate subgraphs via a plain random walk strategy [44,52]
- the BSAS resolution θ assumes linearly spaced values in range $[0, 1]$ with step size 0.1
- $\mu = 5$ and $\lambda = 15$, the parents and offspring population sizes, respectively, for the evolution of Σ
- $\eta = 0.5$, weight between compactness f_{co} and cardinality f_{ca} in Eq. (3)
- maximum number of 20 generations for the evolution of Σ
- $t_{\text{min}} = -10$ and $t_{\text{max}} = 10$, respectively the maximum penalty and the maximum reward values for updating the symbol quality Q_s (see Eq. (4))
- ρ assumes equally spaced values in the range $[0, 1]$ with step size $\frac{1}{20}$ (see Eq. (5))
- $K = 10$, namely the number of candidate alphabets for embedding purposes.

Other than the topology (already introduced in Sect. 5), further parameters for the GNN include:

- $B_s = 30$ is the size of the input batch;
- $\delta = 0.4$ is the dropout rate for the neural network;
- $\alpha = 0.01$ is the learning rate for the Adam optimizer;

- $p = 5$ is the patience delta rate to trigger the early stop criterion;
- lastly, the whole network is trained with a maximum number of 50 epochs.

Computational results are shown in Table 2, in which we show the comparison in terms of accuracy on the test set, averaged across 10 different runs. For Graph E-ABC, we report two separate entries stemming from the sensitivity analysis carried out in [13]: indeed, in our previous work, we identified two critical parameters (W and T, namely the number of subgraphs that agents need to extract before running the clustering procedure and the maximum number of class-related subgraphs to be included in the alphabet, respectively) which underwent a sensitivity analysis by varying their respective values in a predefined range with the final goal of investigating how the performances change as a function of their respective values. Therefore, in Table 2, we report the best point on this 3D grid (i.e., the $\langle T, W\rangle$ pair yielding the maximum accuracy) and the accuracy obtained by averaging the performance on the test set across the entire $T \times W$ grid.

Table 2. Comparison between Graph E-ABC and the GNN for graph classification system in terms of accuracy on the test set. For Graph E-ABC we also report, in brackets, the size of the embedding space (i.e., number of symbols).

Technique	AIDS	GREC	Letter-L	Letter-M	Letter-H
Graph E-ABC (best point)	98.07 (34)	80.09 (1284)	96.89 (2260)	86.27 (4407)	79.78 (2517)
Graph E-ABC (average)	96.95 (222)	74.56 (1748)	92.20 (1460)	81.79 (3879)	74.76 (4592)
GNN	94.25	95.45	93.33	93.33	93.33

In Table 3, we show the running times for both GNN and Graph E-ABC. The GNN does not exploit any dedicated hardware (i.e., GPUs) and all tests have been performed on a Linux CentOS 7 workstation equipped with an Intel® Xeon® Gold 5120 CPU @ 2.20GHz and 192GB of RAM. Results for Graph-EABC include the entire evolutive optimization[4], whereas the GNN has a fixed architecture (determined by means of a simple trial-and-error approach) and does not feature any hyperparameter optimization phase. For Graph E-ABC, we report both the best point (i.e., the running time obtained

Table 3. Comparison between Graph E-ABC and the GNN in terms of running times (training time + testing time). Times are reported in seconds.

Technique	AIDS	GREC	Letter-L	Letter-M	Letter-H
Graph E-ABC (best point)	3244.43	7639.86	5062.73	5495.32	5915.07
Graph E-ABC (average)	7164.30	14316.64	7920.285	9414.85	10680.48
GNN	13.886	6.304	6.556	6.792	6.894

[4] A reasonable estimate for a fair comparison can be obtained by dividing the Graph E-ABC running time reported in Table 3 by 20 – i.e., the number of iterations.

by the $\langle T, W \rangle$-pair that yields the lowest running time) and the average value (i.e., by averaging the running times across the entire $T \times W$ grid[5]).

7 Discussions and Conclusions

In this paper, we have compared Graph E-ABC against a common GNN for graph classification purposes. Computational results on five open-access datasets of fully labelled graphs show interesting performances in terms of accuracy on the test set, counterbalanced by high running times due to the large embedding space dimensionality.

Specifically, by analyzing their respective results on the test set, it can be seen that Graph E-ABC has comparable results with respect to the GNN for 'easy' classification problems (i.e., AIDS and Letter-L), whereas it still shows some difficulties for harder datasets such as GREC, Letter-M and Letter-H. As the running times are concerned, Graph E-ABC features higher running times with respect to a CPU-trained GNN. This aspect is due to the fact that Graph E-ABC features a double optimization procedure, which inevitably slows down the model tuning and training, especially if we consider that agents in the second optimization stage must build several embedding spaces and train as many classifiers to build the ensemble.

References

1. Ahmad, R., Ali, S., Kim, D.H.: A multi-agent system for documents classification. In: 2012 International Conference on Open Source Systems and Technologies, pp. 28–32 (2012). https://doi.org/10.1109/ICOSST.2012.6472823
2. Alamgir, M., von Luxburg, U.: Multi-agent random walks for local clustering on graphs. In: 2010 IEEE International Conference on Data Mining, pp. 18–27 (2010). https://doi.org/10.1109/ICDM.2010.87
3. Apolloni, B., Pedrycz, W., Bassis, S., Malchiodi, D.: The Puzzle of Granular Computing. Springer, Heidelberg (2008)
4. Bacciu, D., Errica, F., Micheli, A., Podda, M.: A gentle introduction to deep learning for graphs. Neural Netw. **129**, 203–221 (2020). https://doi.org/10.1016/j.neunet.2020.06.006
5. Baldini, L.: Granular Graph Embedding Techniques for Graph Classification. Ph.D. thesis, University of Rome "La Sapienza" (2021)
6. Baldini, L., Martino, A., Rizzi, A.: Stochastic information granules extraction for graph embedding and classification. In: Proceedings of the 11th International Joint Conference on Computational Intelligence - NCTA, (IJCCI 2019), pp. 391–402. INSTICC, SciTePress (2019). https://doi.org/10.5220/0008149403910402
7. Baldini, L., Martino, A., Rizzi, A.: Complexity vs. performance in granular embedding spaces for graph classification. In: Proceedings of the 12th International Joint Conference on Computational Intelligence - NCTA, pp. 338–349. INSTICC, SciTePress (2020). https://doi.org/10.5220/0010109503380349
8. Baldini, L., Martino, A., Rizzi, A.: Exploiting cliques for granular computing-based graph classification. In: 2020 International Joint Conference on Neural Networks (IJCNN), pp. 1–9 (2020). https://doi.org/10.1109/IJCNN48605.2020.9206690

[5] Full sensitivity analysis on the running times not shown for brevity.

9. Baldini, L., Martino, A., Rizzi, A.: Relaxed dissimilarity-based symbolic histogram variants for granular graph embedding. In: Proceedings of the 13th International Joint Conference on Computational Intelligence - NCTA, pp. 221–235. INSTICC, SciTePress (2021). https://doi.org/10.5220/0010652500003063
10. Baldini, L., Martino, A., Rizzi, A.: Towards a class-aware information granulation for graph embedding and classification. In: Merelo, J.J., Garibaldi, J., Linares-Barranco, A., Warwick, K., Madani, K. (eds.) IJCCI 2019. SCI, vol. 922, pp. 263–290. Springer, Cham (2021). https://doi.org/10.1007/978-3-030-70594-7_11
11. Baldini, L., Martino, A., Rizzi, A.: A class-specific metric learning approach for graph embedding by information granulation. Appl. Soft Comput. **115**, 108199 (2022). https://doi.org/10.1016/j.asoc.2021.108199
12. Baldini, L., Martino, A., Rizzi, A.: A multi-objective optimization approach for the synthesis of granular computing-based classification systems in the graph domain. SN Comput. Sci. **3**(6), 436 (2022). https://doi.org/10.1007/s42979-022-01260-4
13. Baldini, L., Rizzi, A.: A multi-agent approach for graph classification. In: Proceedings of the 13th International Joint Conference on Computational Intelligence - NCTA, pp. 334–343. INSTICC, SciTePress (2021). https://doi.org/10.5220/0010677300003063
14. Bargiela, A., Pedrycz, W.: Granular computing: an introduction. Kluwer Academic Publishers, Boston (2003)
15. Bargiela, A., Pedrycz, W.: Toward a theory of granular computing for human-centered information processing. IEEE Trans. Fuzzy Syst. **16**(2), 320–330 (2008)
16. Beyer, H.G., Schwefel, H.P.: Evolution strategies - a comprehensive introduction. Nat. Comput. **1**(1), 3–52 (2002). https://doi.org/10.1023/A:1015059928466
17. Bianchi, F.M., Grattarola, D., Livi, L., Alippi, C.: Graph neural networks with convolutional arma filters. IEEE Trans. Pattern Anal. Mach. Intell. **44**(7), 3496–3507 (2022). https://doi.org/10.1109/TPAMI.2021.3054830
18. Bianchi, F.M., Maiorino, E., Livi, L., Rizzi, A., Sadeghian, A.: An agent-based algorithm exploiting multiple local dissimilarities for clusters mining and knowledge discovery. Soft. Comput. **21**(5), 1347–1369 (2015). https://doi.org/10.1007/s00500-015-1876-1
19. Bianchi, F.M., Rizzi, A., Sadeghian, A., Moiso, C.: Identifying user habits through data mining on call data records. Eng. Appl. Artif. Intell. **54**, 49–61 (2016). https://doi.org/10.1016/j.engappai.2016.05.007
20. Bollobás, B.: Modern Graph Theory, vol. 184. Springer (2013)
21. Bronstein, M.M., Bruna, J., LeCun, Y., Szlam, A., Vandergheynst, P.: Geometric deep learning: going beyond euclidean data. IEEE Signal Process. Mag. **34**(4), 18–42 (2017). https://doi.org/10.1109/MSP.2017.2693418
22. Bruna, J., Zaremba, W., Szlam, A., LeCun, Y.: Spectral networks and locally connected networks on graphs. arXiv preprint arXiv:1312.6203 (2013)
23. Cao, W., Yan, Z., He, Z., He, Z.: A comprehensive survey on geometric deep learning. IEEE Access **8**, 35929–35949 (2020)
24. Capillo, A., De Santis, E., Frattale Mascioli, F.M., Rizzi, A.: Mining m-grams by a granular computing approach for text classification. In: Proceedings of the 12th International Joint Conference on Computational Intelligence - NCTA, pp. 350–360. INSTICC, SciTePress (2020). https://doi.org/10.5220/0010109803500360
25. Carvalho, L.F., Barbon, S., de Souza Mendes, L., Proença, M.L.: Unsupervised learning clustering and self-organized agents applied to help network management. Expert Syst. Appl. **54**, 29–47 (2016). https://doi.org/10.1016/j.eswa.2016.01.032
26. Chaimontree, S., Atkinson, K., Coenen, F.: Clustering in a multi-agent data mining environment. In: Cao, L., Bazzan, A.L.C., Gorodetsky, V., Mitkas, P.A., Weiss, G., Yu, P.S. (eds.) ADMI 2010. LNCS (LNAI), vol. 5980, pp. 103–114. Springer, Heidelberg (2010). https://doi.org/10.1007/978-3-642-15420-1_9

27. Chaimontree, S., Atkinson, K., Coenen, F.: A Multi-agent based approach to clustering: harnessing the power of agents. In: Cao, L., Bazzan, A.L.C., Symeonidis, A.L., Gorodetsky, V.I., Weiss, G., Yu, P.S. (eds.) ADMI 2011. LNCS (LNAI), vol. 7103, pp. 16–29. Springer, Heidelberg (2012). https://doi.org/10.1007/978-3-642-27609-5_3
28. Chen, L., Zhang, H., Xiao, J., He, X., Pu, S., Chang, S.F.: Counterfactual critic multi-agent training for scene graph generation. In: 2019 IEEE/CVF International Conference on Computer Vision (ICCV), pp. 4613–4623 (2019). https://doi.org/10.1109/ICCV.2019.00471
29. Chen, X.: Understanding spectral graph neural network. arXiv preprint arXiv:2012.06660 (2020)
30. Corrêa, L., Arantes, L., Sens, P., Inostroza-Ponta, M., Dorn, M.: A dynamic evolutionary multi-agent system to predict the 3d structure of proteins. In: 2020 IEEE Congress on Evolutionary Computation (CEC), pp. 1–8. IEEE (2020). https://doi.org/10.1109/CEC48606.2020.9185761
31. De Santis, E., Livi, L., Sadeghian, A., Rizzi, A.: Modeling and recognition of smart grid faults by a combined approach of dissimilarity learning and one-class classification. Neurocomputing **170**, 368–383 (2015)
32. De Santis, E., Martino, A., Rizzi, A.: On component-wise dissimilarity measures and metric properties in pattern recognition. PeerJ Comput. Sci. **8**, e1106 (2022). https://doi.org/10.7717/peerj-cs.1106
33. Defferrard, M., Bresson, X., Vandergheynst, P.: Convolutional neural networks on graphs with fast localized spectral filtering. Adv. Neural. Inf. Process. Syst. **29**, 3844–3852 (2016). https://doi.org/10.5555/3157382.3157527
34. Di Noia, A., Martino, A., Montanari, P., Rizzi, A.: Supervised machine learning techniques and genetic optimization for occupational diseases risk prediction. Soft. Comput. **24**(6), 4393–4406 (2019). https://doi.org/10.1007/s00500-019-04200-2
35. Ding, W., Lin, C.T., Chen, S., Zhang, X., Hu, B.: Multiagent-consensus-mapreduce-based attribute reduction using co-evolutionary quantum pso for big data applications. Neurocomputing **272**, 136–153 (2018). https://doi.org/10.1016/j.neucom.2017.06.059
36. Dosch, P., Valveny, E.: Report on the second symbol recognition contest. In: Liu, W., Lladós, J. (eds.) GREC 2005. LNCS, vol. 3926, pp. 381–397. Springer, Heidelberg (2006). https://doi.org/10.1007/11767978_35
37. Duin, R.P.W., Pekalska, E.: Open issues in pattern recognition. In: Kurzyński, M., Puchała, E., Woźniak, M., Żołnierek, A. (eds.) Computer Recognition Systems, pp. 27–42. Springer, Heidelberg (2005). https://doi.org/10.1007/3-540-32390-2_3
38. Duin, R.P., Roli, F., de Ridder, D.: A note on core research issues for statistical pattern recognition. Pattern Recogn. Lett. **23**(4), 493–499 (2002)
39. Gärdenfors, P.: Conceptual Spaces: The Geometry of Thought. MIT Press (2004)
40. Giampieri, M., Baldini, L., De Santis, E., Rizzi, A.: Facing big data by an agent-based multimodal evolutionary approach to classification. In: 2020 International Joint Conference on Neural Networks (IJCNN), pp. 1–8 (2020). https://doi.org/10.1109/IJCNN48605.2020.9206966
41. Giampieri, M., De Santis, E., Rizzi, A., Frattale Mascioli, F.M.: A supervised classification system based on evolutive multi-agent clustering for smart grids faults prediction. In: 2018 International Joint Conference on Neural Networks (IJCNN), pp. 1–8 (2018). https://doi.org/10.1109/IJCNN.2018.8489145
42. González-Briones, A., Villarrubia, G., De Paz, J.F., Corchado, J.M.: A multi-agent system for the classification of gender and age from images. Comput. Vis. Image Underst. **172**, 98–106 (2018). https://doi.org/10.1016/j.cviu.2018.01.012
43. Gunning, D., Stefik, M., Choi, J., Miller, T., Stumpf, S., Yang, G.Z.: Xai-explainable artificial intelligence. Sci. Robot. **4**(37), eaay7120 (2019)

44. Göbel, F., Jagers, A.A.: Random walks on graphs. Stochast. Process. Appl. **2**(4), 311–336 (1974). https://doi.org/10.1016/0304-4149(74)90001-5
45. Hajduk, M., Sukop, M., Haun, M.: Cognitive Multi-agent Systems (2019)
46. Herrera, M., Pérez-Hernández, M., Kumar Parlikad, A., Izquierdo, J.: Multi-agent systems and complex networks: review and applications in systems engineering. Processes **8**(3), 312 (2020)
47. İnkaya, T., Kayalıgil, S., Özdemirel, N.E.: Ant colony optimization based clustering methodology. Appl. Soft Comput. **28**, 301–311 (2015). https://doi.org/10.1016/j.asoc.2014.11.060
48. Jennings, N.R.: An agent-based approach for building complex software systems. Commun. ACM **44**(4), 35–41 (2001)
49. Kipf, T.N., Welling, M.: Semi-supervised classification with graph convolutional networks. arXiv preprint arXiv:1609.02907 (2016)
50. Kwapień, J., Drożdż, S.: Physical approach to complex systems. Phys. Rep. **515**(3–4), 115–226 (2012)
51. Lombardo, G., Fornacciari, P., Mordonini, M., Tomaiuolo, M., Poggi, A.: A multi-agent architecture for data analysis. Future Internet **11**(2), 49 (2019). https://doi.org/10.3390/fi11020049
52. Lovász, L.: Random walks on graphs: a survey. Combinatorics **2**, 1–46 (1993)
53. Ma, Y., Wang, S., Aggarwal, C.C., Tang, J.: Graph convolutional networks with eigenpooling. In: Proceedings of the 25th ACM SIGKDD International Conference on Knowledge Discovery & Data Mining (KDD '19), pp. 723–731. Association for Computing Machinery, New York (2019). https://doi.org/10.1145/3292500.3330982
54. Martino, A., Baldini, L., Rizzi, A.: On information granulation via data clustering for granular computing-based pattern recognition: a graph embedding case study. Algorithms **15**(5) (2022). https://doi.org/10.3390/a15050148
55. Martino, A., De Santis, E., Rizzi, A.: An ecology-based index for text embedding and classification. In: 2020 International Joint Conference on Neural Networks (IJCNN), pp. 1–8. IEEE (2020)
56. Martino, A., Frattale Mascioli, F.M., Rizzi, A.: On the optimization of embedding spaces via information granulation for pattern recognition. In: 2020 International Joint Conference on Neural Networks (IJCNN), pp. 1–8 (2020). https://doi.org/10.1109/IJCNN48605.2020.9206830
57. Martino, A., Giampieri, M., Luzi, M., Rizzi, A.: Data mining by evolving agents for clusters discovery and metric learning. In: Esposito, A., Faundez-Zanuy, M., Morabito, F.C., Pasero, E. (eds.) WIRN 2017 2017. SIST, vol. 102, pp. 23–35. Springer, Cham (2019). https://doi.org/10.1007/978-3-319-95098-3_3, italian Workshop on Neural Nets 2017
58. Martino, A., Giuliani, A., Todde, V., Bizzarri, M., Rizzi, A.: Metabolic networks classification and knowledge discovery by information granulation. Comput. Biol. Chem. **84**, 107187 (2020). https://doi.org/10.1016/j.compbiolchem.2019.107187
59. Martino, A., Rizzi, A.: An enhanced filtering-based information granulation procedure for graph embedding and classification. IEEE Access **9**, 15426–15440 (2021). https://doi.org/10.1109/ACCESS.2021.3053085
60. Martino, A., Rizzi, A., Frattale Mascioli, F.M.: Distance matrix pre-caching and distributed computation of internal validation indices in k-medoids clustering. In: 2018 International Joint Conference on Neural Networks (IJCNN), pp. 1–8 (2018). https://doi.org/10.1109/IJCNN.2018.8489101
61. Martino, A., Rizzi, A., Frattale Mascioli, F.M.: Efficient approaches for solving the large-scale *k*-medoids problem: towards structured Data. In: Sabourin, C., Merelo, J.J., Madani, K., Warwick, K. (eds.) IJCCI 2017. SCI, vol. 829, pp. 199–219. Springer, Cham (2019). https://doi.org/10.1007/978-3-030-16469-0_11

62. Modi, P.J., Shen, W.M.: Collaborative multiagent learning for classification tasks. In: Proceedings of the Fifth International Conference on Autonomous Agents (AGENTS '01), pp. 37–38. Association for Computing Machinery, New York (2001). https://doi.org/10.1145/375735.375854
63. Mostafa, J., Ke, W., Fu, Y.: Automated text classification using a multi-agent framework. In: Proceedings of the 5th ACM/IEEE-CS Joint Conference on Digital Libraries (JCDL '05), pp. 157–158 (2005). https://doi.org/10.1145/1065385.1065420
64. Niepert, M., Ahmed, M., Kutzkov, K.: Learning convolutional neural networks for graphs. In: Balcan, M.F., Weinberger, K.Q. (eds.) Proceedings of The 33rd International Conference on Machine Learning, vol. 48, pp. 2014–2023. PMLR (2016)
65. Ogston, E., Overeinder, B., Van Steen, M., Brazier, F.: A method for decentralized clustering in large multi-agent systems. In: Proceedings of the Second International Joint Conference on Autonomous Agents and Multiagent Systems, pp. 789–796. ACM (2003). https://doi.org/10.1145/860575.860702
66. Pan, X., Chen, H.: Multi-agent evolutionary clustering algorithm based on manifold distance. In: Proceedings of the 2012 Eighth International Conference on Computational Intelligence and Security (CIS), pp. 123–127. IEEE (2012)
67. Pan, X., Jiao, L.: A granular agent evolutionary algorithm for classification. Appl. Soft Comput. **11**(3), 3093–3105 (2011). https://doi.org/10.1016/j.asoc.2010.12.012
68. Park, J.E., Oh, K.W.: Multi-agent systems for intelligent clustering. Int. J. Comput. Inf. Eng. **1**(11), 3663–3668 (2007). https://doi.org/10.5281/zenodo.1076836
69. Pedrycz, W., Skowron, A., Kreinovich, V.: Handbook of Granular Computing. John Wiley & Sons (2008)
70. Preuss, M.: Introduction: towards multimodal optimization. In: Multimodal Optimization by Means of Evolutionary Algorithms. NCS, pp. 1–25. Springer, Cham (2015). https://doi.org/10.1007/978-3-319-07407-8_1
71. Riesen, K., Bunke, H.: IAM graph database repository for graph based pattern recognition and machine learning. In: da Vitoria Lobo, N., et al. (eds.) SSPR/SPR 2008. LNCS, vol. 5342, pp. 287–297. Springer, Heidelberg (2008). https://doi.org/10.1007/978-3-540-89689-0_33
72. Schlichtkrull, M., Kipf, T.N., Bloem, P., van den Berg, R., Titov, I., Welling, M.: Modeling relational data with graph convolutional networks. In: Gangemi, A., et al. (eds.) ESWC 2018. LNCS, vol. 10843, pp. 593–607. Springer, Cham (2018). https://doi.org/10.1007/978-3-319-93417-4_38
73. Schneider, M., Somers, M.: Organizations as complex adaptive systems: implications of complexity theory for leadership research. Leadersh. Q. **17**(4), 351–365 (2006)
74. Shea, N.: Representation in Cognitive Science. Oxford University Press (2018)
75. Theodoridis, S., Koutroumbas, K.: Pattern Recognition, 4 edn. Academic Press (2008)
76. Vapnik, V.N., Chervonenkis, A.Y.: On the uniform convergence of relative frequencies of events to their probabilities. Theory Prob. Appl. **16**(2), 264–280 (1971). https://doi.org/10.1137/1116025
77. Ward, I.R., Joyner, J., Lickfold, C., Rowe, S., Guo, Y., Bennamoun, M.: A practical guide to graph neural networks. arXiv preprint arXiv:2010.05234 (2020)
78. Wong, K.: Evolutionary multimodal optimization: a short survey. arXiv preprint arXiv:1508.00457 (2015)
79. Wu, Z., Pan, S., Chen, F., Long, G., Zhang, C., Yu, P.S.: A comprehensive survey on graph neural networks. IEEE Trans. Neural Netw. Learn. Syst. **32**(1), 4–24 (2021). https://doi.org/10.1109/TNNLS.2020.2978386
80. Xu, F., Uszkoreit, H., Du, Y., Fan, W., Zhao, D., Zhu, J.: Explainable AI: a brief survey on history, research areas, approaches and challenges. In: Tang, J., Kan, M.-Y., Zhao, D., Li,

S., Zan, H. (eds.) NLPCC 2019. LNCS (LNAI), vol. 11839, pp. 563–574. Springer, Cham (2019). https://doi.org/10.1007/978-3-030-32236-6_51
81. Xu, K., Hu, W., Leskovec, J., Jegelka, S.: How powerful are graph neural networks? In: International Conference on Learning Representations, pp. 1–17 (2019)
82. Yao, J.T., Vasilakos, A.V., Pedrycz, W.: Granular computing: perspectives and challenges. IEEE Trans. Cybernet. **43**(6), 1977–1989 (2013)
83. Yao, Y.: Perspectives of granular computing. In: 2005 IEEE International Conference on Granular Computing, vol. 1, pp. 85–90. IEEE (2005). https://doi.org/10.1109/GRC.2005.1547239
84. Zadeh, L.A.: Toward a theory of fuzzy information granulation and its centrality in human reasoning and fuzzy logic. Fuzzy Sets Syst. **90**(2), 111–127 (1997). https://doi.org/10.1016/S0165-0114(97)00077-8
85. Zhang, S., Tong, H., Xu, J., Maciejewski, R.: Graph convolutional networks: a comprehensive review. Comput. Soc. Netw. **6**(1), 1–23 (2019). https://doi.org/10.1186/s40649-019-0069-y
86. Zhang, Z., Cui, P., Zhu, W.: Deep learning on graphs: a survey. IEEE Trans. Knowl. Data Eng. 1 (2020). https://doi.org/10.1109/TKDE.2020.2981333
87. Zhou, J., et al.: Graph neural networks: a review of methods and applications. AI Open **1**, 57–81 (2020). https://doi.org/10.1016/j.aiopen.2021.01.001
88. De Santis, E., Rizzi, A.: Prototype theory meets word embedding: a novel approach for text categorization via granular computing. Cogn. Comput. **15**(3), 976–997 (2023). https://doi.org/10.1007/s12559-023-10132-9

One-Shot Identification with Different Neural Network Approaches

Janis Mohr(✉) and Jörg Frochte

Interdisciplinary Institute for Applied Artificial Intelligence and Data Science Ruhr, Bochum University of Applied Sciences, 42579 Heiligenhaus, Germany
{janis.mohr,joerg.frochte}@hs-bochum.de

Abstract. Convolutional neural networks (CNNs) have been widely used in the computer vision community, significantly improving the state-of-the-art. But learning good features often is computationally expensive in machine learning settings and is especially difficult when there is a lack of data. One-shot learning is one such area where only limited data is available. In one-shot learning, predictions have to be made after seeing only one example from one class, which requires special techniques. In this paper we explore different approaches to one-shot identification tasks in different domains including an industrial application and face recognition. We use a special technique with stacked images and use siamese capsule networks. It is encouraging to see that the approach using capsule architecture achieves strong results and exceeds other techniques on a wide range of datasets from industrial application to face recognition benchmarks while being easy to use and optimise.

Keywords: One-shot learning · Convolutional neural networks · Siamese networks · Capsules · Industrial application

1 Introduction

Mammals have the ability to recognise patterns and acquire new skills. In particular, humans can decide, whether two objects are identical or belong to the same category. For example, while humans are capable of perceiving that two different breeds of cats belong to the same category, they can also easily distinguish between a cat and a dog. This skill, more than any other, serves as an essential survival skill, particularly as it relates to identifying food and avoiding dangerous entities. Humans have the ability to quickly and accurately categorise species, even though they may have never seen an individual from the species before - all based on their memories and experience [22]. Little or even no additional data is needed.

Popular machine learning tasks and settings, on the other hand, tend to use very large amounts of data for training but this does not necessarily prevent them from failing when they are shown new and unfamiliar data. Generalisation is a problem of its own.

A challenge of particular interest is the training of machine learning models with limited training data. One-shot learning refers to the idea that an image of a class is only ever seen once. One-shot identification is a special case. Determining whether

J. Garibaldi et al. (Eds.): IJCCI 2021, SCI 1119, pp. 205–222, 2023.
https://doi.org/10.1007/978-3-031-46221-4_10

or not two items belong to the same class is the task at hand. For instance, after being shown images of an object through a series of various manufacturing processes, a neural network would be able to discern the object.

Computer vision methods can be used for a wide variety of tasks. Not only image recognition but also quality assurance. A special use-case in industrial vision is the inspection of produced objects. For industrial tasks, it often is complicated to acquire large amounts of training data because it can be difficult to mount cameras or other sensors during ongoing production. Furthermore convolutional neural networks are used for face recognition tasks. With facial recognition, special attention must always be paid to data protection concerns, which also makes it particularly difficult to obtain larger data sets. This makes these two areas in particular common tasks in the field of one-shot learning, which is under special focus in research and of significant relevance in industrial application.

1.1 Related Work

Overall, research into one-shot learning algorithms has received limited attention by the machine learning community. Despite this, there are a few key papers on this task. Initial work on this topic dates back to the years 2003 and 2006 [12, 13]. The authors developed a variational Bayesian framework for one-shot image classification using the premise that previously learned classes can be leveraged to help forecast future ones when very few examples are available from a given class. In a series of papers, the problem of one-shot learning was approached from the viewpoint of cognitive science, proposing a method called Hierarchical Bayesian Program Learning which focuses on character recognition [25–27]. This method aims at decomposing the image into small pieces and determining a structural explanation for the observed pixels. Some researchers have considered transfer learning approaches for recognising new words from unknown speakers [24] and path planning algorithms [49].

Although the idea of siamese networks has been around for quite some time [5], it has only recently been used for one-shot learning [8]. They have mostly been used for robust face recognition and verification as [9,46] have shown. [21] use siamese neural networks for one-shot image recognition in which they use two identical neural networks that are trained and then fed with two different images. The task is to determine whether those two images belong to the same class. [11] especially used siamese networks to detect defects in steel surfaces during manufacturing somewhat similar to the industrial application used as a benchmark in this paper.

Scientific work focused on a mathematical representation of the human visual cortex led to [15] identifies that some of the processes in the human brain related to vision have similarities to the concept of inverse computer graphics. In inverse computer graphics, an image is analysed and the task is to locate entities in this image in such a way that complex objects are broken down into several simple geometric shapes [16]. Currently used CNNs do not operate in a way that is consistent with this principle [47]. Building on initial experiments [17], so-called capsule networks and dynamic routing in which neurons are grouped into capsules and vectors are shared between layers were proposed [42]. Some additional work was done in different domains to study the performance and possibilities of capsule networks. Rajasegaran et al. [39, 50] analysed performance on key-benchmark datasets like Fashion-MNIST and CIFAR-10. Several papers work out that capsule networks surpass baseline CNNs in image classification tasks across several domains [2,3,14,20,23,30,34,36,38,40,41,44] focused on the qualities regarding explainability which are based on the proximity to the functioning of human vision.

Extensive work has been done to identify objects by their structure. Minderer et al. [33] uses video clips to extract the structure of objects. Using and creating videos is often not possible due to constraints regarding to privacy or because of costs. Abdel et al. [1] studied various computer vision algorithms, including fast Haar transform and Canny edge detector for the detection of cracks and other kinds of damage in concrete. This study was enhanced by Cha et al. [7] which presented an approach to recognise defects on concrete surfaces. Instead of classic computer vision algorithms, they used a deep convolutional neural network and furthermore used a rule-based approach as a baseline. They summarise that it is possible to outperform classic machine vision techniques with neural networks and that these approaches can be very robust in situations where different lighting can occur. Benhimane et al. [4] uses a template management algorithm to track 3D objects in real-time. Brusey et al. and Mahieu et al. [6,32] propose to use RFID chips and vision systems to identify objects during manufacturing processes.

Mohr et al. [35] proposed the approach for merged images used in this paper and also have an in-depth explanation for the industrial application. This paper extends Mohrs' work by looking at another and additional approach, the siamese capsule networks, and by conducting experiments on two other datasets from different domains.

Our contributions in this paper are summarised as follows: (1) We analyse approaches for identifying objects which have only been seen once with capsule networks. (2) We focus on techniques that are economical with data. (3) We show the versatility and high accuracy of the presented approaches and compare several approaches in different domains.

2 Approach

In this section, the three approaches used in this paper including the siamese capsule network are described and explained in detail.

2.1 Classic Convolutional Neural Network with Merged Images

Although deep learning has shown great promise for image recognition tasks, it comes with the cost of large data requirements. Since it is not always possible to gather large amounts of data due to e.g. information privacy, the possible applications for machine learning are still limited. To address this issue, there has been recent interest in the research community to develop neural networks, that can effectively learn from a small amount of data.

One methodology to decrease the quantity of training data required while keeping reliable predictions is one-shot learning. One-shot image recognition's core tenet is that a neural network should be able to determine whether or not examples belong to the same class given a single sample of an image out of that class. Throughout training, the

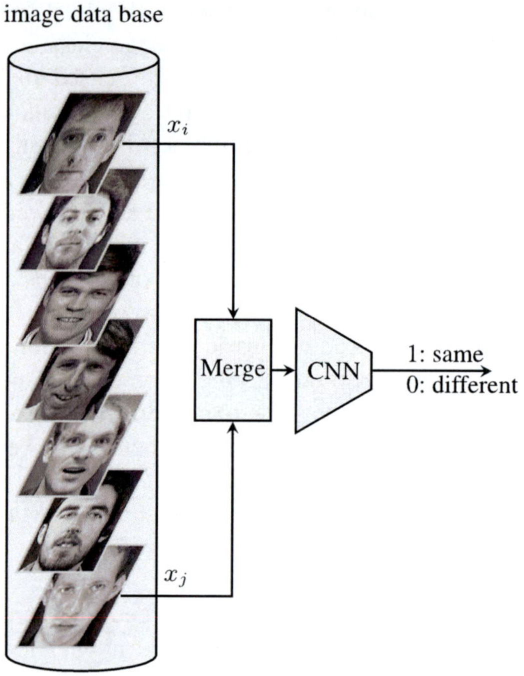

Fig. 1. General approach to one-shot identification using merged images. Two images are merged and then put into a CNN to classify whether they contain the same or different objects. The images on the left are a random selection from the AT&T dataset used in Sect. 3.

network determines which features of the input image pair deviate from one another. Every convolutional neural network may be employed with this method of one-shot identification, depending on the application. Once trained, the model ought to be able to recognise objects despite variations in colour or material properties. The network's output is [same, different]. *Same* denotes the presence of the same object in both photos, whereas *different* denotes the presence of different objects. The solution outlined by [35] seeks to act in the unique situation when only very little training data is provided and comparable images are only seen once.

Figure 1 shows the proposed approach with a data set. A classic convolutional neural network is trained on a data set that consists of images of human faces with different expressions and lighting. The task is to compare two images, which have never been seen before, and to decide if they show the same person or not. The network is trained with merged and labeled images, which show the same or different humans. The meaning of *merged* in practice in this method is explained in Sect. 2.1. Therefore, the neural network learns to discriminate between different objects and can generalise this information to compare images of objects it has never seen before.

The hyperparameter values used for training the network are provided in Table 1. Four convolutional layers with 32, 32, 64, and 64 filters are used, followed by a max pooling [43]. A kernel size of 3×3 is used for convolutions with a stride of 1. The ReLU activation function is used on the output feature maps of each layer. The convolutional layers are followed by two fully connected layers of size 128 and 2 respectively.

Table 1. Hyperparameter values used for training the convolutional neural network. [35].

Parameter	Value
Batch Size	32
Number of Epochs	20
Learning Rate	1e-4
Optimiser	RMSprop [18]

Handover of Two Images. A basic CNN uses one picture at a time as input [28]. Finding a procedure that can take two inputs and compare two images is essential for the comparison of images. This results in two possible alternatives. Two images are merged, and the combined picture is supplied into a CNN for training. The alternative method is based on utilizing two convolution and pooling sequences. The convolutional neural network receives two images as input. Following flattening, the two sequences-each supplied with a distinct image-are concatenated in a particular layer.

Several ways of merging the images are possible. One option is, that the images could be joined horizontally or vertically, as shown in Fig. 2. If the original dimensions are n and m the result is an $2n \times m$ or $n \times 2m$ image.

Fig. 2. Merging two images into a bigger image by joining them horizontally or vertically. [35].

The other way is to stack the images' channels. For an image with colour channels, this leads to a total of six channels. The process is illustrated in Fig. 3, where the images are converted to greyscale for better handling.

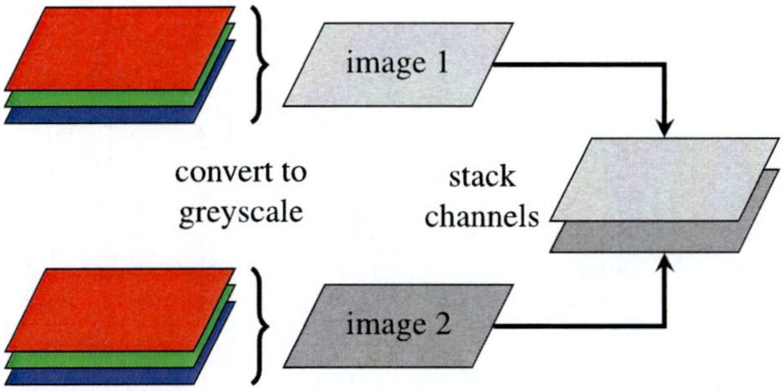

Fig. 3. Merging two images converted to greyscale by stacking them resulting in a two-channel image. [35].

Both approaches lead to an altered tensor, that is delivered to the neural network. Therefore, attention must be paid to the distinctions between the input layer and images. The influence of these two approaches is evaluated. With the same database and the same convolutional neural network the stacking approach of merging leads to a prediction accuracy of 98.36%. When joining the images, as in Fig. 2, they only achieve an accuracy of 61.48%. Because of these results, the approach with joined images will not be investigated any further. Therefore, better results can be expected from a database with stacked images. When the colour channels of the photos are stacked, the important characteristics are in the same relative location. The performance of the CNN may be impacted even if a CNN can achieve a small invariance to translations through parameter sharing. In addition to this, the data are mixed up with the merged images because of convolutions and pooling. With stacked pictures, this is not the case since each operation is applied to each channel independently.

2.2 Siamese Networks

Siamese Networks (SNs) are neural networks that learn relationships between encoded representations of instance pairs that lie on the low dimensional manifold, where a chosen distance function $d\omega$ is used to find the similarity in output space. Below we briefly describe state-of-the-art convolutional SNs which have mainly been used for face verification and face recognition.

Sun et al. [45] presented a joint identification-verification approach for learning face verification with a contrastive loss and face recognition using cross-entropy loss. To balance loss signals for both identification and verification, they investigate the effects of varying weights controlled by λ on the intra-personal and inter-personal variations.

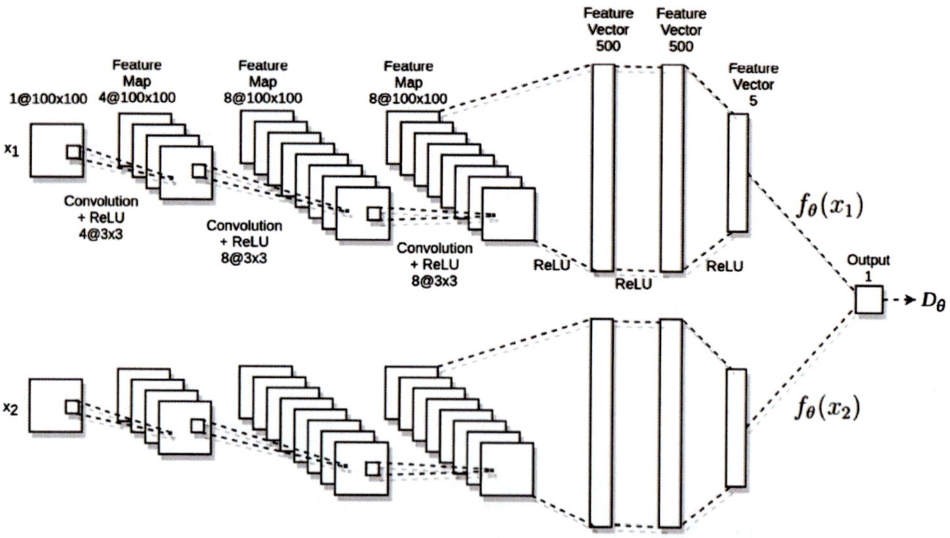

Fig. 4. Architecture of the Siamese Network which is used as a baseline [11].

Wen et al. [48] propose a center loss function to improve discriminative feature learning. The center loss function proposed aims to improve the discriminability between feature representations by minimising the intra-class variation while keeping features from different classes separable.

The center loss is given as

$$\mathcal{L} = -\sum_{i=1}^{m} \log(e^{z})/(\sum_{j=1}^{n} e^{z}) + \lambda_2 \sum_{m} i = 1||x_i - c_{y_i}||_2^2 \tag{1}$$

where $z = W_j^T x_i + b_j$. The c_{y_i} is the centroid of feature representations pertaining to the i_{th} class. This penalises the distance between class centers and minimises the intra-class variation while the softmax keeps the inter-class features separable. The centroids are computed during stochastic gradient descent as full batch updates would not be feasible for large networks.

Liu et al. [31] proposed Sphereface, a hypersphere embedding that uses an angular softmax loss that constrains discrimination on a hypersphere manifold, motivated by the prior that faces lie on a manifold. The model achieves competitive results on several datasets.

The most relevant and notable use of Siamese Networks for face verification is the DeepFace network, introduced by Taigman et al. [46]. The performance obtained was on par with human-level performance on the Faces in the Wild (LFW) dataset and significantly outperformed previous methods. However, it is worth noting this model is trained on a very large dataset from Facebook (SFC), therefore the model can be considered to be performing transfer learning before evaluation. The model also carries out some manual steps for detecting, aligning, and cropping faces from the images. For detecting and aligning the face a 3D model is used. The images are normalized to avoid any differences in illumination values, before creating a 3D model which is created by first identifying 6 fiducial points in the image using a Support Vector Regressor from an LBP histogram image descriptor. Once the faces are cropped based on these points, a further 67 fiducial points are identified for a 3D mesh model, followed by a piece-wise affine transformation for each section of the image. The cropped image is then passed to 3 CNN layers with an initial max-pooling layer followed by two fully-connected layers. Similar to Capsule Networks, the authors refrain from using max pooling at each layer due to information loss.

The above work all achieve comparable state-of-the-art results for face verification using either a single CNN or a combination of various CNNs, some of which are pre-trained on large related datasets. A siamese network as proposed by Deshpande et al. [11] is used as a baseline to compare the results of the other approaches discussed in this paper. The architecture can be seen in Fig. 4. The two networks used are identical and the input to the model is a single channel or grayscale image pairs x_1 and x_2. Each module has three convolutional layers with a number of feature maps as 4, 8, and 8 from left to right respectively of size 100 $\times$ 100 each. The convolutional layers are followed by three fully connected layers of size 500, 500, and 5 respectively. The kernel size of 3 $\times$ 3 is used for convolutions with a stride of 1. The ReLU activation function is used on the output feature maps from each layer. During training, the contrastive loss function is used. The loss function is parameterised by the weights of the neural network θ and the training sample i. The i_{th} training sample from the dataset is a tuple (x_1, x_2, y) where x_1 and x_2 are a pair of images and the label y is equal to 1 if x_1 and x_2 belong to same class and 0 otherwise. The following equation describes the contrastive loss function

$$L(\theta, (x_1, x_2, y)^i) = y\frac{1}{2}D_{\theta,i}^2 + (1-y)\frac{1}{2}(\max\{0, m - D_{\theta,i}\})^2 \quad (2)$$

with $D_{\theta,1} = ||f_\theta(x_1) - f_\theta(x_2)||_{2,i}$. The first term of the right hand side of Eq. 2 imposes cost on the network if the input image pair x_1 and x_2 belongs to the same class ($y = 1$). The second term penalises the input sample if the data belongs to different classes ($y = 0$). A more in-depth explanation of the contrastive loss function can be found at [19].

In contrast, this work looks to use a smaller Capsule Network that is more efficient, and requires little preprocessing steps (no aligning, cropping, etc.) and can learn from less data, which is particularly evident in the fact that it achieves high accuracy on small data sets that are challenging for classical CNNs as shown in Sect. 3.

2.3 Siamese Network with Capsules

One model architecture, which seems especially suitable for one-shot identification tasks, is the capsule network architecture proposed by [42]. Their potential comes from vectors, referred to as capsules, which contain properties that are intelligible to humans. In a specific approach called Dynamic Routing, the information transfer between the capsules is amplified or mitigated. During training, the vector output of the last capsules is masked and then inserted into a decoder to restore the perceived features of the input image which can be used to generate images.

Subsequently, we use the term *CapsNet* for the model that consists of the capsule network itself and also the decoder. The architecture of our CapsNet is illustrated in Fig. 5. It is based on the architecture proposed by [38] with slight alterations. We use a Leaky-ReLU activation function with a leak of $a = 0.01$ is used for the two subsequent convolutional layers.

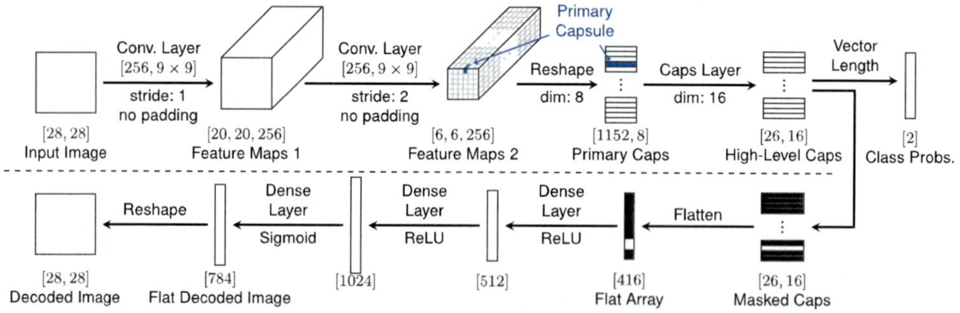

Fig. 5. Architecture of the used CapsNet. [38].

Each primary capsule consists of a group of n_p feature maps. The number of values inside a capsule n_p must be a divisor of the number of feature maps n_m (here: $n_m =$

256), such that $\frac{n_m}{n_p}$ is the number of primary capsules per location. Together with the dimensions $[h_m, w_m, n_m]$ of the feature maps it results in the number N_p of primary capsules [38].

$$N_p = h_m \cdot w_m \cdot \frac{n_m}{n_p} \tag{3}$$

Capsules also have an activation function, similar to the more commonly used neurons. As in [42], we use the squashing function

$$\hat{\mathbf{g}} = \frac{||\mathbf{g}||_2^2}{(1 + ||\mathbf{g}||_2^2)}, \frac{\mathbf{g}}{||\mathbf{g}||_2} \tag{4}$$

where $\mathbf{g}$ is the vector of a capsule. It squashes the length of the output vector $\hat{\mathbf{g}}$ of a capsule between 0 and 1. The resulting length depends non-linearly on $\mathbf{g}$. The squashing function is performed on both primary and subsequent high-level capsules.

Since each primary capsule i and each high-level capsule j are entirely linked to one another by a weight matrix $\mathbf{W}_{ij}$, the values in a capsule may be thought of as neurons. There isn't a bias vector, though. The Dynamic Routing algorithm [42] is executed between the primary capsules and the high-level capsules. An additional coupling coefficient c_{ij} is added between each primary capsule i and each high-level capsule j, which stems from a routing logit b_{ij} by applying a softmax across j. The routing logits b_{ij} are initialised with zeros for each forward pass and updated within the routing iterations. Their values result from the relevance of the prediction of a primary capsule i to the prediction of a high-level capsule j. As a result, the connection between both capsules is amplified or mitigated.

In this paper, a CapsNet as described is used as part of a siamese network as proposed by [37]. Two identical CapsNet's are used in a setup like a siamese network and a contrastive loss function is used during training. This is a straightforward implementation of the original idea of siamese networks with mainly used CNNs replaced by capsule networks.

3 Experimental Results

In the following sections, experiments are conducted on three different data sets. A dataset with images from an industrial application, the smallNORB dataset, and the AT&T database of faces. This provides a large variety in the experiments and tests the approaches in different application domains.

In all three of these experiments, three alternative strategies have been applied. A Siamese network is used as a baseline. A convolutional neural network, which was specially designed for stacked images, and additionally a Siamese network, which is based on capsule networks instead of convolutional neural networks. 10-fold cross-validation was used for the presented accuracies in every experiment except the industrial application.

3.1 Industrial Application

Anodes are produced in an automatic process in aluminum plants. Due to the need for high temperature and force, this process is expensive and time-consuming. One of the production steps during manufacturing an anode is to burn them in a special way in a kiln. During the burning, the anodes lie in irregular and not pre-defined ways on a conveyor belt. For reasons like traceability of defects, quality control, predictive maintenance, and quality assurance it is of high importance for the aluminum plant to be able to identify the anodes between initial jogging and the final electrolysis.

Anodes do undergo some surface transformation during baking. Additionally, burn-off during this procedure has the potential for harm to some locations. Moreover, the substance used to stabilise the anodes during baking might adhere to the surface of the anodes. In both instances, just a portion of the anode is influenced. Other places' features and surface textures are unaffected. In the event that tags are utilised, it is possible for one of the aforementioned processes to destroy the attached tag, rendering the tag useless for object identification and leaving this particular anode unrecognisable.

There is only one instance the anodes go through the manufacturing process, thus there isn't enough data produced to train a neural network to recognise each anode (every anode is a class for itself). Due to just having one picture of each anode accessible for training, a large number of anodes would need to be learned. Because of this, a neural network is to be trained to compare two anodes and make an educated guess, if both anodes are the same or not. This allows to create a neural network, that does not need to be retrained for every anode, thus solving the problem of not having enough training data. Only two images of every anode are required to identify an object. For more detailed information on the dataset kindly refer to Mohr et. al. [35].

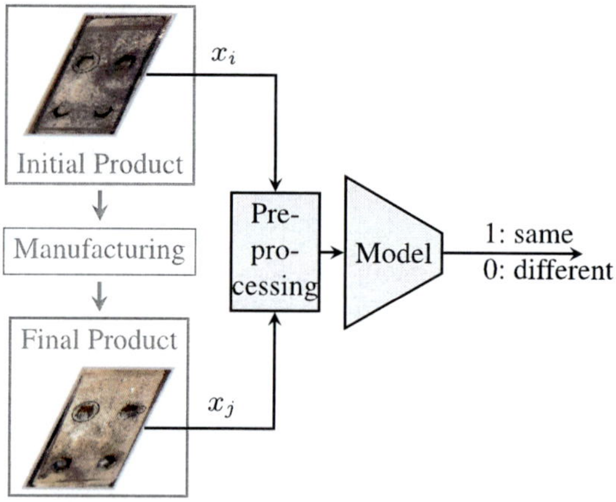

Fig. 6. The workflow for the industrial application. During the manufacturing step several operations are done which alter the appearance of the anodes.

Every anode produced on the conveyor is captured using photographs. A novel picture is created and sent into the neural network each time an anode moves through the furnace. The neural network determines how certain it is that the two anodes presented to the network are identical by comparing this to every other anode. To compare the processed anode to all previously observed anodes, the network must be run through several times because it can only compare two pictures at a time. The two anodes with the highest likelihood are determined after all anodes have been observed. Unlike applications like facial recognition, only a small number of anodes cross the cameras at once. Figure 6 explains the whole process flow.

Fig. 7. A random example from the dataset showing variances in surface texture before (left) and after (right) the anode is baked in the furnace. [35]

Preprocessing and Data Augmentation. During ongoing production, a number of images of anodes were taken. The images were taken before and after the baking in the kiln and various other manufacturing steps. In Fig. 7 an example of these images can be seen. The number of images is limited because the manufacturing of one anode takes a significant amount of time. Therefore images are generated and the changes in the appearance of the anodes are simulated to enhance the dataset and increase its size.

As can be seen in Fig. 7 significant features on the surface of the anode remain after processing. Reviewing the available pictures of anodes before and after the baking process allowed to analyse and identify the most common changes and similarities in an anode during production.

During the green anode's production, the stubs are formed. Because the device used to make the stub is removed after the anode is formed, the alignment of these always varies between two anodes. Between every processing operation, this characteristic is maintained. During the furnace's baking process, the surface's brightness and texture change. The anodes are also added to and altered with spots, markings, and other modifications during processing.

To overcome the problem of limited quantity and limited diversity of data the existing images are manipulated with affine transformations. For simulating the variances all images of unaltered anodes are manipulated. Each image in the data set is rotated randomly about its centre. Brightness and fine contours are changed. Count and position of burn-offs and build-ups are manipulated randomly and simulated with blurred circles, that are layered over the anodes. These images are saved separately and used as part of the training data set. The data set is combined out of simulated and original images. Figure 8 exemplarily shows a simulated image of an anode next to the original image.

Fig. 8. Generated image as it is used in the training data set. This image was generated out of a photo of an anode with data augmentation techniques. [35].

Results. Our experiments have shown, that stacking images achieves high accuracy. Further experiments have displayed, that the convolutional neural network reaches an accuracy of 98.4% when the early stopping terminated the training process. As 9 illustrates this high level is reached in a very stable way. No signs of overfitting can be found as new images are also identified correctly, and the accuracy of the validation data does oscillate strongly or even shrink during the training. Figure 9 shows the learning behaviour based on the accuracy and loss during training.

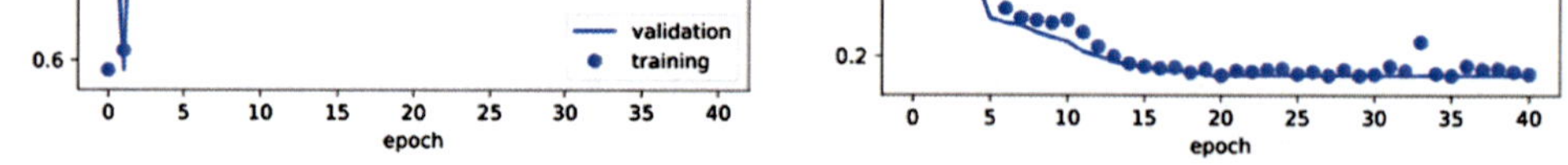

Fig. 9. The accuracy (left) and the loss (right) during training for 40 epochs. The shown development stems from training the approach for merged images on the industrial application dataset.

The siamese network was trained on the same training data and the accuracy was achieved on the same set of test images as used for training the neural network with the merged images. It reaches an accuracy of 96.4%.

The approach with merged images for one-shot identification has the main advantage over state-of-the-art siamese networks. Siamese networks have a more complex architecture regarding the prediction because the images are forwarded through two neural networks instead of just one. In industrial applications with a small batch of objects to identify just once – in opposite to face recognition where siamese networks are very common – pre-processing images through the network and storing outputs and then in an additional step calculating the similarity is less helpful. Therefore, in such industrial applications, they are often slower regarding prediction. Fast predictions are of the utmost importance in industrial applications because in general, they have some kind of real-time requirements.

The siamese CapsNet is tested in two different settings for the industrial application. At first, the same dataset is used which is used for the merged images and the siamese network. On this dataset, the siamese CapsNet achieves an accuracy of 96.4%. Additionally, we used the decoder network of the CapsNet which gives it the ability to generate images after a training phase [36]. The CapsNet is trained on the original images of the anodes for 20 epochs. After this training, the decoder network is used to generate the same amount of images as previously generated manually. A new dataset is composed of the original and the generated images and the siamese CapsNet is trained with the same architecture and parameters achieving an accuracy of 98.5%.

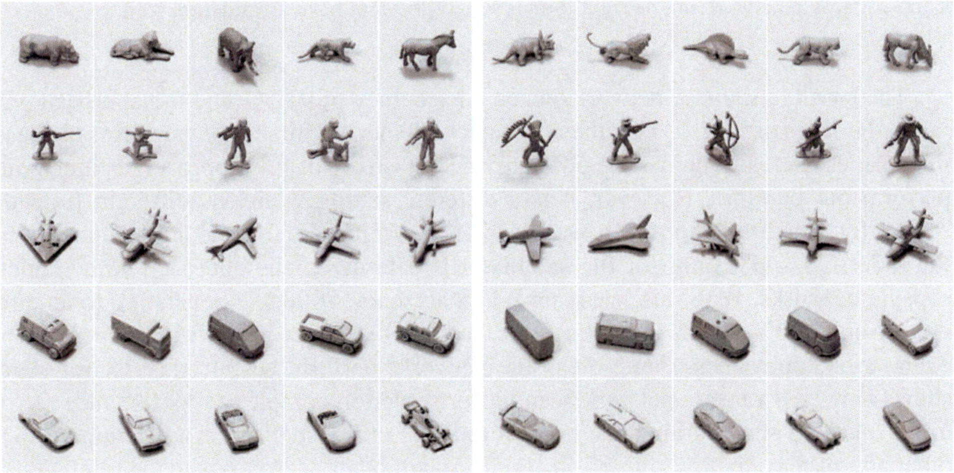

Fig. 10. Random examples from the smallNORB dataset with examples for all 5 categories from the training (left) and test set (right). [29].

3.2 Results on SmallNORB Dataset

The dataset smallNORB [29] is a collection of 48600 stereo, grayscale images (96 × 96 × 2), representing 50 plastic toys belonging to 5 generic categories: humans, airplanes, trucks, cars, and four-legged animals. Each toy was photographed by two cameras under 6 lighting conditions, 9 elevations, and 18 azimuths. A turntable was used to move the toys and they were painted in a uniform green colour. The dataset is split in half; 5 instances of each category for the training and the remaining ones for the testing leading to a few-shot learning task. Examples can be seen in Fig. 10

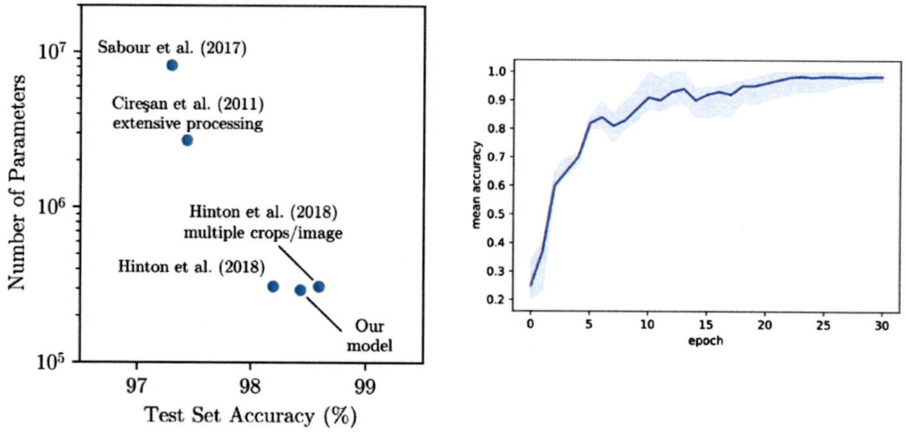

Fig. 11. Test set accuracy and number of parameters of models that have achieved state-of-the-art results on smallNORB visual recognition on the left and development of the accuracy of the siamese capsule network on the right. SmallNORB does not have a validation split.

The baseline siamese networks reach an accuracy of 92.5% and the approach with merged images of 94.7%. That the approach with merged images outperforms the baseline siamese network shows the strength and versatility of this approach as it can outperform our baseline. However, it is a different setting, which is different from the setting for which this approach was developed. We used the same architecture as for the industrial application for the smallNORB dataset for the approach with siamese capsule networks. With this, we received an accuracy of 98.4% translating to an error rate of only 1.6% which is on par with the state-of-the-art [10] with 30 epochs. Out of the used approaches, siamese capsule networks perform best on this dataset which shows how well capsule networks can perform even in complex tasks like face recognition and zero-shot learning tasks. Refer to Fig. 11 for visualisation and comparison to other approaches. In the comparison, a selection of other approaches to the smallNORB dataset can be seen which are sorted based on accuracy and number of parameters. Some of these other approaches have extensively preprocessed or cropped the images.

3.3 Results on AT&T Database of Faces

The AT&T face recognition and verification dataset consists of 40 different subjects with only 10 grey-pixel images per subject in a controlled setting. The pictures have been taken between April 1992 and April 1994. For some subjects, the images were taken at different times, varying the lighting, facial expressions and additional features like glasses. All the images were taken against a dark homogeneous background with the subjects in an upright, frontal position. Examples can be seen in Fig. 1.

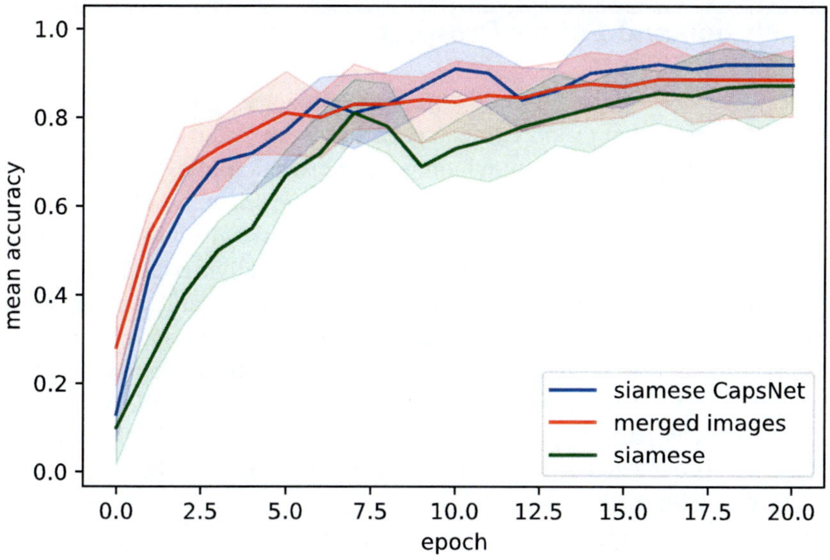

Fig. 12. Mean accuracy after each epoch of training for all three approaches. Shaded area denotes deviation.

This smaller dataset allows us to test how the approaches perform with little data. For testing, we hold out 5 subjects so that we are testing on unseen subjects, as opposed to training on a given viewpoint of a subject and testing on another viewpoint of the same subject. Hence, zero-shot pairwise prediction is performed during testing. The siamese network achieves an accuracy of 87.3% and the CNN with merged images 88.6%, which shows that this approach would require more tuning and adaptation because the face recognition task highly differs from the industrial application and the approach was specifically designed for the requirements of this application without paying attention to generalisation or performance on other tasks. As expected the siamese CapsNet achieves the best accuracy with 90.2% showing as already expected and explained in Sect. 2.3 that CapsNet's perform very well with little data making them suitable for particular tasks and settings like tiny data. Figure 12 shows the development of the accuracy for all three approaches on this dataset.

Table 2. Results for the experiments and approaches in this paper presented as accuracy in percent.

Approach	Industrial Dataset	smallNORB	AT&T faces
merged images	98.4%	94.7%	88.6%
siamese	96.4%	92.5%	87.3%
siamese CapsNet	97.9%	98.4%	90.2%

4 Conclusion and Future Prospects

In summary, we have presented three different approaches to dealing with limited data and one-shot or zero-shot tasks. These approaches were tested on data sets of different sizes and from different domains including an industrial application and a face recognition task. Overall, the approach with the siamese capsule networks achieves the best results in terms of accuracy. This approach is only narrowly beaten by the approach with merged images in the industrial application specifically designed for it.

All experiments in this paper show that the amount of available data has a significant influence on the achievable accuracy and that special approaches have a right to exist for special tasks. It can also be seen that there are some promising approaches for dealing with small data sets and one-shot tasks. Here, the performance of the recently proposed capsule networks is particularly noteworthy. As these are still in the early stages of development, it is expected that the achievable accuracy and number of parameters will improve further with future work.

These capsule networks also provide ideas for further development in the context of one-shot tasks. It has already been shown that the decoder network can be used for continuous learning when training on large data sets. In addition, due to the high comparability of the capsules with elements of human vision, high explanatory power can be achieved. It is valuable to investigate how this explanatory power can also be applied to one-shot and especially face recognition tasks.

Acknowledgements. This work was funded by the federal state of North Rhine-Westphalia and the European Regional Development Fund FKZ: EFRE-040021. The industrial application was supported by Trimet Aluminium SE.

References

1. Abdel-Qader, I., Abudayyeh, O., Kelly, M.: Analysis of edge-detection techniques for crack identification in bridges. J. Comput. Civil Eng. (2003)
2. Abeysinghe, C., Perera, I., Meedeniya, D.: Capsule networks for character recognition in low resource languages. In: Machine Vision Inspection Systems, vol. 2: Machine Learning-Based Approaches, pp. 23–46 (2021)
3. Afshar, P., Mohammadi, A., Plataniotis, K.N.: Brain tumor type classification via capsule networks. In: 25th IEEE International Conference on Image Processing ICIP (2018)
4. Benhimane, S., Najafi, H., Grundmann, M., Malis, E., Genc, Y., Navab, N.: Real-time object detection and tracking for industrial applications. In: VISAPP 2008: Third International Conference on Compter Vision Theory and Applications, vol. 2 (Jan 2008)
5. Bromley, J., Guyon, I., LeCun, Y., Säckinger, E., Shah, R.: Signature verification using a "siamese" time delay neural network. In: Proceedings of the 6th International Conference on Neural Information Processing Systems. NIPS 1993 (1993)
6. Brusey, J., Mcfarlane, D.C.: Effective rfid-based object tracking for manufacturing. Inter. J. Compu. Integrated Manufact. 638–647 (2009)
7. Cha, Y.J., Choi, W., Buyukozturk, O.: Deep learning-based crack damage detection using convolutional neural networks. Comput. Aided Civil Infrastructure Eng. **32**, 361–378 (2017)
8. Chicco, D.: Siamese Neural Networks: An Overview. Methods in molecular biology (2020)

9. Chopra, S., Hadsell, R., Lecun, Y.: Learning a similarity metric discriminatively, with application to face verification. In: Proceedings of Computer Vision and Pattern Recognition (July 2005)
10. Ciresan, D.C., Meier, U., Masci, J., Gambardella, L.M., Schmidhuber, J.: High-performance neural networks for visual object classification. Technical Report No. IDSIA-01-11 (2011)
11. Deshpande, A.M., Minai, A.A., Kumar, M.: One-shot recognition of manufacturing defects in steel surfaces. In: 48th SME North American Manufacturing Research Conference, NAMRC, vol. 48 (2020)
12. Fei-Fei, L., Fergus, R., Perona, P.: A bayesian approach to unsupervised one-shot learning of object categories. In: Proceedings of the Ninth IEEE International Conference on Computer Vision, ICCV 2003, vol. 2, p. 1134. ,IEEE Computer Society (2003)
13. Fei-Fei, L., Fergus, R., Perona, P.: One-shot learning of object categories. IEEE Trans. Pattern Anal. Mach. Intell. (2006)
14. Gagana, B., Athri, H.U., Natarajan, S.: Activation function optimizations for capsule networks. In: 2018 International Conference on Advances in Computing, Communications and Informatics (ICACCI), pp. 1172–1178 (2018)
15. Hinton, G.E.: A parallel computation that assigns canonical object-based frames of reference. In: Seventh International Joint Conference on Artificial Intelligence (1981)
16. Hinton, G.E.: Shape representation in parallel systems. In: Seventh International Joint Conference on Artificial Intelligence (1981)
17. Hinton, G.E., Krizhevsky, A., Wang, S.: Transforming auto-encoders. In: International Conference on Artificial Neural Networks (2011)
18. Graves, A.: Generating sequences with recurrent neural networks. ArXiv (2013)
19. Hadsell, R., Chopra, S., LeCun, Y.: Dimensionality reduction by learning an invariant mapping. In: 2006 IEEE Computer Society Conference on Computer Vision and Pattern Recognition (CVPR 2006) (2006)
20. Iesmantas, T., Alzbutas, R.: Convolutional capsule network for classification of breast cancer histology images (2018)
21. Koch, G., Zemel, R., Salakhutdinov, R.: Siamese neural networks for one-shot image recognition. In: 32nd International Conference on Machine Learning (2015)
22. Kühl, N., Goutier, M., Baier, L., Wolff, C., Martin, D.: Human vs. supervised machine learning: who learns patterns faster? ArXiv (2020)
23. Kumar, A.D.: novel deep learning model for traffic sign detection using capsule networks. Inter. J. Pure Appli. Math. **118**(20) (2018)
24. Lake, B.M., ying Lee, C., Glass, J.R., Tenenbaum, J.B.: One-shot learning of generative speech concepts. Cogn. Sci. **36** (2014)
25. Lake, B.M., Salakhutdinov, R., Gross, J., Tenenbaum, J.B.: One shot learning of simple visual concepts. Cogn. Sci. **33** (2011)
26. Lake, B.M., Salakhutdinov, R., Tenenbaum, J.B.: Concept learning as motor program induction: a large-scale empirical study. Cogn. Sci. **34** (2012)
27. Lake, B.M., Salakhutdinov, R., Tenenbaum, J.B.: One-shot learning by inverting a compositional causal process. In: NIPS (2013)
28. LeCun, Y., Bottou, L., Bengio, Y., Haffner, P.: Gradient-based learning applied to document recognition. In: Proceedings of the IEEE (1998)
29. Lecun, Y., Huang, F., Bottou, L.: Learning methods for generic object recognition with invariance to pose and lighting. In: IEEE Conference on Computer Vision and Pattern Recognition, CVPR (2004)
30. Li, C., Quan, C., Peng, L., Qi, Y., Deng, Y., Wu, L.: A capsule network for recommendation and explaining what you like and dislike. In: Proceedings of the 42nd International ACM SIGIR Conference on Research and Development in Information Retrieval (2019)

31. Liu, W., Wen, Y., Yu, Z., Li, M., Raj, B., Song, L.: Sphereface: deep hypersphere embedding for face recognition. In: IEEE Conference on Computer Vision and Pattern Recognition (CVPR) (2017)
32. Mahieu, P., Genin, X., Bouché, C., Brismalein, D., Pédroli, H.: Carbon block tracking package based on vision technology. Light Metals (2019)
33. Minderer, M., Sun, C., Villegas, R., Cole, F., Murphy, K., Lee, H.: Unsupervised learning of object structure and dynamics from videos. In: 33rd Conference on Neural Information Processing Systems (2019)
34. Mobiny, A., van Nguyen, H.: Fast CapsNet for Lung Cancer Screening (2018)
35. Mohr, J., Breidenbach, F., Frochte, J.: An approach to one-shot identification with neural networks. IJCCI (2021)
36. Mohr, J., Tousside, B., Schmidt, M., Frochte, J.: Explainability and continuous learning with capsule networks. In: Proceedings of the 13th International Joint Conference on Knowledge Discovery, Knowledge Engineering and Knowledge Management (10 2021)
37. Neill, J.O.: Siamese capsule networks (2018)
38. Quetscher, F., Kaufmann, C., Frochte, J.: Investigation of capsule networks regarding their potential of explainability and image rankings. In: ICAART (2022)
39. Rajasegaran, J., Jayasundara, V., Jayasekara, S., Jayasekara, H., Seneviratne, S., Rodrigo, R.: DeepCaps: Going deeper with capsule networks. In: IEEE/CVF Conference on Computer Vision and Pattern Recognition CVPR (2019)
40. Renkens, V., van hamme, H.: Capsule networks for low resource spoken language understanding. In: Proceedings of Interspeech 2018 (2018)
41. Renzulli, R., Tartaglione, E., Grangetto, M.: Rem: routing entropy minimization for capsule networks (2022)
42. Sabour, S., Frosst, N., Hinton, G.E.: Dynamic routing between capsules. In: Proceedings of the 31st International Conference on Neural Information Processing Systems, NIPS 2017 (2017)
43. Scherer, D., Müller, A., Behnke, S.: Evaluation of pooling operations in convolutional architectures for object recognition. In: 20th International Conference on Artificial Neural Networks (ICANN), pp. 92–101 (Jan 2010)
44. Shahroudnejad, A., Mohammadi, A., Plataniotis, K.N.: Improved explainability of capsule networks: Relevance path by agreement. In: IEEE Global Conference on Signal and Information Processing (GlobalSIP) (2018)
45. Sun, Y., Wang, X., Tang, X.: Deep learning face representation by joint identification-verification. In: Neural Information Processing NIPS (2014)
46. Taigman, Y., Yang, M., Ranzato, M., Wolf, L.: Deepface: closing the gap to human-level performance in face verification. In: Proceedings of the IEEE Computer Society Conference on Computer Vision and Pattern Recognition (Nov 2014)
47. Tielenman, T.: Optimizing Neural Networks That Generate Images. Dissertation, University of Toronto (2014)
48. Wen, Y., Zhang, K., Li, Z., Qiao, Yu.: A discriminative feature learning approach for deep face recognition. In: Leibe, B., Matas, J., Sebe, N., Welling, M. (eds.) ECCV 2016. LNCS, vol. 9911, pp. 499–515. Springer, Cham (2016). https://doi.org/10.1007/978-3-319-46478-7_31
49. Wu, D., Zhu, F., Shao, L.: One shot learning gesture recognition from rgbd images. In: 2012 IEEE Computer Society Conference on Computer Vision and Pattern Recognition Workshops (2012)
50. Xi, E., Bing, S., Jin, Y.: Capsule network performance on complex data. In: International Joint Conference on Neural Networks (IJCNN) (2017)

Check for updates

Evaluation of Gated Recurrent Neural Networks for Embedded Systems Applications

Jean-Baptiste Chaudron(✉) and Arnaud Dion

ISAE-SUPAERO, Université de Toulouse, Toulouse, France
jean-baptiste.chaudron@isae-supaero.fr,
arnaud.dion@isae-supaero.fr

Abstract. Artificial Neural Networks (ANNs), based on the concept of neuron cells, are widely used nowadays for multiple applications. Recurrent Neural Networks (RNNs) are special types of ANNs capable of capturing temporal relationship into a set of data and offer interesting features. However, RNN architectures are complex and their integration on embedded systems implies a proper understanding of inherent structures and algorithms. This paper describes and analyses four types of gated RNN cells: Long-Short Term Memory (LSTM), Gated Recurrent Unit (GRU), Minimal Gated Unit (MGU) and STAckable Recurrent cell (STAR). We explain the algorithmic details and how these cells can be combined and integrated to build complex structures. We illustrate our approach with experimental results using different architectures and implementations on dedicated test cases.

Keywords: RNN · GRU · LSTM · MGU · STAR · Embedded systems

1 Introduction

An ANN can be considered as a complex Artificial Intelligence (AI) system combining and interconnecting a large number of nonlinear processing units called *neurons* [16]. The perceptron algorithm was the first attempt to model numerically the behavior of a biologic neuron [47] and was initially used as a binary classifier. In the early days, the processing machines were not powerful enough to compute large amount of data and to enable the capability of *learning* [36]. Over the years, following the increase of processing power, the initial structure proposed for the original perception model has been extended to create new complex architectures with more neuron entities integrated in multiple layers starting from the three layers perceptron [19]. These updated structures associated to efficient learning algorithm were enabling scalability and extending possibilities to support different problems and applications, to model complex systems and provide more sophisticated outputs than simple binary classification. Therefore, ANN technology has grown exponentially and is, nowadays, at the heart of scientific concerns for several applications in multiple areas where research communities design, combine and develop new types and variants of ANNs to extend their features. However, all these ANNs and their variants can be classified in three main types:

J. Garibaldi et al. (Eds.): IJCCI 2021, SCI 1119, pp. 223–244, 2023.
https://doi.org/10.1007/978-3-031-46221-4_11

1. The Multi-Layer Perceptrons (MLPs), or Fully Connected (FC) networks, are the extended version of the original perceptron [19] that contains three or more linear layers. These are usually integrated as sub-component in other architectures.
2. The Convolutional Neural Networks (CNNs) are neural networks specialized mainly in image processing applications. These are based on matrix convolution operations commonly used in image processing (such as the well-known Sobel filter [23]). Integrating neural networks learning feature into convolution filters has push CNNs to outperform traditional techniques in image classification and speech recognition [44].
3. The RNNs contain some internal memory entities enabling a feedback loop. This principle allows a temporal relationship in the processed data. Therefore, the output of a cell is not only influenced by its input but also by the previous computations. This is the type of ANNs analyzed in this paper.

For future embedded systems, ANNs (associated with other AI methods) are enabling new possibilities such as enhancing avionic systems [52], extend autonomous driving for automotive systems [13], answer new challenges for space systems [50] or optimize autopilot features for drone systems [15]. In particular, the capacity of RNNs to handle temporal relationships into data, make these type of ANNs very interesting for various embedded systems applications [27,46]. However, this feature comes with the cost of complexity and the integration of RNNs on embedded systems, having a limited amount of memory and processing power [45], needs to be carefully analyzed. In an initial work [38], we investigated the use of two RNN architectures (i.e. LSTM and GRU) for Inertial Measurement Unit (IMU) sensors fusion and we have shown interesting first results. The work presented in this paper is an extension of this initial effort and propose several improvements. First, we introduce two other basic RNN cells: the MGU and the STAR that are described together with the LSTM and the GRU presented in previous paper. We describe the characteristics of all these gated RNN cells. Additionally, we investigate how to use these basic cells to build complex structures. We have updated our open-source implementation accordingly, tested it on a another similar data-set and compared our implementation with well-known Pytorch framework [42]. Finally, an updated bibliography is presented along this document to argument and illustrate our work. This paper is organized as follow:

- Section 2 explains the background with the emergence of RNNs and their applications in embedded systems context.
- Section 3 outlines the architectural aspects and characteristics of LSTMs, GRUs, STARs and MGUs basic cells.
- Section 4 investigates design and implementation considerations when implementing complex structures based on these basic cells.
- Section 5 presents test case application, shows our experimental results and finally Sect. 6 concludes and offers some possible perspectives to pursue this work.

2 State of the Art

2.1 Emergence of RNNs

A basic ANN can be seen as a set of mathematical operations to generate an output $Y = (y_1, ..., y_M)$ from an input $X = (x_1, ..., x_N)$ as depicted in Fig. 1 and described in expression 1 (note that X and Y are noted here as linear vectors for sake of clarity).

$$Y = \Phi(W \times X + b) \tag{1}$$

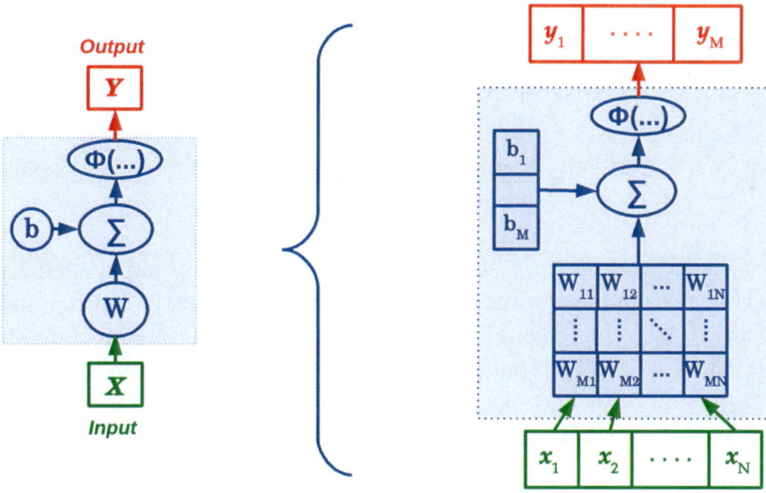

Fig. 1. Illustration of a classic MLP.

More precisely, the inputs are weighted, summed and passed into a mathematical function noted $\Phi(...)$. This function, called the activation function, is the non-linear processing function of the neuron and is key for the behavior of a neural network [7]. In this work, two well known logarithm based activation functions will be considered, the sigmoid (noted σ) and the hyperbolic tangent (noted τ). To resume, mathematically, each element y_j from output vector Y can be calculated using expression 2.

$$y_j = \Phi(\sum_{i=1}^{N} W_i^j \times x_i + b_j) \tag{2}$$

For some applications, this basic architecture has a major drawback inherent to its structure: only the inputs can be used to predict the output. Therefore, it can not capture a dependency between the precedent processed output and the next output because there is no form of *memory* (i.e. previous executions results are not stored). To overcome this issue, a first variant of the basic ANN structure has been proposed in 1986 [21] where

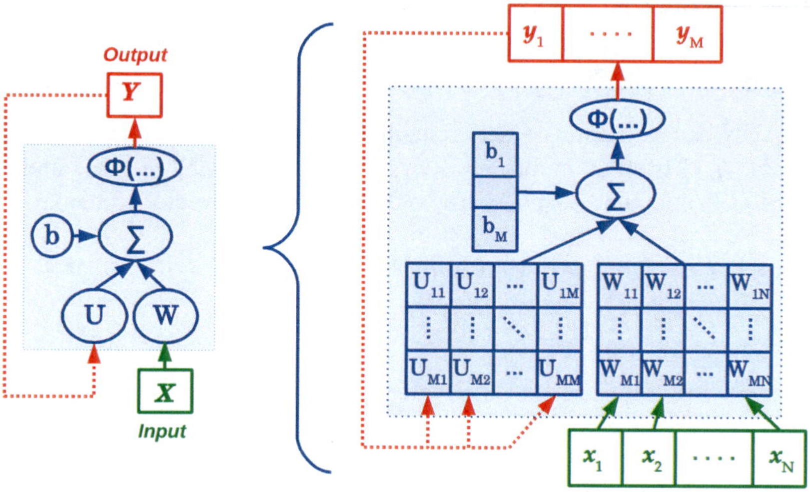

Fig. 2. Illustration of a classic RNN.

extra inputs (called states) were added and connected to the output of the previous execution. This recurrent evolution, containing a feedback loop, is depicted in Fig. 2.

Considering t as the increment for the execution steps in this architecture, the output Y_t is generated based on the input X_t for the current execution step and, also, on the output from previous execution step Y_{t-1}. Accordingly, expressions 1 and 2 can be adapted to expressions 3 and 4.

$$Y_t = \Phi(W \times X_t + U \times Y_{t-1} + b) \tag{3}$$

$$y_j^t = \Phi(\sum_{i=1}^{N} W_j^i \times x_i^t + \sum_{j=1}^{M} U_j^j \times y_j^{t-1} + b_j) \tag{4}$$

This recurrent architecture principle has been extended to multiple cells and layers [8]. In the literature, other types of RNNs have also been proposed such as Time-Delay Neural Networks (TDNN) [54], Narx Network [29], Echo State Network (ESN) [20], clock work NN [26] or gated architectures such as LSTM [11] and GRU [4] studied in this paper.

2.2 Training with Back-Propagation

The heart of all ANN algorithms is the ability to *learn* from experience that has been enabled by the first *back-propagation* algorithm [49] which has been then generalized and enhanced over the years using different techniques or mathematical improvements [56,59,63]. The name back-propagation comes from the main purpose of these algorithms to propagate the error signal generated at the output (with respect to desired output i.e. label) back across the network. Therefore, the contributions (gradients) of each parameter to this error are calculated in order to correct the weights and the biases. For RNNs, the back-propagation variant is called Back-Propagation Through Time (BPTT) which consists of unfolding the considered sequence dependencies in the neural network (unrolling the sequence) as depicted in Fig. 3 where a 3 steps sequence is considered. In BPTT, the error is propagated from step to step, calculating the variations caused by the error from previous iteration using the chain rule. The goal is to determine the error produced by the output with respect to the inherent sequence relationship and to determine the corresponding gradients. The gradients for the weights are accumulated over the considered steps. Therefore, the weights remains shared across the sequence (in Fig. 3 U, W weights and b bias components are the same for each time $t-2$, $t-1$ and t).

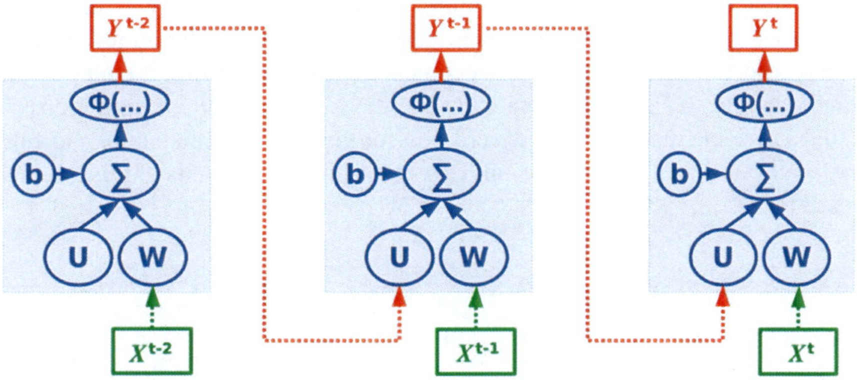

Fig. 3. Example of unfolding 3 steps sequence in a RNN.

For long sequence relations, due to this gradient accumulation, standard RNNs are usually suffering from the vanishing gradient problem during training with back-propagation [48]. To tackle this issue, new specific RNN gated architectures have been proposed which implement separate gates mechanisms to add and, also, remove information from the recurrent relation (i.e. the past states). Nowadays, the two most popular RNNs architectures are the LSTM [11] and the GRU [4] which outperform the standard (also called vanilla) RNN [5,41] and we have chosen them for our work. Two others gated architecture are considered as well: an resource optimized GRU variant called MGU [61] and the recent STAR [53] architecture. The mathematical aspects of the back-propagation algorithms are not described in this work, interested readers may refers to our previous work [38] for a more detailed description.

2.3 Applications of RNNs in Embedded Systems (ESs)

In the context of the PRISE project (French acronym for Platform for Embedded Systems Research and Engineering), we are always investigating new concepts and techniques for embedded systems. ANNs and AI methods are offering new interesting opportunities and challenges for aircraft avionic [52], automotive [13], space [50] and drone systems [15]. In particular, the special feature of RNNs encouraged investigations for their integration to enhance flight control systems with adaptive control [2,35,60] or optimize the design of robust flight controllers [39]. RNNs have been also used for in-flight parameters estimation [34], abnormal detection scheme [58] or actuators fault detection [9]. IMU sensors fusion using RNNs have also been investigated [31,55] similarly to what we presented in our previous paper [38] and, also, to this work (see Sect. 5.1).

3 Basic RNN Cells Description

3.1 LSTM Cell

A LSTM cell has a complex structure [11], described in Fig. 4, storing information in two state vectors: a cell state (C_t) captures the *long-term* memory of the cell and an output state (Y_t) handles the *short-term* memory and renders the output of the cell at each step. The activation of the state vectors is done by four gates: the output gate (O), the candidate gate (G), the input gate (I) and the forget gate (F). Expressions (5), (6), (7) and (8) give the mathematical description for these gates. The output and cell states are calculated according to expressions (9) and (10). Some variants of this baseline have proposed [12,22].

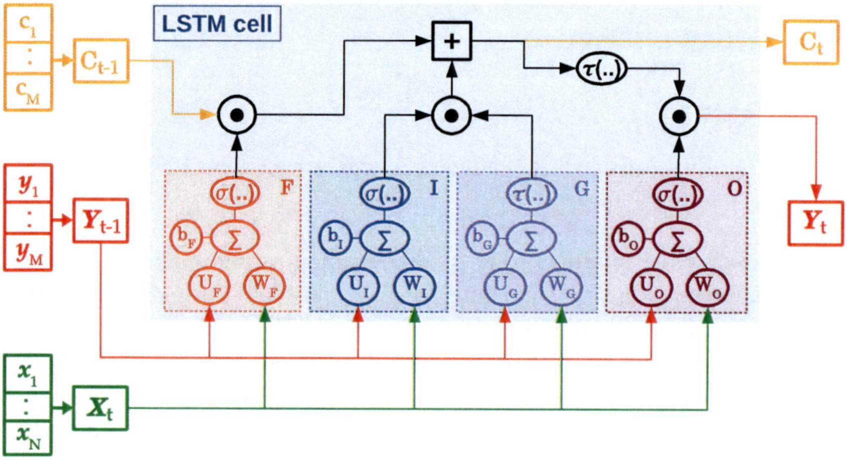

Fig. 4. Anatomy of a LSTM cell.

$$O_t = \sigma(W_O \times X_t + U_O \times Y_{t-1} + b_O) \tag{5}$$

$$G_t = \tau(W_G \times X_t + U_G \times Y_{t-1} + b_G) \tag{6}$$

$$I_t = \sigma(W_I \times X_t + U_I \times Y_{t-1} + b_I) \tag{7}$$

$$F_t = \sigma(W_F \times X_t + U_F \times Y_{t-1} + b_F) \tag{8}$$

$$C_t = F_t \cdot C_{t-1} + I_t \cdot G_t \tag{9}$$

$$Y_t = O_t \cdot \tau(C_t) \tag{10}$$

3.2 GRU Cell

The architecture of a GRU cell, described in Fig. 5, is composed of three gates [4]: the reset gate (R), the update gate (Z) and the candidate gate (H). Expressions (11), (12) and (13) provide the mathematical descriptions for these gates and the calculation of output based on these gates is described in (14) (where **1** represents a vector of ones). Similar to LSTM cell, other variants have been proposed [6,22].

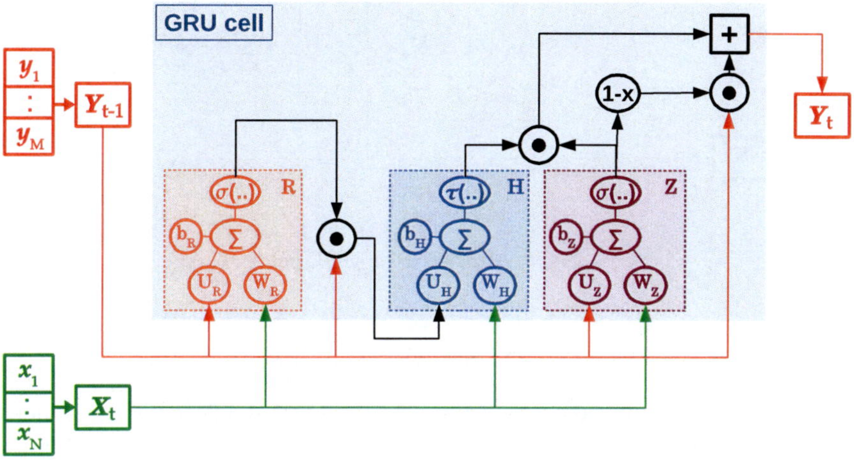

Fig. 5. Anatomy of a GRU cell.

$$R_t = \sigma(W_R \times X_t + U_R \times Y_{t-1} + b_R) \tag{11}$$

$$Z_t = \sigma(W_Z \times X_t + U_Z \times Y_{t-1} + b_Z) \tag{12}$$

$$H_t = \tau(W_H \times X_t + U_H \times (R_t \cdot Y_{t-1}) + b_H) \tag{13}$$

$$Y_t = (\mathbf{1} - Z_t) \cdot Y_{t-1} + Z_t \cdot H_t \tag{14}$$

3.3 MGU Cell

The architecture of a MGU cell is a simplified version of the GRU cell [61] and is illustrated in Fig. 6. The reset gate (R) has been removed and the two other gates remain, namely: the update gate, renamed (F), and the candidate gate (H). The mathematical descriptions for these gates are given in expressions (15) and (16) as well as the output expression (14) (where **1** represents a vector of ones). Other versions, such as simplified MGU [17], have also been proposed.

$$F_t = \sigma(W_F \times X_t + U_F \times Y_{t-1} + b_F) \tag{15}$$
$$H_t = \tau(W_H \times X_t + U_H \times (F_t \cdot Y_{t-1}) + b_H) \tag{16}$$
$$Y_t = (\mathbf{1} - F_t) \cdot Y_{t-1} + F_t \cdot H_t \tag{17}$$

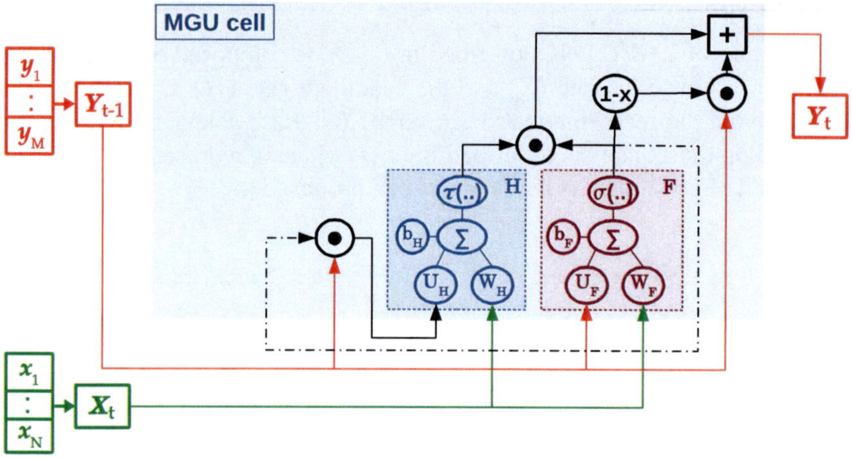

Fig. 6. Anatomy of a MGU cell.

3.4 STAR Cell

A recent work has introduced a new gated RNN cell called STAR [53]. This new architecture is illustrated in Fig. 7 and has two gates. The K gate is the analogue of the forget gate of LSTM cell and controls how the information from previous hidden state and the new input are fused into a new hidden state. The Z gate is simply considering the new inputs. It has to be noted that the output state is fed into an hyperbolic tangent activation function which limits the output to interval $[-1; 1]$. Expressions (18) and (19) provide the mathematical descriptions for these gates and the output calculation is given in expression (20).

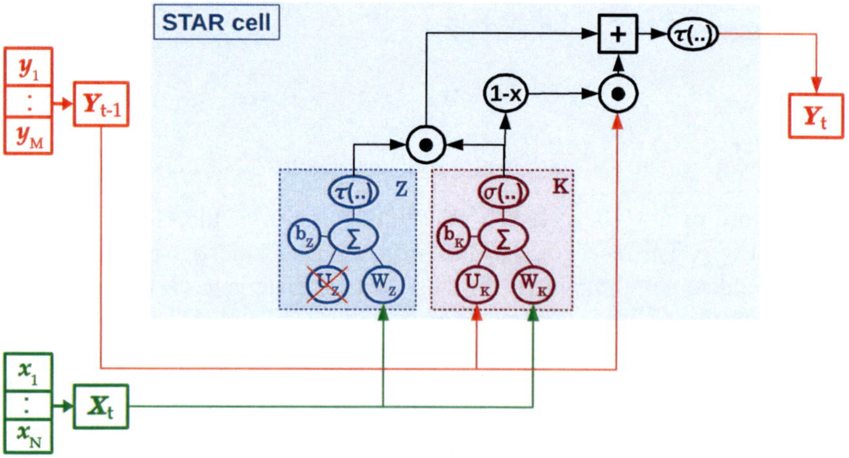

Fig. 7. Anatomy of a STAR cell.

$$K_t = \sigma(W_K \times X_t + U_K \times Y_{t-1} + b_K) \tag{18}$$

$$Z_t = \tau(W_Z \times X_t + b_Z) \tag{19}$$

$$Y_t = \tau((\mathbf{1} - K_t) \cdot Y_{t-1} + K_t \cdot Z_t) \tag{20}$$

4 Building Deep RNN Structures

4.1 Discussion on Basic Cells Complexity

The four types of RNN cells, presented in Sect. 3, are complex but we can highlight some relevant parameters for integration. Considering that input size of a cell is N, the output size is M and the sequence size is T (i.e. number of considered steps), we can calculate the total size of weight memory, the number of floating point operations and activation functions calls for each cell (for inference only i.e. no back-propagation). These parameters are provided in Table 1.

Table 1. Basic cells characteristics.

Cell type	Weights size	FLOPs	τ calls	σ calls
LSTM	$4(NM + M^2 + M)$	$4(2TNM + 2TM^2) + 4TM$	$2TM$	$3TM$
GRU	$3(NM + M^2 + M)$	$3(2TNM + 2TM^2) + 4TM$	TM	$2TM$
MGU	$2(NM + M^2 + M)$	$2(2TNM + 2TM^2) + 5TM$	TM	TM
STAR	$2NM + M^2 + 2M$	$4TNM + 2TM^2 + 4TM$	$2TM$	TM

4.2 Bi-Directional Variants

Regular RNN cells can be extended to a bidirectional version called Bi-Directional Recurrent Network (BRNN) cells. The BRNN cells are working simultaneously in positive ($[t-2] \rightarrow [t-1] \rightarrow [t]$) and negative ($[t] \rightarrow [t-1] \rightarrow [t-1]$) step directions [51]. For example bi-directionnal LSTM have shown better results than traditional LSTM for activity recognition [1]. The idea is to keep information from the sequence dependencies in the two directions having twice the trainable weight parameters to capture the relevant information. The calculation from these parallel and bi-directional computation lines are then aggregated using a merging operator (such as standard mean function or even an activation function). This type of architecture is illustrated in Fig. 8 as unfolded.

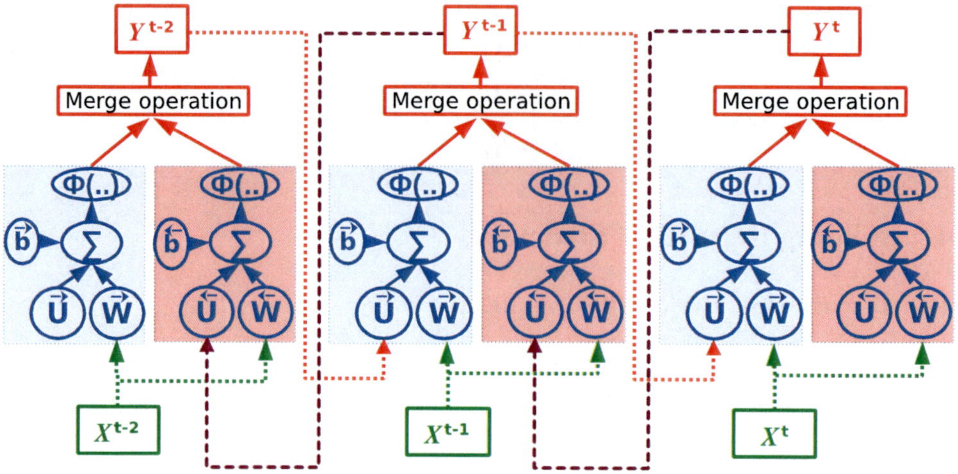

Fig. 8. Unfolded Bi-directionnal RNN illustration.

4.3 Stacking Recurrent Cells

The idea of stacking recurrent neural network came in the early days [18] and has been later extended [40]. The goal is to have more trainable parameters in order to capture more relevant and fine information during training phase. The resulting RNN architecture is more performant for complex task due to its big number of trained parameters. As an illustration, for complex avionics system, these architecture outperform single cell RNN architectures [39]. The basic idea here is to stack the cells, each new one on the top of the other. The output of one cell becoming the input of the other one in a sequential manner as depicted in Fig. 9. This type of architecture is considered in this work (see Sect. 5).

5 Experiments and Results

5.1 Test Cases Overview

An IMU sensors fusion problem has been considered here. The inputs considered are the measurements taken from an IMU containing accelerometers, gyroscopes, and magnetometers to predict attitude angles with a stacked RNN architecture. The attitude angles, to predict by the RNN, are the phi angle (i.e. roll), theta angle (i.e. pitch) and psi angle (i.e. yaw) as described in Fig. 10. Compare to our original work [38], we decided to integrate another test case provided from [37] which has the data-set with 6 logs (D1-6) publicly available thanks to the authors. The choice of selecting another test case came from the statement (based on our experiments) that our original data-set is not fully suitable to properly train and test our implementations for sensors fusion (lot of sensors noise in IMU measurements, very inconsistencies in flights scenarios, ...). Also, with this data-set, we are using each log independently of each other spilling the first 70% of each log as training data and the remaining 30% as testing data. For all the tests

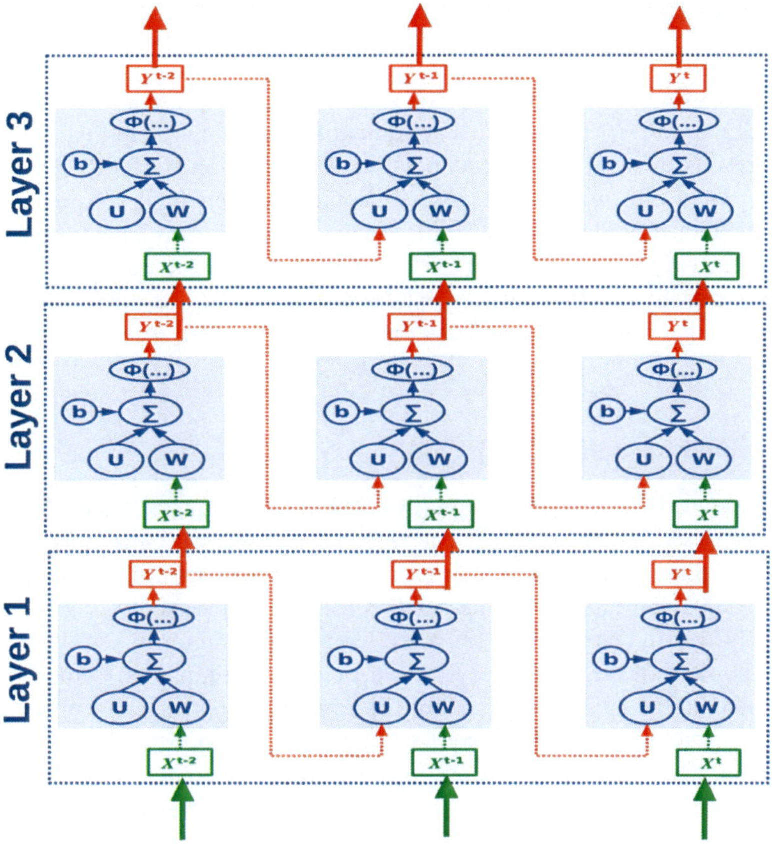

Fig. 9. Stacking RNN cells illustration

presented here, we selected an arbitrary sequence of 5 steps for temporal dependency consideration (i.e. $T = 5$).

Table 2. Pytorch implementation characteristics.

ID	Type	Layers size	Prediction
LSTM-P1	Stacked RNN	(9,20,20,20,3)	all angles
LSTM-P2	Stacked RNN	(9,20,20,20,1)	one angle
GRU-P1	Stacked RNN	(9,20,20,20,3)	all angles
GRU-P2	Stacked RNN	(9,20,20,20,1)	one angle

5.2 Pytorch Implementation

The test case has been first implemented with the well-known Pytorch framework [42]. Our Pytorch implementation is integrating 4 types of RNNs as detailed in Table 2. These

RNNs were trained using Adam optimizer [25] with learning rate $lr = 0.0005$ and momentum parameters $\beta_1 = 0.99$ and $\beta_2 = 0.999$.

The losses for each considered RNN architectures are presented in Fig. 11. We can see that all these RNN architectures are quickly converging. Note that, LSTM-P1 and GRU-P1 results are noted with an *_all* notation. LSTM-P2 and GRU-P2 results are noted with *_phi*, *_theta* and *_psi* because these are applied to each angle separately (one RNN for each predicted angle). The resulting outputs compared to labels are presented in Fig. 12 for LSTM and in Fig. 13 for GRU. We can observe that the version dedicated to one angle (here, LSTM-P2 and GRU-P2 for phi angle) are less noisy than their equivalent predicting all angles. It confirms the behavior we had observed in our previous work [38].

5.3 Our Implementation

We have extended our C++ framework initiated in [38] with stacked RNNs support for all the cells described in Sect. 3. For testing, we choose 4 architectures as detailed in Table 3 trained using two different optimizers: (1) Adam optimizer [25] with learning rate $lr = 0.001$ and momentum parameters as $\beta_1 = 0.99$ and $\beta_2 = 0.999$; (2) SGD optimizer [43] with learning rate $lr = 0.0005$ and momentum $\mu = 0.9$. Note that we changed the sizes of the stacked layers to decrease from inputs to outputs. The losses are presented in Fig. 14 showing that our RNN architectures are converging. The losses starts higher than in Pytorch and we believe this is due to different weights and bias initialization. The resulting predicted phi angle is compared to labels in Fig. 15 for GRU and MGU and in Fig. 16 for LSTM and STAR. Curves are showing relatively good estimations per our implementations which seems less noisy than their Pytorch equivalents.

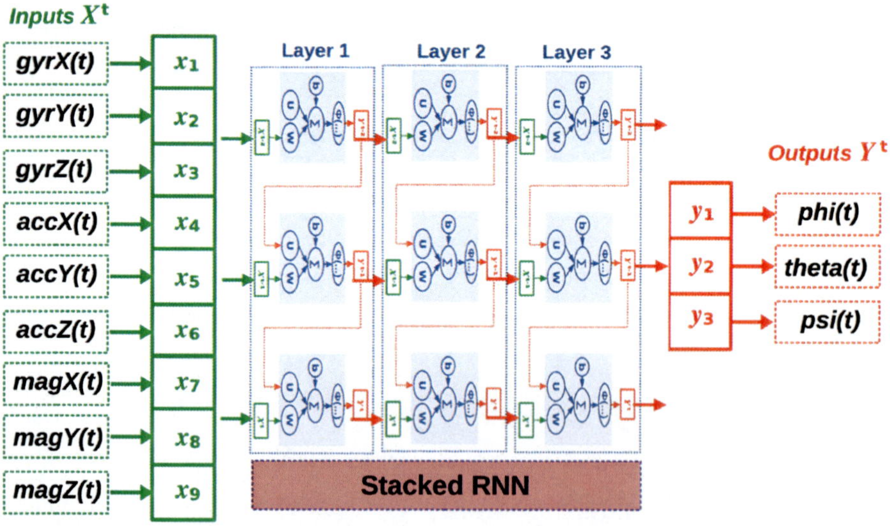

Fig. 10. Test case illustration.

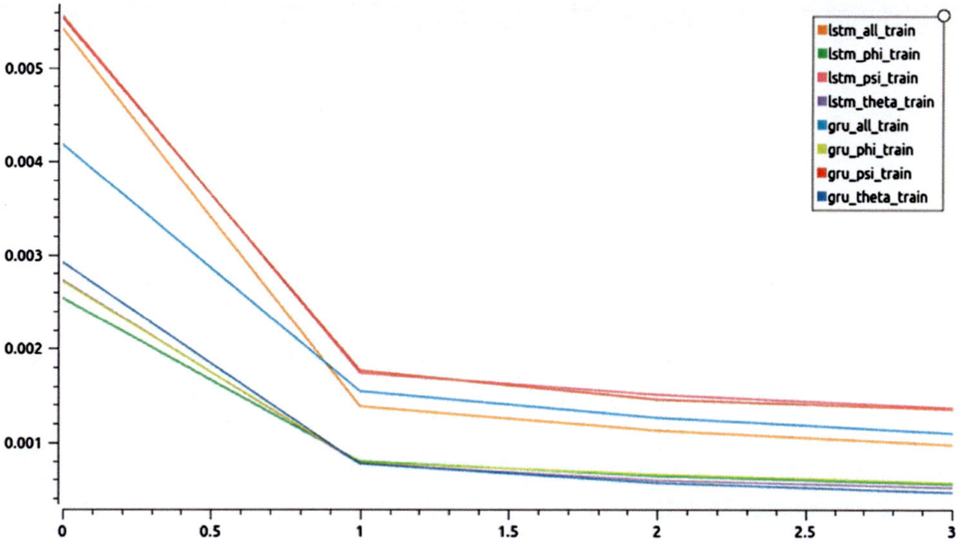

Fig. 11. Losses for our Pytorch RNNs (D1).

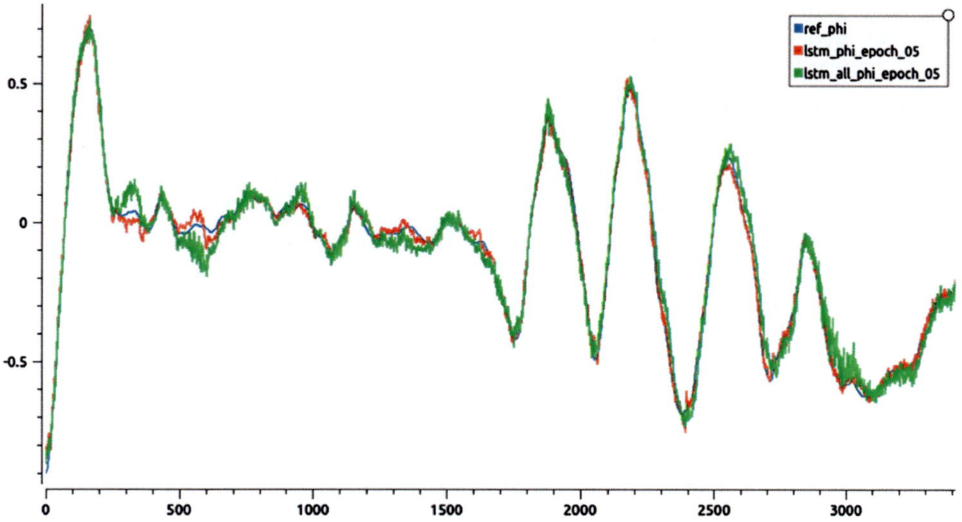

Fig. 12. Test after 5 epochs: Phi angle predictions for LSTM-P1 and LSTM-P2 (D1).

Table 3. C++ implementation characteristics.

ID	Type	Layers size	Prediction	Optimizer
LSTM-C1	Stacked RNN	(9,20,10,5,3)	all angles	ADAM
LSTM-C2	Stacked RNN	(9,20,20,20,3)	all angles	SGD
GRU-C1	Stacked RNN	(9,20,10,5,3)	all angles	ADAM
GRU-C2	Stacked RNN	(9,20,20,20,3)	all angles	SGD
STAR-C1	Stacked RNN	(9,20,10,5,3)	all angles	ADAM
STAR-C2	Stacked RNN	(9,20,20,20,3)	all angles	SGD
MGU-C1	Stacked RNN	(9,20,10,5,3)	all angles	ADAM
MGU-C2	Stacked RNN	(9,20,20,20,3)	all angles	SGD

5.4 Performance Results

Such as in our previous work [38], our C++ prototypes have been tested on 4 embedded devices with different software configurations:

1. **bbai:** *Beaglebone AI* with Linux Debian (kernel 4.14) and gcc 6.3.0
   – https://beagleboard.org/
2. **rpi4:** *Raspberry 4* with Linux Raspy (kernel 5.10) and gcc 8.3.0
   – https://www.raspberrypi.org/
3. **nano:** *Jetson Nano* with Linux Ubuntu (kernel 4.9) and gcc 7.5.0
   – https://developer.nvidia.com/embedded/
4. **pynq:** *Pynq Z2* with PetaLinux (kernel 4.19) and gcc 7.3.0
   – http://www.pynq.io/

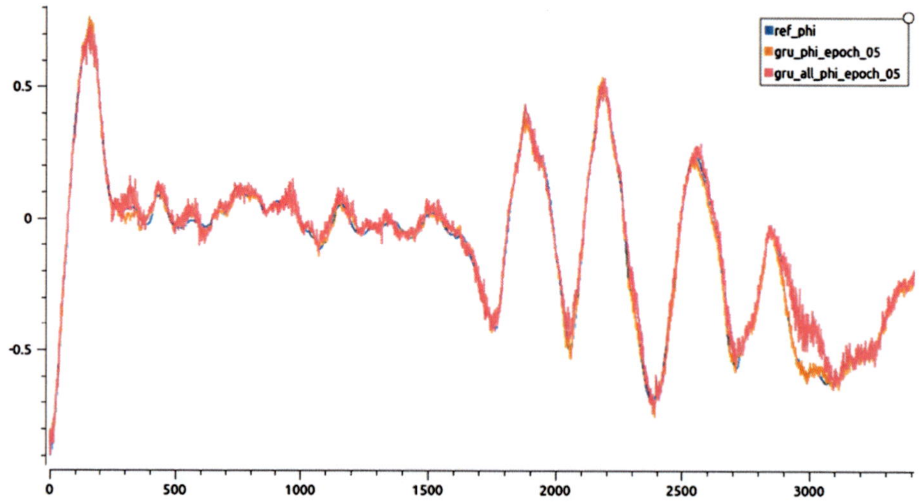

Fig. 13. Test after 5 epochs: Phi angle predictions for GRU-P1 and GRU-P2 (D1).

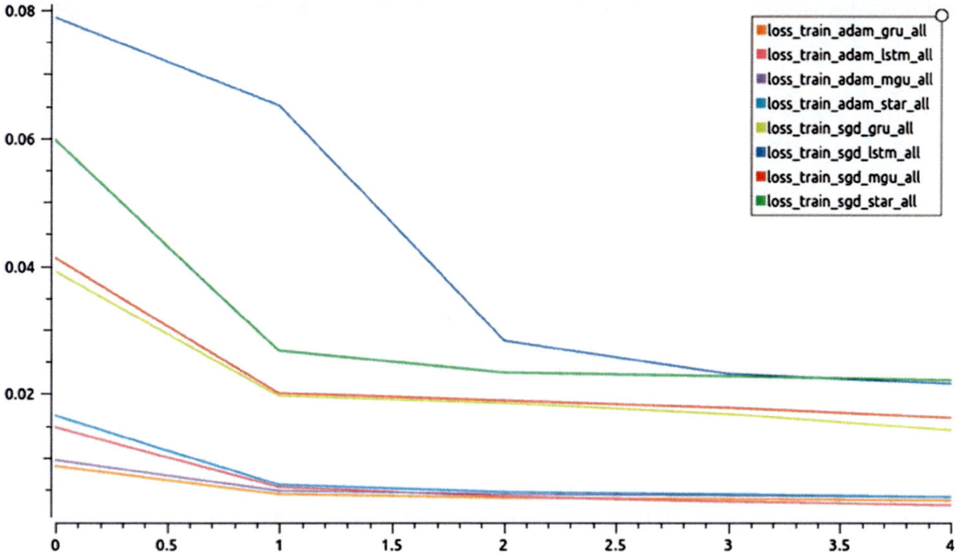

Fig. 14. Losses for our C++ RNNs (D1).

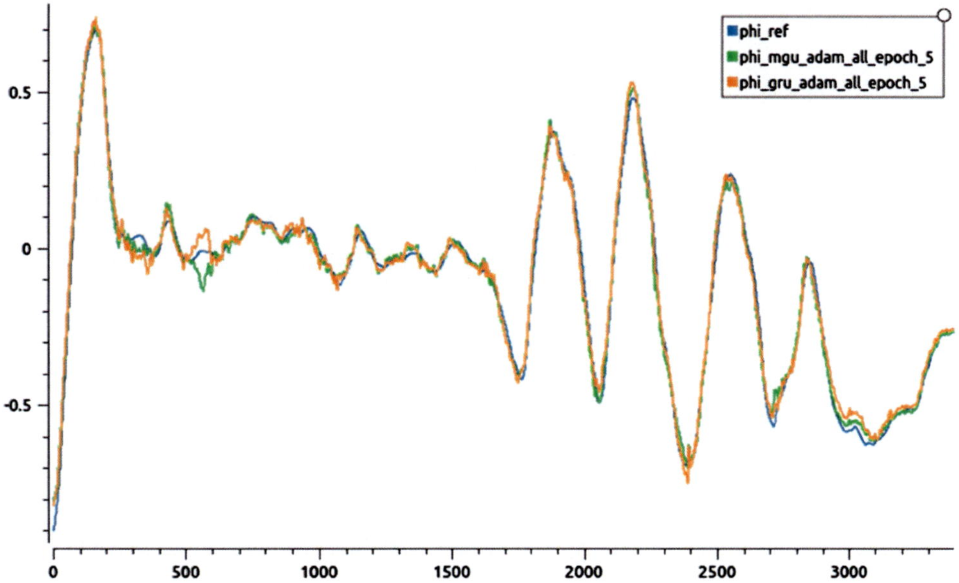

Fig. 15. Test after 5 epochs: Phi angle predictions for GRU-C1 and MGU-C1 (D1).

Performance measurements obtained are presented in Tables 4, 5, 6 and 7. Execution times are expressed in milliseconds (ms) and represent the duration of **one** execution step for a feed-forward process (noted *ff only*) and **one** execution step for a complete training process (feed-forward and back-propagation steps). The back-propagation measurements include performance for Adam optimizer (noted *bp adam*, used in C1

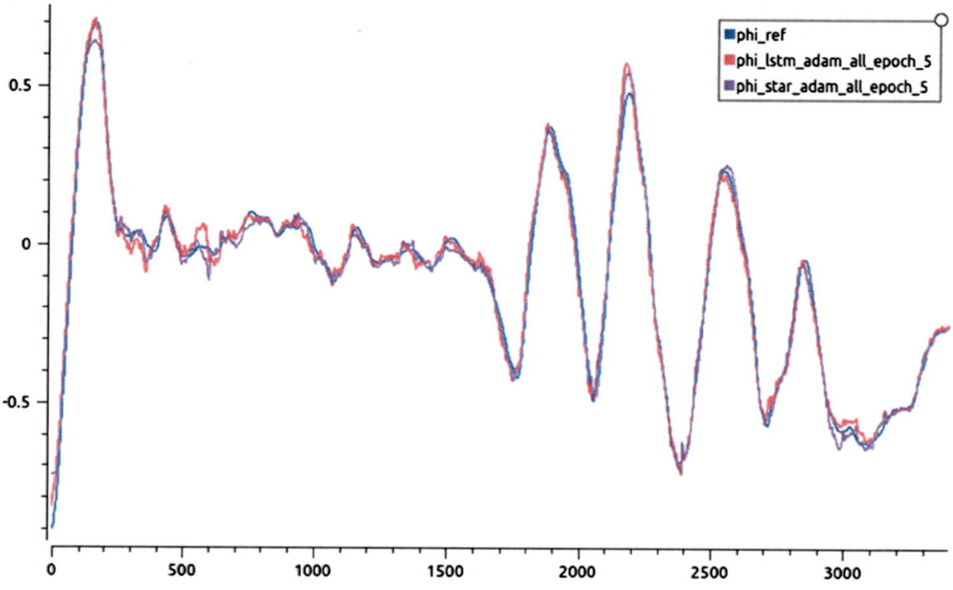

Fig. 16. Test after 5 epochs: Phi angle predictions for LSTM-C1 and STAR-C1 (D1).

versions) and the SGD optimizer (noted *bp sgd*, used in C2 versions). These tables are also including measurements using normal logarithmic activation functions from standard C++ library (noted N) and using interpolation tables such as in [10] (noted I). We can observe that the Raspberry 4 and the Jetson Nano re much more efficient than the Beaglebone AI and the Pynq Z2. The use of interpolation tables for activation function calculations offers slightly better performances and overall, stacked STAR architectures are the most efficient ones.

Table 4. LSTM-C1/C2 (9,20,10,5,3) performances on devices.

Boards		bbai (ms)		rpi4 (ms)		pynq (ms)		nano (ms)	
		N	I	N	I	N	I	N	I
ff only	min	4.216	3.988	0.8678	0.837	5.347	5.258	0.926	0.875
	max	6.984	6.755	1.265	1.245	7.993	7.915	1.435	1.334
	mean	4.239	4.004	0.875	0.844	5.382	5.291	0.941	0.886
bp sgd	min	14.292	14.073	2.992	2.937	17.436	17.336	3.186	3.109
	max	18.214	15.610	3.551	3.376	19.204	19.043	3.538	3.407
	mean	14.347	14.120	3.018	2.958	17.484	17.320	3.211	3.123
bp adam	min	17.290	17.100	3.612	3.582	20.663	20.552	3.836	3.776
	max	19.628	18.197	5.477	5.436	32.610	24.236	5.226	4.531
	mean	17.389	17.180	3.652	3.622	20.725	20.613	3.888	3.707

Table 5. GRU-C1/C2 (9,20,10,5,3) performances on devices.

Boards		bbai (ms)		rpi4 (ms)		pynq (ms)		nano (ms)	
		N	I	N	I	N	I	N	I
ff only	min	3.512	3.373	0.706	0.692	4.409	4.383	0.754	0.718
	max	6.293	6.235	1.128	1.117	7.085	7.060	1.250	1.221
	mean	3.568	3.423	0.714	0.701	4.444	4.404	0.761	0.726
bp sgd	min	11.359	11.122	2.342	2.215	13.837	13.790	2.481	2.459
	max	15.767	11.639	3.224	2.951	17.202	14.909	3.733	2.655
	mean	11.423	11.178	2.365	2.318	13.880	13.808	2.497	2.473
bp adam	min	13.379	13.242	2.789	2.623	16.314	16.268	2.950	2.923
	max	16.894	13.793	3.328	3.055	18.673	18.338	4.187	3.289
	mean	13.443	13.292	2.816	2.748	16.362	16.311	2.979	2.953

Table 6. MGU-C1/C2 (9,20,10,5,3) performances on devices.

Boards		bbai (ms)		rpi4 (ms)		pynq (ms)		nano (ms)	
		N	I	N	I	N	I	N	I
ff only	min	2.403	2.316	0.474	0.457	2.961	2.950	0.501	0.480
	max	5.240	5.153	0.918	0.902	5.613	5.596	1.005	0.979
	mean	2.446	2.354	0.482	0.465	2.980	2.913	0.507	0.484
bp sgd	min	8.434	8.37	1.717	1.695	10.072	10.041	1.801	1.778
	max	12.753	8.878	1.836	1.818	11.177	11.007	2.114	1.883
	mean	8.466	8.395	1.738	1.715	10.288	10.089	1.814	1.789
bp adam	min	9.807	9.608	2.070	1.983	11.707	11.700	2.128	2.107
	max	13.970	10.347	2.131	2.058	12.729	12.054	3.370	2.242
	mean	9.879	9.762	2.094	2.007	11.736	1.723	2.153	2.122

Table 7. STAR-C1/C2 (9,20,10,5,3) performances on devices.

Boards		bbai (ms)		rpi4 (ms)		pynq (ms)		nano (ms)	
		N	I	N	I	N	I	N	I
ff only	min	1.946	1.773	0.379	0.358	2.384	2.280	0.411	0.370
	max	4.505	4.310	0.724	0.703	4.945	4.846	1.636	0.826
	mean	1.982	1.810	0.385	0.361	2.401	2.299	0.417	0.374
bp sgd	min	6.072	5.887	1.241	1.207	7.227	7.121	1.294	1.258
	max	7.782	6.127	1.659	1.540	8.314	8.141	3.078	1.340
	mean	6.114	5.923	1.257	1.218	7.261	7.159	1.305	1.265
bp adam	min	7.023	6.864	1.432	1.408	8.360	8.255	1.531	1.487
	max	8.784	7.362	1.787	1.634	9.425	9.109	3.995	1.593
	mean	7.095	6.916	1.452	1.422	8.406	8.311	1.545	1.498

6 Conclusions and Perspectives

In this work, we have successfully pursued our investigations on the use of RNNs for real-time embedded applications. This work has required the mastering on multiple aspects from the full understating of mathematical aspects in RNNs to their implementations. Our experiments have shown that our stacked RNN architectures are efficient and can be used on CPU based embedded systems. However, for IMU sensors fusion, RNNs seem greedy in term of CPU usage compared to dedicated efficient mathematical based algorithms such as Kalman, Madwick or Mahony algorithms [32] which take few microseconds to process the same task on the selected embedded devices. Therefore, the use of RNNs needs justification in order to improve some aspects that are not possible per other traditional approaches such improving an EKF filter to tolerate GPS outage [30]. A part of our work has been released as an open-source package and, therefore, is accessible and results are reproducible[1]. Also, we have confirmed that our implementation is correct by comparing to the well known Pytorch framework. There are several interesting perspectives to pursue this study. We can continue to optimize our RNN implemented architectures by integrating other features such as refined gates mechanisms [14] or regularization to prevent over-fitting [33]. We started investigations for a proper initialization of a RNN i.e. how to trim their initial states [62]. Additionally, we are also working on the integration of other RNN architectures such as ESN [20] or Simple Recurrent Unit (SRU) [28]. Besides this, we consider the online training aspect even if, currently, the training phase seems too greedy in term of computing resources to allow an *online* training capacity for embedded devices. In this senses, we are thinking to use of our flight simulator [3] to reproduce similar online learning scenarios for flight controller as the ones described in [24]. In parallel, we started working on MPSoC systems containing GPU and/or FPGA as processing accelerators [57] to enhance online training efficiency for on-board systems.

References

1. Alawneh, L., Mohsen, B., Al-Zinati, M., Shatnawi, A., Al-Ayyoub, M.: A comparison of unidirectional and bidirectional LSTM networks for human activity recognition. In: 2020 IEEE International Conference on Pervasive Computing and Communications Workshops (PerCom Workshops), pp. 1–6 (2020). https://doi.org/10.1109/PerComWorkshops48775.2020.9156264
2. Chang, X., Rong, L., Chen, K., Fu, W.: LSTM-based output-constrained adaptive fault-tolerant control for fixed-wing UAV with high dynamic disturbances and actuator faults. Math. Prob. Eng. **2021**, 1–18 (2021). https://doi.org/10.1155/2021/8882312
3. Chaudron, J.B., Joksimović, A., Siron, P., Vingerhoeds, R., Carbonneau, X.: Running large-scale and hybrid real-time aircraft simulations in an HLA framework. In: Cerone, A., et al. Software Engineering and Formal Methods. SEFM 2021 Collocated Workshops. SEFM 2021. LNCS, vol. 13230, pp. 221–237. Springer, Cham (2022). https://doi.org/10.1007/978-3-031-12429-7_16

[1] https://github.com/ISAE-PRISE/rnn4ap (only LSTM and GRU).

4. Cho, K., et al.: Learning phrase representations using RNN encoder-decoder for statistical machine translation. In: Moschitti, A., Pang, B., Daelemans, W. (eds.) Proceedings of the 2014 Conference on Empirical Methods in Natural Language Processing, EMNLP 2014, 25–29 October 2014, Doha, Qatar, A meeting of SIGDAT, a Special Interest Group of the ACL, pp. 1724–1734. ACL (2014). https://doi.org/10.3115/v1/d14-1179
5. Chung, J., Gulcehre, C., Cho, K., Bengio, Y.: Empirical evaluation of gated recurrent neural networks on sequence modeling, pp. 1–9 (2014). http://arxiv.org/abs/1412.3555
6. Dey, R., Salem, F.M.: Gate-variants of gated recurrent unit (GRU) neural networks. In: 2017 IEEE 60th International Midwest Symposium on Circuits and Systems (MWSCAS), pp. 1597–1600 (2017). https://doi.org/10.1109/MWSCAS.2017.8053243
7. Dubey, S.R., Singh, S.K., Chaudhuri, B.B.: A comprehensive survey and performance analysis of activation functions in deep learning. CoRR abs/2109.14545 (2021). https://arxiv.org/abs/2109.14545
8. Elman, J.L.: Finding structure in time. Cogn. Sci. **14**(2), 179–211 (1990). https://doi.org/10.1016/0364-0213(90)90002-E
9. Eroglu, B., Sahin, M.C., Ure, N.K.: Autolanding control system design with deep learning based fault estimation. Aerosp. Sci. Technol. **102**, 105855 (2020). https://doi.org/10.1016/j.ast.2020.105855
10. Faiedh, H., Gafsi, Z., Besbes, K.: Digital hardware implementation of sigmoid function and its derivative for artificial neural networks. In: ICM 2001 Proceedings. The 13th International Conference on Microelectronics, pp. 189–192 (2001). https://doi.org/10.1109/ICM.2001.997519
11. Gers, F.A., Schmidhuber, J.A., Cummins, F.A.: Learning to forget: continual prediction with LSTM. Neural Comput. **12**(10), 2451–2471 (2000). https://doi.org/10.1049/cp:19991218
12. Greff, K., Srivastava, R.K., Koutník, J., Steunebrink, B.R., Schmidhuber, J.: LSTM: a search space odyssey. IEEE Trans. Neural Netw. Learn. Syst. **28**(10), 2222–2232 (2017). https://doi.org/10.1109/TNNLS.2016.2582924
13. Grigorescu, S.M., Trasnea, B., Cocias, T.T., Macesanu, G.: A survey of deep learning techniques for autonomous driving. J. Field Robot. **37**, 362–386 (2020). https://doi.org/10.1002/rob.21918
14. Gu, A., Gulcehre, C., Paine, T., Hoffman, M., Pascanu, R.: Improving the gating mechanism of recurrent neural networks. In: Proceedings of the 37th International Conference on Machine Learning, ICML2020, JMLR.org (2020)
15. Gu, W., Valavanis, K.P., Rutherford, M.J., Rizzo, A.: UAV model-based flight control with artificial neural networks: a survey. J. Intell. Robot. Syst. **100**, 1469–1491 (2020). https://doi.org/10.1007/s10846-020-01227-8
16. Haykin, S.: Neural Networks: A Comprehensive Foundation. 2nd edn, Prentice Hall, Upper Saddle River, NJ (1999)
17. Heck, J., Salem, F.: Simplified minimal gated unit variations for recurrent neural networks. In: 2017 IEEE 60th International Midwest Symposium on Circuits and Systems, pp. 1593–1596, August 2017. https://doi.org/10.1109/MWSCAS.2017.8053242
18. Hihi, S., Bengio, Y.: Hierarchical recurrent neural networks for long-term dependencies. In: Touretzky, D., Mozer, M., Hasselmo, M. (eds.) Advances in Neural Information Processing Systems, vol. 8. MIT Press (1995). https://proceedings.neurips.cc/paper/1995/file/c667d53acd899a97a85de0c201ba99be-Paper.pdf
19. Irie, M.: Capabilities of three-layered perceptrons. In: IEEE 1988 International Conference on Neural Networks, vol. 1, pp. 641–648 (1988). https://doi.org/10.1109/ICNN.1988.23901
20. Jaeger, H., Haas, H.: Harnessing nonlinearity: predicting chaotic systems and saving energy in wireless communication. Science **304**(5667), 78–80 (2004). https://doi.org/10.1126/science.1091277, https://www.science.org/doi/abs/10.1126/science.1091277

21. Jordan, M.I.: Serial order: a parallel distributed processing approach. Technical Report, June 1985-march 1986 https://www.osti.gov/biblio/6910294
22. Jozefowicz, R., Zaremba, W., Sutskever, I.: An empirical exploration of recurrent network architectures. In: Proceedings of the 32nd International Conference on International Conference on Machine Learning, ICML 2015, vol. 37, pp. 2342–2350. JMLR.org (2015)
23. Kanopoulos, N., Vasanthavada, N., Baker, R.: Design of an image edge detection filter using the sobel operator. IEEE J. Solid-State Circuits **23**(2), 358–367 (1988). https://doi.org/10.1109/4.996
24. Kim, B.S., Calise, A., Kam, M.: Nonlinear flight control using neural networks and feedback linearization. In: Proceedings. The First IEEE Regional Conference on Aerospace Control Systems, pp. 176–181 (1993). https://doi.org/10.1109/AEROCS.1993.720919
25. Kingma, D.P., Ba, J.: Adam: a method for stochastic optimization. In: Bengio, Y., LeCun, Y. (eds.) 3rd International Conference on Learning Representations, ICLR 2015, San Diego, CA, USA, 7–9 May 2015, Conference Track Proceedings (2015)
26. Koutnik, J., Greff, K., Gomez, F., Schmidhuber, J.: A clockwork RNN. In: Xing, E.P., Jebara, T. (eds.) Proceedings of the 31st International Conference on Machine Learning. Proceedings of Machine Learning Research, 22–24 June 2014, vol. 32, pp. 1863–1871. PMLR, Bejing, China (2014). http://proceedings.mlr.press/v32/koutnik14.pdf
27. Lalapura, V.S., Amudha, J., Satheesh, H.S.: Recurrent neural networks for edge intelligence: a survey. ACM Comput. Surv. **54**(4) (2021). https://doi.org/10.1145/3448974
28. Lei, T., Zhang, Y., Wang, S.I., Dai, H., Artzi, Y.: Simple recurrent units for highly parallelizable recurrence. In: Proceedings of the 2018 Conference on Empirical Methods in Natural Language Processing, pp. 4470–4481. Association for Computational Linguistics, Brussels, Belgium, November 2018. https://doi.org/10.18653/v1/D18-1477, https://aclanthology.org/D18-1477
29. Lin, T., Horne, B., Tino, P., Giles, C.: Learning long-term dependencies in NARX recurrent neural networks. IEEE Trans. Neural Networks **7**(6), 1329–1338 (1996). https://doi.org/10.1109/72.548162
30. Liu, J., Guo, G.: Vehicle localization during GPS outages with extended kalman filter and deep learning. IEEE Trans. Instrum. Meas. **70**, 1–10 (2021). https://doi.org/10.1109/TIM.2021.3097401
31. Liu, Y., Zhou, Y., Zhang, Y.: A novel hybrid attitude fusion method based on LSTM neural network for unmanned aerial vehicle. In: 2021 IEEE International Conference on Robotics and Biomimetics (ROBIO), pp. 1630–1635 (2021). https://doi.org/10.1109/ROBIO54168.2021.9739330
32. Ludwig, S.A., Burnham, K.D.: Comparison of Euler estimate using extended kalman filter, Madgwick and Mahony on quadcopter flight data. In: 2018 International Conference on Unmanned Aircraft Systems (ICUAS), pp. 1236–1241 (2018). https://doi.org/10.1109/ICUAS.2018.8453465
33. Merity, S., Keskar, N.S., Socher, R.: Regularizing and optimizing LSTM language models (2017). https://doi.org/10.48550/ARXIV.1708.02182, https://arxiv.org/abs/1708.02182
34. Mersha, B.W., Jansen, D.N., Ma, H.: Angle of attack prediction using recurrent neural networks in flight conditions with faulty sensors in the case of f-16 fighter jet. Complex Intell. Syst. (2021). https://doi.org/10.1007/s40747-021-00612-6
35. Mingzhong, L., Fuli, W.: Adaptive control of black-box nonlinear systems using recurrent neural networks. In: Proceedings of the 36th IEEE Conference on Decision and Control, vol. 5, pp. 4165–4170 (1997). https://doi.org/10.1109/CDC.1997.649486
36. Minsky, M., Papert, S.A.: Perceptrons: An Introduction to Computational Geometry. The MIT Press, Cambridge (1969). https://doi.org/10.7551/mitpress/11301.001.0001

37. Narkhede, P., Walambe, R., Poddar, S., Kotecha, K.: Incremental learning of LSTM framework for sensor fusion in attitude estimation. PeerJ Comput. Sci. **7** (2021). https://doi.org/10.7717/peerj-cs.662
38. Neves, G., Chaudron, J.B., Dion, A.: Recurrent neural networks analysis for embedded systems. In: Proceedings of the 13th International Joint Conference on Computational Intelligence - Volume 1: NCTA, pp. 374–383. INSTICC, SciTePress (2021). https://doi.org/10.5220/0010715700003063
39. Nivison, S.A., Khargonekar, P.P.: Development of a robust deep recurrent neural network controller for flight applications. In: 2017 American Control Conference (ACC), pp. 5336–5342 (2017). https://doi.org/10.23919/ACC.2017.7963784
40. Pascanu, R., Gulcehre, C., Cho, K., Bengio, Y.: How to construct deep recurrent neural networks (2013). https://doi.org/10.48550/ARXIV.1312.6026, https://arxiv.org/abs/1312.6026
41. Pascanu, R., Mikolov, T., Bengio, Y.: On the difficulty of training recurrent neural networks. In: Proceedings of the 30th International Conference on International Conference on Machine Learning, ICML2013, vol. 28, pp. III-1310-III-1318. JMLR.org (2013)
42. Paszke, A., et al.: PyTorch: an imperative style, high-performance deep learning library. CoRR abs/1912.01703 (2019). http://arxiv.org/abs/1912.01703
43. Qian, N.: On the momentum term in gradient descent learning algorithms. Neural Netw. **12**(1), 145–151 (1999). https://doi.org/10.1016/S0893-6080(98)00116-6
44. Rawat, W., Wang, Z.: Deep convolutional neural networks for image classification: a comprehensive review. Neural Comput. **29**(9), 2352–2449 (2017)
45. Rezk, N.M., Purnaprajna, M., Nordstrom, T., Ul-Abdin, Z.: Recurrent neural networks: an embedded computing perspective. IEEE Access **8**, 57967–57996 (2020). https://doi.org/10.1109/access.2020.2982416, http://dx.doi.org/10.1109/ACCESS.2020.2982416
46. Rezk, N.M., Purnaprajna, M., Nordström, T., Ul-Abdin, Z.: Recurrent neural networks: an embedded computing perspective. IEEE Access **8**, 57967–57996 (2020). https://doi.org/10.1109/ACCESS.2020.2982416
47. Rosenblatt, F.: The Perceptron - A Perceiving and Recognizing Automaton. Technical Report, 85–460-1, Cornell Aeronautical Laboratory (1957)
48. Rosindell, J., Wong, Y.: Biodiversity, the tree of life, and science communication. Phylogenet. Divers. Appl. Challenges Biodivers. Sci. **2**, 41–71 (2018). https://doi.org/10.1007/978-3-319-93145-6_3
49. Rumelhart, D.E., McClelland, J.L.: Learning internal representations by error propagation, pp. 318–362 (1987)
50. Russo, A., Lax, G.: Using artificial intelligence for space challenges: a survey. Appl. Sci. **12**(10) (2022). https://doi.org/10.3390/app12105106
51. Schuster, M., Paliwal, K.: Bidirectional recurrent neural networks. IEEE Trans. Signal Process. **45**(11), 2673–2681 (1997). https://doi.org/10.1109/78.650093
52. Schweiger, A., et al.: Classification for avionics capabilities enabled by artificial intelligence. In: 2021 IEEE/AIAA 40th Digital Avionics Systems Conference (DASC), pp. 1–10 (2021). https://doi.org/10.1109/DASC52595.2021.9594364
53. Turkoglu, M.O., D'Aronco, S., Wegner, J.: Gating revisited: deep multi-layer RNNs that can be trained. In: IEEE Transactions on Pattern Analysis and Machine Intelligence, p. 1, March 2021. https://doi.org/10.1109/TPAMI.2021.3064878
54. Waibel, A., Hanazawa, T., Hinton, G., Shikano, K., Lang, K.: Phoneme recognition using time-delay neural networks. IEEE Trans. Acoust. Speech Signal Process. **37**(3), 328–339 (1989). https://doi.org/10.1109/29.21701
55. Weber, D., Gühmann, C., Seel, T.: Neural networks versus conventional filters for inertial-sensor-based attitude estimation. In: 2020 IEEE 23rd International Conference on Information Fusion (FUSION), pp. 1–8 (2020). https://doi.org/10.23919/FUSION45008.2020.9190634

56. Werbos, P.J.: Backpropagation through time: what it does and how to do it. Proc. IEEE **78**(10), 1550–1560 (1990). https://doi.org/10.1109/5.58337
57. Wu, R., Guo, X., Du, J., Li, J.: Accelerating neural network inference on FPGA-based platforms - a survey. Electronics **10**(9) (2021). https://doi.org/10.3390/electronics10091025
58. Xiao, K., Zhao, J., He, Y., Li, C., Cheng, W.: Abnormal behavior detection scheme of UAV using recurrent neural networks. IEEE Access **7**, 110293–110305 (2019). https://doi.org/10.1109/ACCESS.2019.2934188
59. Yam, Y., Chow, T.: Extended backpropagation algorithm. Electron. Lett. **29**(1), 1701–1702 (1993)
60. Youlal, H., Kada, A.: A class of recurrent neural networks for adaptive control of nonlinear systems. IFAC Proc. Volumes **27**(8), 1393–1398 (1994). https://doi.org/10.1016/S1474-6670(17)47905-5, https://www.sciencedirect.com/science/article/pii/S1474667017479055, iFAC Symposium on System Identification (SYSID'94), Copenhagen, Denmark, 4-6 July
61. Zhou, G.B., Wu, J., Zhang, C.L., Zhou, Z.H.: Minimal gated unit for recurrent neural networks. Int. J. Autom. Comput. **13** (2016). https://doi.org/10.1007/s11633-016-1006-2
62. Zimmermann, H.-G., Tietz, C., Grothmann, R.: Forecasting with recurrent neural networks: 12 tricks. In: Montavon, G., Orr, G.B., Müller, K.-R. (eds.) Neural Networks: Tricks of the Trade. LNCS, vol. 7700, pp. 687–707. Springer, Heidelberg (2012). https://doi.org/10.1007/978-3-642-35289-8_37
63. Zweiri, Y., Whidborne, J., Althoefer, K., Seneviratne, L.: A new three-term backpropagation algorithm with convergence analysis. In: Proceedings 2002 IEEE International Conference on Robotics and Automation (Cat. No.02CH37292), vol. 4, pp. 3882–3887 (2002). https://doi.org/10.1109/ROBOT.2002.1014327

Author Index

J. Garibaldi et al. (Eds.): IJCCI 2021, SCI 1119, p. 245, 2023.
https://doi.org/10.1007/978-3-031-46221-4